SOLUTIONS MANUAL TO ACCOMPANY GAME THEORY

SOLUTIONS MANUAL TO ACCOMPANY GAME THEORY

An Introduction

SECOND EDITION

E.N. BARRON
Loyola University Chicago
Chicago, Illinois

To Christina

Contents

Foreword

This solutions manual is a companion to *Game Theory: An Introduction, Second Edition*. It provides the statements and solutions of all the problems in the book. An Appendix contains the main definitions and theorems in the book. References in the problems to theorems or equations can be found in the Appendix, but references to examples are found in the book. The problem numbers correspond exactly to those in the book.

I thank Susanne Steitz-Filler for supporting this work at Wiley and Amy Hendrickson for her help with some Latex problems. I am especially grateful to the National Science Foundation that provided partial support during the writing of this work under DMS-1008602.

I would appreciate notification of errors at ebarron@luc.edu. All errors and software associated with the book will be posted on my website:

www.math.luc.edu/~enb

CHAPTER ONE

Matrix Two-Person Games

1.1 The Basics

Problems

1.1 There are 100 bankers lined up in each of 100 rows. Pick the richest banker in each row. Javier is the poorest of those. Pick the poorest banker in each column. Raoul is the richest of those. Who is richer: Javier or Raoul?

1.1 Answer: Think of this as a 100×100 game matrix and we are looking for the upper and lower values except that we are really doing it for the transpose of the matrix.

If we take the maximum in each row and then Javier is the minimum maximum, Javier is v^{+}. If we take the minimum in each column, and Raoul is the maximum of those, then Raoul is the maximum minimum, or v^{-}. Thus, Javier is richer.

Another way to think of this is that the poorest rich guy is wealthier than the richest poor guy. Common sense.

1.2 In a Nim game start with 4 pennies. Each player may take 1 or 2 pennies from the pile. Suppose player I moves first. The game ends when there are no pennies left and the player who took the last penny pays 1 to the other player.

(a) Draw the game as we did in 2×2 Nim.

1.2.a Answer: The game tree is

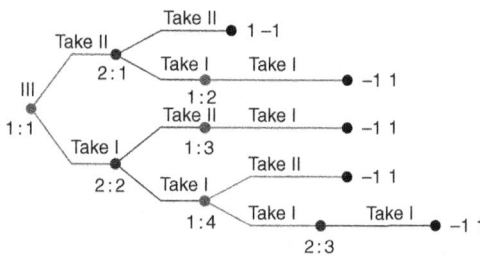

(b) Write down all the strategies for each player and then the game matrix.

Solutions Manual to Accompany Game Theory: An Introduction, Second Edition. E.N. Barron.
© 2013 John Wiley & Sons, Inc. Published 2013 by John Wiley & Sons, Inc.

1.2.b Answer:
Using the notation from the figure, we may list the strategies for player I as follows:
1. Go to $2:1$; if at $1:2$ take 1. [Same as Take 2. (there are no more choices for I after that)]
2. Go to $2:2$; if at $1:3$ take 1; if at $1:4$ take 2. [Same as Take 1, then if 2 are left, take 1.]
3. Go to $2:2$; if at $1:3$ take 1; if at $1:4$ take 1. [Same as Take 1, then if 2 are left, take 2.]

For player II, the strategies are as follows:
1. If at $2:1$ take 2; if at $2:2$ take 2. [Same as If there are 3 left, take 2.]
2. If at $2:1$ take 2; if at $2:2$ take 1.
3. If at $2:1$ take 1; if at $2:2$ take 2.
4. If at $2:1$ take 1; if at $2:2$ take 1; if at $2:3$ take 1.

The game matrix is

I/II	1	2	3	4
1	1	1	−1	−1
2	−1	−1	−1	−1
3	−1	1	−1	1

(c) Find v^+, v^-. Would you rather be player I or player II?

1.2.c Answer: Since $v^+ = -1$, $v^- = -1$, this game has a value of -1. Player II can always win by playing as follows: If player I takes 2 pennies, then II should take 1. If player I takes 1 penny, then II should take 2 pennies. No matter what player I does, player II wins.

1.3 In the game rock-paper-scissors both players select one of these objects simultaneously. The rules are as follows: paper beats rock, rock beats scissors, and scissors beats paper. The losing player pays the winner $1 after each choice of object. If both choose the same object the payoff is 0.

(a) What is the game matrix?

1.3.a Answer: The rock-paper-scissors game matrix with the rules of the problem is

I/II	Rock	Paper	Scissors
Rock	0	−1	1
Paper	1	0	−1
Scissors	−1	1	0

(b) Find v^+ and v^- and determine whether a saddle point exists in pure strategies, and if so, find it.

1.3.b Answer: $v^+ = 1$, $v^- = -1$. No saddle point in pure strategies since $v^+ > v^-$.

1.4 Each of two players must choose a number between 1 and 5. If a player's choice = opposing player's choice $+1$, she loses $2; if a player's choice $\geq$ opposing player's choice $+2$, she wins $1. If both players choose the same number the game is a draw.

(a) What is the game matrix?

1.4.a Answer: The game matrix is

I/II	1	2	3	4	5
1	0	2	−1	−1	−1
2	−2	0	2	−1	−1
3	1	−2	0	2	−1
4	1	1	−2	0	2
5	1	1	1	−2	0

(b) Find v^+ and v^- and determine whether a saddle point exists in pure strategies, and if so, find it.

1.4.b Answer: $v^+ = 1$, $v^- = -1$, no pure saddle point.

1.5 Each player displays either one or two fingers and simultaneously guesses how many fingers the opposing player will show. If both players guess either correctly or incorrectly, the game is a draw. If only one guesses correctly, he wins an amount equal to the total number of fingers shown by both players. Each pure strategy has two components: the number of fingers to show, the number of fingers to guess. Find the game matrix, v^+, v^-, and optimal pure strategies if they exist.

1.5 Answer: For each player we let (i, j) be the pure strategy in which the player shows i fingers, and guesses the other player will show j fingers. The matrix is

I/II	(1, 1)	(1, 2)	(2, 1)	(2, 2)
(1, 1)	0	2	−3	0
(1, 2)	−2	0	0	3
(2, 1)	3	0	0	−4
(2, 2)	0	−3	4	0

Since $v^+ = 2, v^- = -2$, there are no pure optimal strategies.

1.6 In the Russian roulette Example 1.5 suppose that if player I spins and survives and player II decides to pass, then the net gain to I is \$1000 and so I gets all of the additional money that II had to put into the pot in order to pass. Draw the game tree and find the game matrix. What are the upper and lower values? Find the saddle point in pure strategies.

1.6 Answer: The game tree stays the same but the payoff at the end of the Spin-Safe-Pass branch becomes 1, instead of $\frac{1}{2}$.

Here is the game tree:

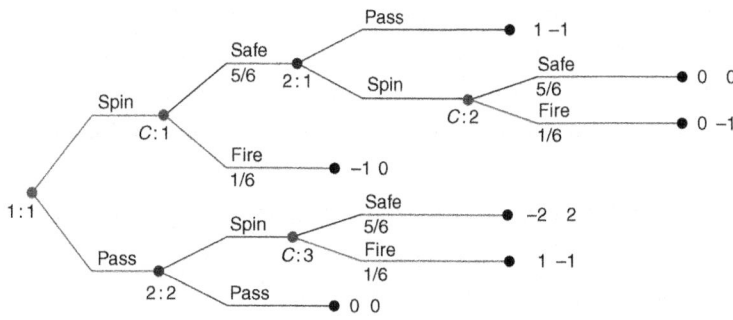

This is a Gambit generated tree. Any node labeled C is a Chance node. A node labeled with a number indicates that player is making a move. The numbers below the branches in chance moves are the probability that branch is taken. The payoffs at the end of the tree are the payoffs to each player. In a zero sum game, the payoff to player II is the negative of the payoff to player I.

Going through the same calculations as before, we get the game matrix

$$A = \begin{bmatrix} \frac{2}{3} & \frac{2}{3}, & -\frac{1}{36} & -\frac{1}{36} \\ -\frac{3}{2} & 0 & -\frac{3}{2} & 0 \end{bmatrix}.$$

The upper and lower values are $v^- = -\frac{1}{36} = v^+$. There is a pure saddle at row 1, column 3. Even if player I takes the entire pot there will be a saddle at row 1, column 3, both players should spin.

1.7 Let x be an unknown number and consider the matrices

$$A = \begin{bmatrix} 0 & x \\ 1 & 2 \end{bmatrix}, \quad B = \begin{bmatrix} 2 & 1 \\ x & 0 \end{bmatrix}.$$

Show that no matter what x is, each matrix has a pure saddle point.

1.7 Answer: Consider first the game with A. To calculate v^-, we take the minimum of x and 0, written $\min\{x, 0\}$, and the minimum of 1, 2 which is 1. Then

$$v^- = \max\{\min\{x, 0\}, 1\} = 1$$

since $\min\{x, 0\} \le 0$ no matter what x is. Similarly,

$$v^+ = \min\{1, \max\{x, 2\}\} = 1 \text{ since } \max\{x, 2\} \ge 2.$$

Thus, $v^- = v^+ = 1$ and there is a pure saddle at row 2, column 1.

For matrix B, we have $v^- = \max\{1, \min\{x, 0\}\} = 1$, $v^+ = \min\{\max\{2, x\}, 1\} = 1$, and there is a pure saddle at row 1, column 2, no matter what x is.

1.8 If we have a game with matrix A and we modify the game by adding a constant C to every element of A, call the new matrix $A + C$, is it true that $v^+(A + C) = v^+(A) + C$?

1.8 Answer: This is true since

$$v^+(A + C) = \min_{1 \le j \le m} \max_{1 \le i \le n} (a_{ij} + C) = \min_{1 \le j \le m} \max_{1 \le i \le n} a_{ij} + C = v^+(A) + C.$$

It is also true that $v^-(A + C) = v^-(A) + C$.

(a) If it happens that $v^-(A + C) = v^+(A + C)$, will it be true that $v^-(A) = v^+(A)$, and conversely?

1.8.a Answer: It is true that $v^-(A + C) = v^+(A + C) \Leftrightarrow v^-(A) = v^+(A)$. This follows from the first part.

(b) What can you say about the optimal pure strategies for $A + C$ compared to the game for just A?

1.8.b Answer: The previous parts of this problem imply that $A + C$ has a pure saddle point if and only if A has a saddle point. Since

$$a_{ij^*} + C \leq a_{i^*j^*} + C \leq a_{i^*j} + C \Leftrightarrow a_{ij^*} \leq a_{i^*j^*} \leq a_{i^*j},$$

we conclude that $A + C$ has a saddle point at (i^*, j^*) if and only if A has a saddle point at (i^*, j^*).

1.9 Consider the square game matrix $A = (a_{ij})$ where $a_{ij} = i - j$ with $i = 1, 2, \ldots, n$, and $j = 1, 2, \ldots, n$. Show that A has a saddle point in pure strategies. Find them and find $v(A)$.

1.9 Answer: We have

$$v^+(A) = \min_{1 \leq j \leq n} \max_{1 \leq i \leq n} (i - j) = \min_{1 \leq j \leq n} (n - j) = n - n = 0$$

and

$$v^-(A) = \max_{1 \leq i \leq n} \min_{1 \leq j \leq n} (i - j) = \max_{1 \leq j \leq n} (i - n) = n - n = 0.$$

Thus, $v = 0$ with a pure saddle point at (n, n).

1.10 Player I chooses 1, 2, or 3 and player II guesses which number I has chosen. The payoff to I is |I's number − II's guess|. Find the game matrix. Find v^- and v^+.

1.10 Answer: The game matrix is

I/II	1	2	3
1	0	1	2
2	1	0	1
3	2	1	0

It is easy to see that $v^- = 0$, $v^+ = 1$ and we have no pure saddle point.

1.11 In the Cat versus Rat game, determine v^+ and v^- without actually writing out the matrix. It is a 16×16 matrix.

1.11 Answer: $v^+ = 1$, $v^- = 0$ because there is always at least one 1 and one 0 in each row and column.

1.12 In a football game, the offense has two strategies: run or pass. The defense also has two strategies: defend against the run, or defend against the pass. A possible game matrix is

$$A = \begin{bmatrix} 3 & 6 \\ x & 0 \end{bmatrix}.$$

This is the game matrix with the offense as the row player I. The numbers represent the number of yards gained on each play. The first row is run, the second is pass. The first column is defend the run and the second column is defend the pass. Assuming that $x > 0$, find the value of x so that this game has a saddle point in pure strategies.

1.12 Answer: Since $v^- = \max\{3, \min\{x, 0\}\} = 3$ and $v^+ = \min\{\max\{x, 3\}, 6\}$. In order to have a pure saddle, we need $v^+ = 3$ that requires $\max\{x, 3\} = 3$ and so $0 < x \leq 3$. Thus, any $0 < x \leq 3$ will produce a pure saddle at row 1, column 1.

1.13 Suppose A is a 2×3 matrix and A has a saddle point in pure strategies. Show that it must be true that either one column dominates another, or one row dominates the other, or both. Then find a matrix A that is 3×3 and has a saddle point in pure strategies, but no row dominates another and no column dominates another.

1.13 Answer: Denote the game matrix as

$$A = \begin{bmatrix} a_{11} & a_{12} & a_{13} \\ a_{21} & a_{22} & a_{23} \end{bmatrix}.$$

Without loss of generality, we may as well assume that the saddle is at row 1, column 1, $v^- = v^+ = a_{11}$. Since $(1, 1)$ is a saddle, we have

$$a_{i1} \leq a_{11} \leq a_{1j}, i = 2, j = 2, 3.$$

In particular, $a_{21} \leq a_{11}$. If it is also true that $a_{22} \leq a_{12}$ and $a_{23} \leq a_{13}$ then row 1 dominates row 2 and we are done. Thus, we need only to suppose that $a_{22} > a_{12}$. Then, from the saddle point inequalities,

$$a_{22} > a_{12} \geq a_{11} \geq a_{21}.$$

But then $a_{12} \geq a_{11}$ and $a_{22} > a_{21}$ says that column 2 is dominated by column 1.

Now consider the 3×3 matrix

$$A = \begin{bmatrix} 4 & -2 & 0 \\ 3 & 1 & 1 \\ 0 & 2 & \frac{1}{2} \end{bmatrix}.$$

Then $v^- = v^+ = 1$ and there is a saddle at row 2, column 3, but no row or column dominates another.

1.2 The von Neumann Minimax Theorem

Problems

1.14 Let $f(x, y) = x^2 + y^2$, $C = D = [-1, 1]$. Find $v^+ = \min_{y \in D} \max_{x \in C} f(x, y)$ and $v^- = \max_{x \in C} \min_{y \in D} f(x, y)$.

1.14 Answer: We have

$$v^+ = \min_{-1 \leq y \leq 1} \max_{-1 \leq x \leq 1} (x^2 + y^2) = \min_{-1 \leq y \leq 1} (1 + y^2) = 1,$$

and

$$v^- = \max_{-1 \leq x \leq 1} \min_{-1 \leq y \leq 1} (x^2 + y^2) = \max_{-1 \leq x \leq 1} (x^2 + 0) = 1.$$

1.15 Let $f(x, y) = y^2 - x^2$, $C = D = [-1, 1]$.

(a) Find $v^+ = \min_{y \in D} \max_{x \in C} f(x, y)$ and $v^- = \max_{x \in C} \min_{y \in D} f(x, y)$.

1.15.a Answer: The upper value is

$$v^+ = \min_{-1 \le y \le 1} \max_{-1 \le x \le 1} (y^2 - x^2) = \min_{-1 \le y \le 1} y^2 = 0,$$

and the lower value is

$$v^- = \max_{-1 \le x \le 1} \min_{-1 \le y \le 1} (y^2 - x^2) = \max_{-1 \le x \le 1} (0) = 0,$$

since if y chooses first, y can choose $y = x$. Thus, $v^+ = 0, v^- = 0$.

(b) Show that $(0, 0)$ is a pure saddle point for $f(x, y)$.

1.15.b Answer: We have $f(0, 0) = 0$, and

$$f(x, 0) = -x^2 \le f(0, 0) = 0 \le f(0, y) = y^2, \quad \forall \ -1 \le x, y \le 1.$$

1.16 Let $f(x, y) = (x - y)^2, C = D = [-1, 1]$. Find $v^+ = \min_{y \in D} \max_{x \in C} f(x, y)$ and $v^- = \max_{x \in C} \min_{y \in D} f(x, y)$.

1.16 Answer: $v^+ = 1, v^- = 0$. Here is why. For $v^- = \max_x \min_y (x - y)^2$, y can be chosen to be $y = x$ to get a minimum of zero. For $v^+ = \min_y \max_x (x - y)^2$, x wants to be as far away from y as possible. So, if $y < 0$, then $x = 1$, and if $y > 0$, then $x = -1$, so

$$\max_{-1 \le x \le 1} (x - y)^2 = \begin{cases} (1 + y)^2 & \text{if } y > 0; \\ (1 - y)^2 & \text{if } y \le 0. \end{cases}$$

The minimum of this over $y \in [-1, 1]$ is 1, so $v^+ = 1$. You can see this with the Maple commands

```
> f:=y->piecewise(y<0,(1-y)^2,y>=0,(1+y)^2);
> plot(f(y),y=-1..1,view=[-1..1,0..3]);
```

Observe that the function $f(x, y)$ is not concave–convex.

1.17 Show that for any matrix $A_{n \times m}$, the function $f : \mathbb{R}^n \times \mathbb{R}^m \to \mathbb{R}$ defined by $f(x, y) = \sum_{i=1}^n \sum_{j=1}^m a_{ij} x_i y_j = x A y^T$ is convex in $y = (y_1, \ldots, y_m)$ and concave in $x = (x_1, \ldots, x_n)$. In fact, it is **bilinear**.

1.17 Answer: Let $x, \xi \in \mathbb{R}^n$, and $\alpha, \beta \in \mathbb{R}$. Then

$$\begin{aligned} f(\alpha x + \beta \xi, y) &= (\alpha x + \beta \xi) A y^T \\ &= \alpha (x A y^T) + \beta (\xi A y^T) \\ &= \alpha f(x, y) + \beta f(\xi, y). \end{aligned}$$

This proves $x \mapsto f(x, y)$ is linear. Similarly, $y \mapsto f(x, y)$ is also linear.

1.18 Show that for any real-valued function $f = f(x, y), x \in C, y \in D$, where C and D are any old sets, it is always true that

$$\max_{x \in C} \min_{y \in D} f(x, y) \le \min_{y \in D} \max_{x \in C} f(x, y).$$

1.18 Answer: We have for any $x \in C$,

$$\min_{y \in D} f(x, y) \le f(x, y) \Rightarrow \max_{x \in C} \min_{y \in D} f(x, y) \le \max_{x \in C} f(x, y).$$

The right-hand side is a function of y. The left-hand side is a fixed number, v^-, always below the right-hand side for any y. Thus, the minimum of the right-hand side is

$$\max_{x \in C} \min_{y \in D} f(x, y) = v^- \le \min_{y \in D} \max_{x \in C} f(x, y) = v^+.$$

1.19 Verify that if there is $x^* \in C$ and $y^* \in D$ and a real number v so that

$$f(x^*, y) \ge v, \ \forall y \in D, \ \text{ and } \ f(x, y^*) \le v, \ \forall x \in C,$$

then

$$v = f(x^*, y^*) = \max_{x \in C} \min_{y \in D} f(x, y) = \min_{y \in D} \max_{x \in C} f(x, y).$$

1.19 Answer: Under the assumptions,

$$f(x^*, y) \ge v, \ \forall y \in D, \Rightarrow \min_{y \in D} f(x^*, y) \ge v.$$

Then

$$\max_{x \in C} \min_{y \in D} f(x, y) \ge \min_{y \in D} f(x^*, y) \ge v.$$

Similarly,

$$f(x, y^*) \le v, \ \forall \, x \in C \Rightarrow \max_{x \in C} f(x, y^*) \le v,$$

and then

$$\min_{y \in D} \max_{x \in C} f(x, y) \le \max_{x \in C} f(x, y^*) \le v.$$

Since min max $\ge$ max min, we have from

$$\min_{y \in D} \max_{x \in C} f(x, y) \le v \le \max_{x \in C} \min_{y \in D} f(x, y)$$

that we must have equality throughout.

1.20 Suppose that $f : [0, 1] \times [0, 1] \to \mathbb{R}$ is strictly concave in $x \in [0, 1]$ and strictly convex in $y \in [0, 1]$ and continuous. Then there is a point (x^*, y^*) so that

$$\min_{y \in [0,1]} \max_{x \in [0,1]} f(x, y) = f(x^*, y^*) = \max_{x \in [0,1]} \min_{y \in [0,1]} f(x, y).$$

In fact, define $y = \varphi(x)$ as the function so that $f(x, \varphi(x)) = \min_y f(x, y)$. This function is well defined and continuous by the assumptions. Also define the function

$x = \psi(y)$ by $f(\psi(y), y) = \max_x f(x, y)$. The new function $g(x) = \psi(\varphi(x))$ is then a continuous function taking points in $[0, 1]$ and resulting in points in $[0, 1]$. There is a theorem, called the **Brouwer fixed-point theorem**, which now guarantees that there is a point $x^* \in [0, 1]$ so that $g(x^*) = x^*$. Set $y^* = \varphi(x^*)$. Verify that (x^*, y^*) satisfies the requirements of a saddle point for f.

1.20 Answer: Use the definitions of $y^* = \varphi(x^*)$ and $x^* = \psi(y^*)$. We have

$$f(x^*, y^*) = f(\psi(\varphi(x^*)), \varphi(x^*)) = \max_x f(x, \varphi(x^*)) \geq f(x, \varphi(x^*)) = f(x, y^*),$$

for all $x \in [0, 1]$, and

$$f(x^*, y^*) = f(\psi(\varphi(x^*)), \varphi(x^*)) = \min_y f(\psi(\varphi(x^*)), y) \leq f(\psi(\varphi(x^*)), y)$$
$$= f(x^*, y),$$

for all $y \in [0, 1]$. Putting these together we have $f(x, y^*) \leq f(x^*, y^*) \leq f(x^*, y)$ for all $x, y \in [0, 1]$.

1.4 Solving 2 × 2 Games Graphically

Problems

1.21 Following the same procedure as that for player I, look at $E(i, Y), i = 1, 2$ with $Y = (y, 1 - y)$. Graph the lines $E(1, Y) = y + 4(1 - y)$ and $E(2, Y) = 3y + 2(1 - y), 0 \leq y \leq 1$. Now, how does player II analyze the graph to find Y^*?

1.21 Answer: Since player II wants to guarantee that player I gets the smallest maximum, look at the line segments that are the highest and then choose the y^* that gives the smallest maximum payoff. The point of intersection of the two lines is where the y^* will be located and the corresponding horizontal coordinate will be the value of the game.

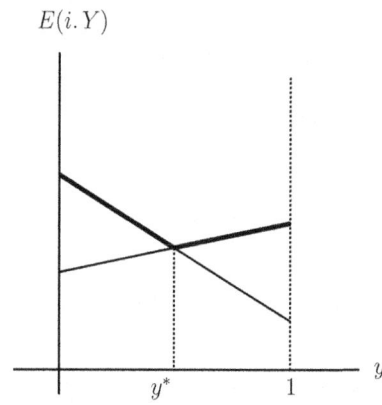

The result for this problem is $3y + 2(1 - y) = y + 4(1 - y)$ which implies $y^* = \frac{1}{2}$ and $v(A) = \frac{5}{2}$.

1.22 Find the value and optimal X^* for the games with matrices

$$\text{(a)} \begin{bmatrix} 1 & 0 \\ -1 & 2 \end{bmatrix} \qquad \text{(b)} \begin{bmatrix} 3 & 1 \\ 5 & 7 \end{bmatrix}$$

What, if anything, goes wrong with (b) if you use the graphical method?

1.22 Answer: For (a), the lines for player I cross where $x - (1 - x) = 2(1 - x)$, which gives $x^* = \frac{3}{4}$. For player II, the lines cross where $y = -y + 2(1 - y)$ which gives $y^* = \frac{1}{2}$. Therefore, the solution of the game in mixed strategies is

$$X^* = \left(\frac{3}{4}, \frac{1}{4}\right), \quad Y^* = \left(\frac{1}{2}, \frac{1}{2}\right), \quad value(A) = \frac{1}{2}.$$

For part (b), the matrix has a saddle point at row 2, column 1, so the optimal strategies won't be mixed strategies. If we didn't spot the pure saddle point and applied the graphical method anyway, we would get for player II, the two lines cross where $3y + 1 - y = 5y + 7(1 - y)$, which gives $y^* = \frac{3}{2} > 1$. The second line lies above the first line for the range $0 \le y \le 1$.

1.23 Curly has two safes, one at home and one at the office. The safe at home is a piece of cake to crack and any thief can get into it. The safe at the office is hard to crack and a thief has only a 15% chance of doing it. Curly has to decide where to place his gold bar (worth 1). On the other hand, if the thief hits the wrong place he gets caught (worth -1 to the thief and $+1$ to Curly). Formulate this as a two-person zero sum matrix game and solve it using the graphical method.

1.23 Answer: Let Curly be the row player and the thief be the column player. The matrix is

I/II	Home	Office
Home	-1	1
Office	1	0.7

The payoff of 0.7 to Curly if he puts the gold bar at the office and the thief hits the office is obtained by calculating $(+1) \times 0.85 + (-1) \times 0.15 = 0.7$. The two lines for the expected payoffs of player I cross where $E((x, 1 - x), 1) = E((x, 1 - x), 2)$, which become $-x + (1 - x) = x + 0.7(1 - x)$. The solution is $x^* = \frac{3}{23}$.

Similarly for player II, the lines cross where $E(1, (y, 1 - y)) = E(2, (y, 1 - y))$, or $-y + (1 - y) = y + 0.7(1 - y)$.

The mixed strategy solution is $X^* = Y^* = \left(\frac{3}{23}, \frac{20}{23}\right)$, $v = \frac{17}{23}$.

1.24 Let z be an unknown number and consider the matrices

$$A = \begin{bmatrix} 0 & z \\ 1 & 2 \end{bmatrix} \quad \text{and} \quad B = \begin{bmatrix} 2 & 1 \\ z & 0 \end{bmatrix}.$$

(a) Find $v(A)$ and $v(B)$ for any z.

1.24.a Answer: No matter what z is the lower value is $v^-(A) = \max\{1, \min\{z, 0\}\} = 1$ and the upper value is $v^+(A) = \min\{1, \max\{2, z\}\} = 1$, so there a saddle at row 2, column 1, and $v(A) = 1$.

Similarly, $v^-(B) = \max\{1, \min\{z, 0\}\} = 1$, and $v^+(B) = \min\{\max\{2, z\}, 1\} = 1$, so $v(B) = 1$ and there is a pure saddle at row 1, column 2.

(b) Now consider the game with matrix $A + B$. Find a value of z so that $v(A + B) < v(A) + v(B)$ and a value of z so that $v(A + B) > v(A) + v(B)$. Find the values of $A + B$ using the graphical method. This problem shows that the value is not a linear function of the matrix.

1.24.b Answer: We know $v(A) = v(B) = 1$ and so $v(A) + v(B) = 2$.

Now pick $z = 3$. Then $A + B = \begin{bmatrix} 2 & 4 \\ 4 & 2 \end{bmatrix}$ and the graphical method gives $X^* = Y^* = \left(\frac{1}{2}, \frac{1}{2}\right)$, and $v(A + B) = 3 > v(A) + v(B)$.

Next pick $z = -1$. Then $A + B = \begin{bmatrix} 2 & 0 \\ 0 & 2 \end{bmatrix}$, and the graphical method gives $X^* = Y^* = \left(\frac{1}{2}, \frac{1}{2}\right)$, and $v(A + B) = 1 < v(A) + v(B)$.

1.25 Suppose that we have the game matrix

$$A = \begin{bmatrix} 13 & 29 & 8 \\ 18 & 22 & 31 \\ 23 & 22 & 19 \end{bmatrix}.$$

Why can this be reduced to $B = \begin{bmatrix} 18 & 31 \\ 23 & 19 \end{bmatrix}$? Now solve the game graphically.

1.25 Answer: Column 2 may be eliminated by dominance: Any $\frac{9}{13} \leq \lambda \leq \frac{3}{4}$ will make

$$13\lambda + 8(1 - \lambda) \leq 29,$$
$$18\lambda + 31(1 - \lambda) \leq 22,$$
$$23\lambda + 19(1 - \lambda) \leq 22.$$

Once column 2 is gone, row 1 may be dropped. Then we apply the graphical method to get $X^* = \left(0, \frac{4}{17}, \frac{13}{17}\right)$ and $Y^* = \left(\frac{12}{17}, 0, \frac{5}{17}\right)$. The value of the game is $v = \frac{37}{17}$.

1.26 Two brothers, Curly and Shemp, inherit a car worth 8000 dollars. Since only one of them can actually have the car, they agree they will present sealed bids to buy the car from the other brother. The brother that puts in the highest sealed bid gets the car. They must bid in 1000 dollar units. If the bids happen to be the same, then they flip a coin to determine ownership and no money changes hands. Curly can bid only up to 5000, while Shemp can bid up to 8000.

Find the payoff matrix with Curly as the row player and the payoffs the expected net gain (since the car is worth 8000). Find v^-, v^+ and use dominance to solve the game.

1.26 Answer: The matrix is 6×9 with the bids $0, 1, \ldots, 5$ for Curly and $0, 1, \ldots, 8$ for Shemp.

Curly/Shemp	0	1	2	3	4	5	6	7	8
0	4	1	2	3	4	5	6	7	8
1	7	4	2	3	4	5	6	7	8
2	6	6	4	3	4	5	6	7	8
3	5	5	5	4	4	5	6	7	8
4	4	4	4	4	4	5	6	7	8
5	3	3	3	3	3	4	6	7	8

For example, if they bid exactly the same amount, the expected payoff to Curly is $\frac{1}{2}8000 = 4000$. If Curly bids 3 and Shemp bids 6 then Shemp gets the car and pays Curly 6000. Shemp's net gain is 2000. This can be thought of as a constant sum game.

It is immediate that $v^- = v^+ = 4$ and the saddle point is Curly should bid 3000 and Shemp should bid 3000, leading to an expected payoff to Curly of 4000. Of course, Shemp also has an expected payoff of 4000.

1.5 Graphical Solution of 2 × m and n × 2 Games

Problems

1.27 In the 2×2 Nim game, we saw that $v^+ = v^- = -1$. Reduce the game matrix using dominance.

1.27 Answer: The game matrix is

Player I/player II	1	2	3	4	5	6
1	1	1	−1	1	1	−1
2	−1	1	−1	−1	1	−1
3	−1	−1	−1	1	1	1

Column 3 immediately dominates every other column. Then it doesn't matter what row player I chooses because the payoff is always -1.

1.28 Consider the matrix game

$$A = \begin{bmatrix} 2 & 0 \\ 0 & 2 \end{bmatrix}.$$

(a) Find $v(A)$ and the optimal strategies.

1.28.a Answer: The graphical method gives $2x = 2(1 - x) \Rightarrow x^* = \frac{1}{2}$. Thus, $v(A) = 1$, $X^* = Y^* = \left(\frac{1}{2}, \frac{1}{2}\right)$.

(b) Show that $X^* = \left(\frac{1}{2}, \frac{1}{2}\right)$, $Y^* = (1, 0)$ is not a saddle point for the game even though it does happen that $E(X^*, Y^*) = v(A)$.

1.28.b Answer: A direct calculation gives $v(A) = 1 = E(X^*, Y^*)$. However, $E(X, Y^*) = 2x$, where $X = (x, 1 - x), 0 \le x \le 1$, and it is not true that $2x < v(A) = 1$ for all x in that range. This means $Y^* = (1, 0)$ is not optimal.

1.29 Use the methods of this section to solve the games:

$$\textbf{(a)} \begin{bmatrix} 4 & -3 \\ -9 & 6 \end{bmatrix}, \qquad \textbf{(b)} \begin{bmatrix} 4 & 9 \\ 6 & 2 \end{bmatrix}, \qquad \textbf{(c)} \begin{bmatrix} -3 & -4 \\ -7 & 2 \end{bmatrix}.$$

1.29 Answer: (a) $X^* = \left(\frac{15}{22}, \frac{7}{22}\right)$; (b) $Y^* = \left(\frac{7}{9}, \frac{2}{9}\right)$; (c) $Y^* = \left(\frac{6}{10}, \frac{4}{10}\right)$.

To see where these come from since they are all 2×2 games without pure saddle points, simply find where the two payoff lines cross for each player.

For (a) we must solve $4x - 9(1 - x) = -3x + 6(1 - x) \Rightarrow x = \frac{15}{22}$, and $4y - 3(1 - y) = -9y + 6(1 - y)$ that gives $y = \frac{9}{22}$. Then plugging in to either payoff line, we get $v = -\frac{3}{22}$. The other parts are similar.

1.30 Use (convex) dominance and the graphical method to solve the game with matrix

$$A = \begin{bmatrix} 0 & 5 \\ 1 & 4 \\ 3 & 0 \\ 2 & 2 \end{bmatrix}.$$

1.30 Answer: We may drop row 4 since it is (weakly) dominated by a convex combination of rows 2 and 3. In fact, $2 \le \frac{1}{2} \cdot 1 + \frac{1}{2} \cdot 3$, and $2 \le \frac{1}{2} \cdot 4 + \frac{1}{2} \cdot 0$.

The lines corresponding to rows 2 and 3 intersect at $y^* = \frac{2}{3}$. That is, $E(2, Y) = y + 4(1 - y) = E(3, Y) = 3y$ and so $y^* = \frac{2}{3}$. The value of the game is $v = 2$. Since we use only rows 2 and 3, it is easy to calculate from the graph that the saddle point for player I is $X^* = \left(0, \frac{1}{2}, \frac{1}{2}, 0\right)$.

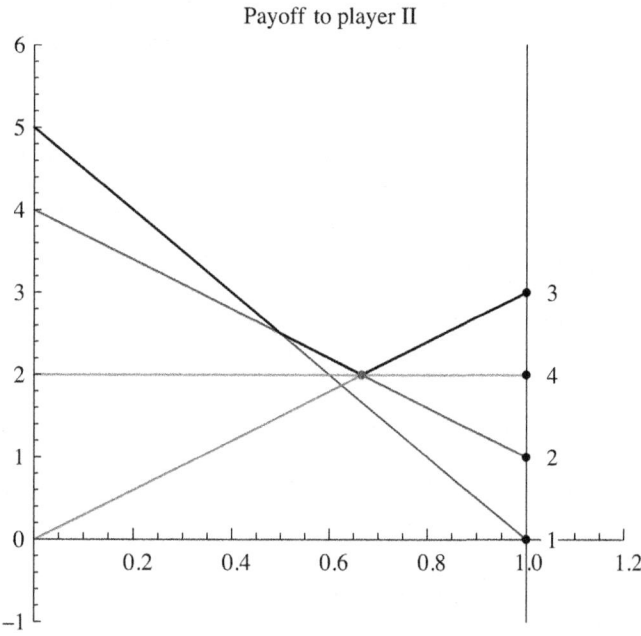

Payoff to player II

1.31 The third column of the matrix

$$A = \begin{bmatrix} 0 & 8 & 5 \\ 8 & 4 & 6 \\ 12 & -4 & 3 \end{bmatrix}$$

is dominated by a convex combination. Reduce the matrix and solve the game.

1.31 Answer: Any $\frac{3}{8} \leq \lambda \leq \frac{7}{16}$ will work for a convex combination of columns 2 and 1. To see why

$$0 \cdot \lambda + 8 \cdot (1 - \lambda) \leq 5 \Rightarrow \frac{3}{8} \leq \lambda,$$

$$8 \cdot \lambda + 4 \cdot (1 - \lambda) \leq 6 \Rightarrow \lambda \leq \frac{1}{2},$$

$$12 \cdot \lambda + (-4) \cdot (1 - \lambda) \leq 3 \Rightarrow \lambda \leq \frac{7}{16}.$$

The reduced matrix is $\begin{bmatrix} 0 & 8 \\ 8 & 4 \\ 12 & -4 \end{bmatrix}$. The graph for this matrix for player II is

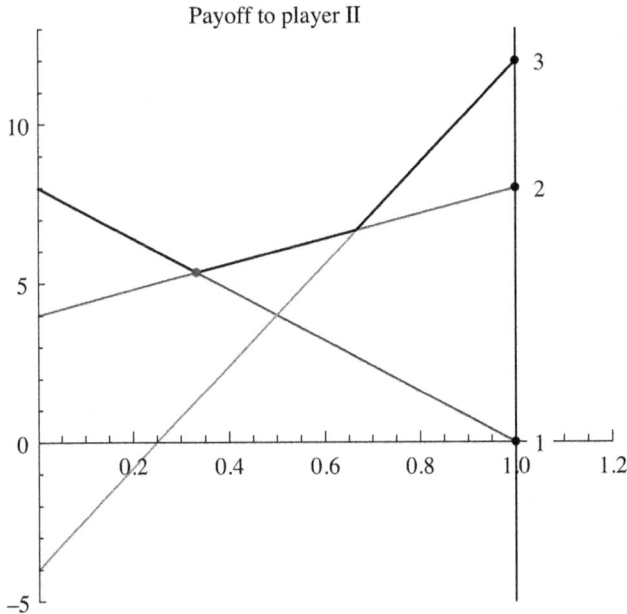

Payoff to player II

The solution of the original game is, therefore, $X^* = \left(\frac{1}{3}, \frac{2}{3}, 0\right)$, $Y^* = \left(\frac{1}{3}, \frac{2}{3}, 0\right)$, and $v(A) = \frac{16}{3}$.

1.32 Four army divisions attack a town along two possible roads. The town has three divisions defending it. A defending division is dug in and hence equivalent to two attacking divisions. Even one division attacking an undefended road captures the town. Each commander must decide how many divisions to attack or defend each road. If the attacking commander captures a road to the town, the town falls. Score 1 to the attacker if the town falls and -1 if it doesn't.

(a) Find the payoff matrix with payoff the attacker's probability of winning the town.

1.32.a Answer: Call the two roads to the town a, b. Each force has strategy comprised of two numbers (i, j), where i is equal to the number of divisions assigned to a and j is equal to the number of divisions assigned to b. The game matrix is then

Attack/Defend	(3, 0)	(2, 1)	(1, 2)	(0, 3)
(4, 0)	-1	-1	1	1
(3, 1)	1	-1	1	1
(2, 2)	1	-1	-1	1
(1, 3)	1	1	-1	1
(0, 4)	1	1	-1	-1

(b) Find the value of the game and the optimal saddle point.

1.32.b Answer: We may reduce the matrix by dominance to the 2×2 game

Attack/Defend	(2, 1)	(1, 2)
(3, 1)	-1	1
(1, 3)	1	-1

Clearly, each row and column should be played with probability $\frac{1}{2}$. The solution of the original game is then

$$X^* = \left(0, \frac{1}{2}, 0, \frac{1}{2}, 0\right), \quad Y^* = \left(0, \frac{1}{2}, \frac{1}{2}, 0\right), \quad v = 0.$$

1.33 Consider the matrix game $A = \begin{bmatrix} a_4 & a_3 & a_3 \\ a_1 & a_6 & a_5 \\ a_2 & a_4 & a_3 \end{bmatrix}$, where $a_1 < a_2 < \cdots < a_5 < a_6$.

Use dominance to solve the game.

1.33 Answer: Column 2 is dominated by column 3, then row 3 is dominated by row 1. The reduced matrix is $\begin{bmatrix} a_4 & a_3 \\ a_1 & a_5 \end{bmatrix}$, which does not have a pure saddle. The resulting solution is obtained where the payoff lines cross; for example, $a_4 x + a_1(1-x) = a_3 x + a_5(1-x) \Rightarrow x^* = \frac{a_5 - a_1}{a_4 + a_5 - a_1 - a_3}$. We have,

$$v(A) = \frac{a_4 a_5 - a_1 a_3}{g}, \quad X^* = \left(\frac{a_5 - a_1}{g}, \frac{a_4 - a_3}{g}, 0\right),$$

$$Y^* = \left(\frac{a_5 - a_3}{g}, 0, \frac{a_4 - a_1}{g}\right),$$

where $g = a_4 + a_5 - a_1 - a_3$.

1.34 Aggie and Baggie are fighting a duel each with one lemon meringue pie starting at 20 paces. They can each choose to throw the pie at either 20 paces, 10 paces, or

0 paces. The probability either player hits the other at 20 paces is $\frac{1}{3}$; at 10 paces it is $\frac{3}{4}$ and at 0 paces it is 1. If they both hit or both miss at the same number of paces, the game is a draw. If a player gets a pie in the face, the score is -1, the player throwing the pie gets $+1$.

Set this up as a matrix game and solve it.

1.34 Answer: The strategies for each player are the number of paces at which to fire the pie. Here is the game matrix.

Aggie/Baggie	20	10	0
20	0	$-\frac{1}{6}$	$-\frac{1}{3}$
10	$\frac{1}{6}$	0	$\frac{1}{2}$
0	$\frac{1}{3}$	$-\frac{1}{2}$	0

For example, suppose Aggie decides to hurl at 10, while Baggie decides she will wait until 0. The expected payoff to Aggie is

Prob(Aggie hits at 10)$(+1)$ + *Prob*(Aggie misses at 10 and Baggie hits at 0)(-1)

$$= \frac{3}{4} - \frac{1}{4} = \frac{1}{2}.$$

Now it is easy to calculate that $v^- = v^+ = 0$, and this zero is achieved at row 2, column 2. Thus, both players should hurl the pie at 10 paces, and the game will be a draw. It is an important part of this problem that both players will choose simultaneously the paces at which they will fire. It would be a different game if they chose by turns.

1.35 Consider the game with matrix

$$A = \begin{bmatrix} -2 & 3 & 5 & -2 \\ 3 & -4 & 1 & -6 \\ -5 & 3 & 2 & -1 \\ -1 & -3 & 2 & 2 \end{bmatrix}.$$

Someone claims that the strategies $X^* = \left(\frac{1}{9}, 0, \frac{8}{9}, 0\right)$ and $Y^* = \left(0, \frac{7}{9}, \frac{2}{9}, 0\right)$ are optimal.

(a) Is that correct? Why or why not?

1.35.a Answer: The given strategies are not optimal because $\max_i E(i, Y) = \frac{31}{9}$ and $\min_j E(X, j) = -\frac{42}{9}$. Another way to see it is to note that since rows 1 and 3 are used with positive probability, Theorem A.8 tells us that if X^* is optimal we must have $E(1, Y^*) = E(3, Y^*) = v$, which you can check easily is not true. Similarly, since columns 2 and 3 are used with positive probability, it must be true that $E(X^*, 2) = E(X^*, 3)$ for X^* to be optimal. But that fails also. Neither X^* nor Y^* are optimal.

(b) If $X^* = \left(\frac{13}{33}, \frac{5}{33}, 0, \frac{15}{33}\right)$ is optimal and $v(A) = -\frac{26}{33}$, find Y^*.

1.35.b Answer: The optimal Y^* is $Y^* = \left(\frac{52}{99}, \frac{8}{33}, 0, \frac{23}{99}\right)$. This is obtained from solving the equations:

$$E(1, Y^*) = -2y_1 + 3y_2 + 5y_3 - 2y_4 = -\frac{26}{33},$$

$$E(2, Y^*) = 3y_1 - 4y_2 + y_3 - 6y_4 = -\frac{26}{33},$$

$$E(4, Y^*) = -y_1 - 3y_2 + 2y_3 + 2y_4 = -\frac{26}{33},$$

$$y_1 + y_2 + y_3 + y_4 = 1.$$

We use the fact that $E(i, Y^*) = v$ if $x_i^* > 0$.

1.36 In the baseball game Example 1.8, it turns out that an optimal strategy for player I, the batter, is given by $X^* = (x_1, x_2, x_3) = \left(\frac{2}{7}, 0, \frac{5}{7}\right)$ and the value of the game is $v = \frac{2}{7}$. It is amazing that the batter should never expect a curveball with these payoffs under this optimal strategy. What is the pitcher's optimal strategy Y^*?

1.36 Answer: The optimal strategy for the pitcher is $Y^* = \left(\frac{5}{7}, \frac{2}{7}, 0\right)$. To see why, since $x_1 > 0, x_3 > 0$, we must have $E(1, Y^*) = E(3, Y^*) = v = \frac{2}{7}$. This leads to the system of equations:

$$0.3y_1 + 0.25y_2 + 0.2y_3 = \frac{2}{7},$$

$$0.28y_1 + 0.3y_2 + 0.33y_3 = \frac{2}{7},$$

$$y_1 + y_2 + y_3 = 1.$$

This system has solution $y_1 = \frac{5}{7}, y_2 = \frac{2}{7}, y_3 = 0$. The pitcher should never throw a slider. The batter will get a hit with probability $\frac{2}{7}$.

1.37 In a football game, we use the matrix $A = \begin{bmatrix} 3 & 6 \\ 8 & 0 \end{bmatrix}$. The offense is the row player. The first row is Run, the second is Pass. The first column is Defend against Run, the second is Defend against the Pass.

(a) Use the graphical method to solve this game.

1.37.a Answer: $X^* = \left(\frac{8}{11}, \frac{3}{11}\right), \quad Y^* = \left(\frac{6}{11}, \frac{5}{11}\right), \quad v(A) = \frac{48}{11}.$

(b) Now suppose the offense gets a better quarterback so the matrix becomes $A = \begin{bmatrix} 3 & 6 \\ 12 & 0 \end{bmatrix}$. What happens?

1.37.b Answer: Solving for the offense, we see that $3x + 12(1 - x) = 6x$, which gives $x^* = \frac{4}{5}$, so the optimal strategy is $X^* = \left(\frac{4}{5}, \frac{1}{5}\right)$. The value of the game is $v(A) = \frac{24}{5}$ and the optimal strategy for the defense is $Y^* = \left(\frac{2}{5}, \frac{3}{5}\right)$. If the offense gets a better quarterback, the team should Run more!

1.38 Two players Reinhard and Carla play a number game. Carla writes down a number 1, 2, or 3. Reinhard chooses a number (again 1, 2, or 3) and guesses that Carla has written down that number. If Reinhard guesses right he wins $1 from Carla; if he guesses wrong, Carla tells him if his number is higher or lower and he gets to guess again. If he is right, no money changes hands but if he guesses wrong he pays Carla $1.

(a) Find the game matrix with Reinhard as the row player and find the upper and lower values. A strategy for Reinhard is of the form

[first guess, guess if low, guess if high].

1.38.a Answer: Carla's strategies are: write down 1, 2, or 3. Reinhard's strategies will be written as $[a, b, c]$, where a is equal to the first number guessed, b is the number guessed if Carla says "lower," c is the number guessed if Carla says "higher." There are 27 strategies, some of which would be stupid, like repeating a number. Eliminating dominated strategies leaves Reinhard with the strategies

$$[1, -, 2], [1, -, 3], [2, 1, 3], [3, 1, -], [3, 2, -].$$

For instance, $[1, -, 2]$ means that Reinhard first guesses 1, Carla will not respond "lower," and then Reinhard responds with 2. The matrix, with Reinhard as the row player, is

I/II	1	2	3
$[1, -, 2]$	1	0	-1
$[1, -, 3]$	1	-1	0
$[2, 1, 3]$	0	1	0
$[3, 1, -]$	0	-1	1
$[3, 2, -]$	-1	0	1

It is easy to calculate the upper value is $v^+ = 1$ and the lower value is $v^- = 0$.

(b) Find the value of the game by first noticing that Carla's strategy 1 and 3 are symmetric as are $[1, -, 2]$, $[3, 2, -]$ and $[1, -, 3]$, $[3, 1, -]$ for Reinhard. Then modify the graphical method slightly to solve.

1.38.b Answer: To find the value of the game reason as follows. Both $[1, -, 2]$ $[3, 2, -]$ $[1, -, 3]$, $[3, 1, -]$ are essentially the same strategy and will, by symmetry, have the same probability of use. Furthermore, 1 and 3 should have the same probability of use by Carla. That means we are looking for optimal strategies of the form $X^* = (x_1, x_2, x_3, x_2, x_1)$ and $Y^* = (y_1, y_2, y_1)$. This requires that

$$2x_1 + 2x_2 + x_3 = 1 \quad \text{and} \quad 2y_1 + y_2 = 1.$$

Since Carla has only two unknowns, we could use the graphical method to solve. In fact, we graph the lines

$$A \cdot \begin{bmatrix} \dfrac{1 - y_2}{2} \\ y_2 \\ \dfrac{1-y_2}{2} \end{bmatrix} = \begin{bmatrix} 0 \\ \dfrac{1 - 3y_2}{2} \\ y_2 \\ \dfrac{1-3y_2}{2} \\ 0 \end{bmatrix}$$

with $0 \le y_2 \le 1$. We see that the intersection of the two lines $y_2 = \frac{1-3y_2}{2}$ gives the optimal strategy for Carla. Since $y_2 = \frac{1}{5}$, we get $Y^* = \left(\frac{2}{5}, \frac{1}{5}, \frac{2}{5}\right)$.

To find the optimal strategies for Reinhard, observe that the optimal strategy for Carla was obtained by using the second and third row of the matrix (or row 3 and row 4). That means we may drop the other rows to find X^*, noting that row 2 and row 4 would be played with the same probability. Automatically, $x_1 = 0$.

We look at the lines for $0 \le x_2 \le 1$,

$$\left(x_2 \quad \frac{1-x_2}{2}\right) \begin{bmatrix} 1 & -1 & 0 \\ 0 & 1 & 0 \end{bmatrix} = \left(x_2 \quad -x_2 + \frac{1-x_2}{2}\right)$$

and they cross at $x_2 = \frac{1}{5}$. This means $x_3 = \frac{3}{5}$ and $X^* = (0, \frac{1}{5}, \frac{3}{5}, \frac{1}{5}, 0)$. The value of the game is $v = \frac{1}{5}$ to Reinhard.

1.39 We have an infinite sequence of numbers $0 < a_1 \le a_2 \le a_3 \le \cdots$. Each of two players chooses an integer independent of the other player. If they both happen to choose the same number k, then player I receives a_k dollars from player II. Otherwise, no money changes hand. Assume that $\sum_{k=1}^{\infty} \frac{1}{a_k} < \infty$.

(a) Find the game matrix (it will be infinite). Find v^+, v^-.

1.39.a Answer: The game matrix is

$$A = \begin{bmatrix} a_1 & 0 & 0 & \cdots \\ 0 & a_2 & 0 & \cdots \\ 0 & 0 & a_3 & \cdots \\ \vdots & \vdots & \vdots & \vdots \end{bmatrix}.$$

Then since there is a 0 in every row and all the $a_k's > 0$, the maximum minimum must be $v^- = 0$. Since there is an $a_k > 0$ in every column, the maximum in every column is a_k. The minimum of those is a_1 and so $v^+ = a_1$.

(b) Find the value of the game if mixed strategies are allowed and find the saddle point in mixed strategies. Use Theorem A.8.

1.39.b Answer: We use $E(i, Y^*) \le v \le E(X^*, j)$, $\forall i, j = 1, 2, \ldots$. We get for $X^* = (x_1, x_2, \ldots)$ and $E(X^*, j) = a_j x_j \ge v \Rightarrow x_j \ge \frac{v}{a_j}$. Similarly, for $Y^* = (y_1, y_2, \ldots)$, implies $y_j \le \frac{v}{a_j}$. Adding these inequalities results in

$$1 = \sum_i x_i \ge v \sum_i \frac{1}{a_i} \quad \text{and} \quad 1 = \sum_j y_j \le v \sum_j \frac{1}{a_j},$$

and we conclude that $v = \frac{1}{\sum_i \frac{1}{a_i}}$. We needed the facts that $\sum_i \frac{1}{a_i} < \infty$, and the fact $\sum_i \frac{1}{a_i} \neq 0$, since $a_k > 0$ for all $k = 1, 2, \ldots$.

Next $x_i \ge \frac{v}{a_i} > 0$, and

$$1 = \sum_{i=1}^{\infty} x_i \ge v \sum_{i=1}^{\infty} \frac{1}{a_i} = 1 \Rightarrow \sum_{i=1}^{\infty} \left[x_i - \frac{v}{a_i} \right] = 0$$

which means it must be true, since each term is nonnegative, $x_i = \frac{v}{a_i}, i = 1, 2, \ldots$. Similarly, $X^* = Y^*$, and the components of both optimal strategies are $\frac{v}{a_i}, i = 1, 2, \ldots$.

(c) Assume next that $\sum_{k=1}^{\infty} \frac{1}{a_k} = \infty$. Show that the value of the game is $v = 0$ and every mixed strategy for player I is optimal, but there is no optimal strategy for player II.

1.39.c Answer: Just as in the second part, we have for any integer $n > 1$, $a_j x_j \geq v$ implies

$$1 \geq \sum_{j=1}^{n} x_j \geq v \sum_{j=1}^{n} \frac{1}{a_j}.$$

Then

$$\frac{1}{\sum_{j=1}^{n} \frac{1}{a_j}} \geq v \geq 0.$$

Sending $n \to \infty$ on the left side and using the fact $\sum_{j=1}^{\infty} \frac{1}{a_j} = \infty$, we get $v = 0$. Note that we know ahead of time that $v \geq 0$ since

$$v \geq v^- = \max_{X \in S_\infty} \min_{j=1,2,\ldots} E(X, j) = \max_{X \in S_\infty} \min_{j=1,2,\ldots} x_j a_j \geq 0.$$

Let $X = (x_1, x_2, \ldots)$ be any mixed strategy for player I. Then, it is always true that $E(X, j) = x_j a_j \geq v = 0$ for any column j. By Theorem A.8 this says that X is optimal for player I. On the other hand, if Y^* is optimal for player II, then $E(i, Y^*) = a_i y_i \leq v = 0, i = 1, 2, \ldots$. Since $a_i > 0$, this implies $y_i = 0$ for every $i = 1, 2, \ldots$. But then $Y = (0, 0, \ldots)$ is not a strategy and we conclude player II does not have an optimal strategy. Since the space of strategies in an infinite sequence space is not closed and bounded, we are not guaranteed that an optimal mixed strategy exists by the minimax theorem.

1.40 Show that for any strategy $X = (x_1, \ldots, x_n) \in S_n$ and any numbers $b_1, \ldots, b_n$, it must be that

$$\max_{X \in S_n} \sum_{i=1}^{n} x_i b_i = \max_{1 \leq i \leq n} b_i \quad \text{and} \quad \min_{X \in S_n} \sum_{i=1}^{n} x_i b_i = \min_{1 \leq i \leq n} b_i.$$

1.40 Answer: Let $\max_i b_i = b_k$. Then $\sum_i x_i b_i - b_k = \sum_i x_i (b_i - b_k) = z$ since $\sum_i x_i = 1$. Now $b_i \leq b_k$ for each i, so $z \leq 0$. Its maximum value is achieved by taking $x_k = 1$ and $x_i = 0, i \neq k$. Hence, $\max_X \sum_i x_i b_i - b_k = 0$, which says $\max_X \sum_i x_i b_i = b_k = \max_i b_i$.

1.41 The properties of optimal strategies (Section A.2) show that $X^* \in S_n$ and $Y^* \in S_m$ are optimal if and only if $\min_j E(X^*, j) = \max_i E(i, Y^*)$. The common value will be the value of the game. Verify this.

1.41 Answer: Using Problem 1.40, we have

$$v = \min_{Y \in S_m} \max_{X \in S_n} E(X, Y) = \min_{Y \in S_m} \max_{X \in S_n} \sum_{i=1}^{n} x_i E(i, Y)$$

$$= \min_{Y \in S_m} \max_{1 \le i \le n} E(i, Y)$$

and

$$v = \max_{X \in S_n} \min_{Y \in S_m} E(X, Y) = \max_{X \in S_n} \min_{Y \in S_m} \sum_{j=1}^{m} y_j E(X, j)$$

$$= \max_{X \in S_n} \min_{1 \le j \le m} E(X, j).$$

Now, if (X^*, Y^*) is a saddle then $v = E(X^*, Y^*)$ and

$$\min_{Y \in S_m} \max_{1 \le i \le n} E(i, Y) \le \max_{1 \le i \le n} E(i, Y^*) \le v$$

$$\le \min_{1 \le j \le m} E(X^*, j) \le \max_{X \in S_n} \min_{1 \le j \le m} E(X^*, j).$$

But we have seen that the two ends of this long inequality are the same. We conclude that

$$v = \max_{1 \le i \le n} E(i, Y^*) = \min_{1 \le j \le m} E(X^*, j).$$

Conversely, if $\max_{1 \le i \le n} E(i, Y^*) = \min_{1 \le j \le m} E(X^*, j) = a$, then we have the inequalities

$$E(i, Y^*) \le a \le E(X^*, j), \quad \forall\, i = 1, 2, \ldots, n, \quad j = 1, 2, \ldots, m.$$

This immediately implies that (X^*, Y^*) is a saddle and $a = E(X^*, Y^*) = $ value of the game.

1.42 Show that if (X^*, Y^*) and (X^0, Y^0) are both saddle points for the game with matrix A, then so is (X^*, Y^0) and (X^0, Y^*). In fact, show that (X_λ, Y_β) where $X_\lambda = \lambda X^* + (1 - \lambda)X^0$, $Y_\beta = \beta Y^* + (1 - \beta)Y^0$ and λ, β any numbers in $[0, 1]$, is also a saddle point. Thus if there are two saddle points, there are an infinite number.

1.42 Answer: By definition of saddle

$$E(X^0, Y^*) \le E(X^*, Y^*) \le E(X^*, Y^0)$$

and

$$E(X^*, Y^0) \le E(X^0, Y^0) \le E(X^0, Y^*).$$

Now put them together to get

$$E(X^*, Y^0) \le E(X^0, Y^0) \le E(X^0, Y^*) \le E(X^*, Y^*) \le E(X^*, Y^0)$$

and so all of them are equal. This implies, for example, that (X^*, Y^0) is also a saddle point since

$$E(X, Y^0) \le E(X^0, Y^0) = E(X^*, Y^0) = E(X^*, Y^*) \le E(X^*, Y), \quad \forall X, Y.$$

It is similar to see that (X^*, Y_β) and (X_λ, Y^*) are also saddle points.

Let (X, Y) be arbitrary strategies. Then using the bilinearity of $E(X, Y)$ and what we just showed,

$$E(X_\lambda, Y_\beta) = \lambda E(X^*, Y_\beta) + (1 - \lambda) E(X^0, Y_\beta)$$
$$\le \lambda E(X^*, Y) + (1 - \lambda) E(X^0, Y) = E(X_\lambda, Y), \quad \forall Y \in S_m$$

and

$$E(X_\lambda, Y_\beta) = \beta E(X_\lambda, Y^*) + (1 - \beta) E(X_\lambda, Y^0)$$
$$\ge \beta E(X, Y^*) + (1 - \beta) E(X, Y^0) = E(X, Y_\beta), \quad \forall X \in S_n$$

1.6 Best Response Strategies

Problems

1.43 Consider the game with matrix

$$\begin{bmatrix} 3 & -2 & 4 & 7 \\ -2 & 8 & 4 & 0 \end{bmatrix}.$$

(a) Solve the game.

1.43.a Answer: We will use the graphical method. The graph is

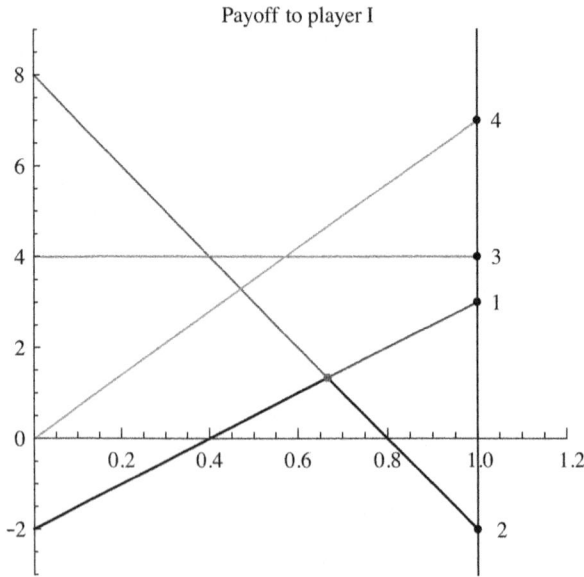

The figure indicates that the two lines determining the optimal strategies will cross where $-2x + 3(1-x) = 8x + (-2)(1-x)$. This tells us that $x^* = \frac{2}{3}$ and then $v = -2\frac{2}{3} + 3\left(1 - \frac{2}{3}\right) = \frac{4}{3}$. We have $X^* = \left(\frac{2}{3}, \frac{1}{3}\right)$.

Next, since the two lines giving the optimal X^* come from columns 1 and 2. we may drop the remaining columns and solve for Y^* using the matrix $\begin{bmatrix} 3 & -2 \\ -2 & 8 \end{bmatrix}$. The two lines for player II cross where $3y - 2(1-y) = -2y + 8(1-y)$, or $y^* = \frac{2}{3}$. Thus, $Y^* = \left(\frac{2}{3}, \frac{1}{3}, 0, 0\right)$.

(b) Find the best response for player I to the strategy $Y = (\frac{1}{4}, \frac{1}{2}, \frac{1}{8}, \frac{1}{8})$.

1.43.b Answer: The best response for player I to $Y = \left(\frac{1}{4}, \frac{1}{2}, \frac{1}{8}, \frac{1}{8}\right)$ is $X = (0, 1)$. The reason is because if we look at the payoff for $X = (x, 1-x)$, we get

$$E(X, Y^*) = X \, AY^{*T} = 4 - \frac{23}{8}x.$$

The maximum of this over $0 \le x \le 1$ occurs when $x = 0$, which means the best response strategy for player I is $X^* = (0, 1)$, with expected payoff 4.

(c) What is II's best response to I's best response?

1.43.c Answer: Player I's best response is $X^* = (0, 1)$, which means always play row II. Player II's best response to that is always play column 1, $Y^* = (1, 0, 0, 0)$, resulting in a payoff to player I of -2.

1.44 An entrepreneur, named Victor, outside Laguna beach can sell 500 umbrellas when it rains and 100 when the sun is out along with 1000 pairs of sunglasses. Umbrellas cost him \$5 each and sell for \$10. Sunglasses wholesale for 2\$ and sell for \$5. The vendor has \$2500 to buy the goods. Whatever he doesn't sell is lost as worthless at the end of the day.

(a) Assume Victor's opponent is the weather set up a payoff matrix with the elements of the matrix representing his net profit.

1.44.a Answer: Weather has two strategies: Rain and Sun. Victor has two strategies: Assume Rain and Assume Sun.

If Victor buys for Rain, he buys 500 umbrellas at \$5 for an investment of 2500. If it does rain, he sells everything and earns $500 \cdot 10 = 5000$ for a net profit of 2500. The remaining numbers are calculated in a similar way.

The payoff matrix to Victor is then

V/W	Rain	Sun
Buy for Rain	2500	-1500
Buy for Sun	-1500	3500

Then using the graphical method we easily see that $X^* = \left(\frac{5}{9}, \frac{4}{9}\right)$, $v(A) = \frac{650}{9}$, and the optimal strategy for the weather is also $Y^* = \left(\frac{5}{9}, \frac{4}{9}\right)$.

(b) Suppose Victor hears the weather forecast and there is a 30% chance of rain. What should he do?

1.44.b Answer: Victor should play his best response strategy to $Y^0 = (0.3, 0.7)$, which is $X^* = (0, 1)$. He should assume that the sun will be out and buy for those conditions, giving him a net expected profit of $-0.3 \times 1500 + 0.7 \times 3000 = \$2000$.

1.45 You're in a bar and a stranger comes to you with a new pickup strategy.[1] The stranger proposes that you each call Heads or Tails. If you both call Heads, the stranger pays you \$3. If both call Tails, the stranger pays you \$1. If the calls aren't a match, then you pay the stranger \$2.

(a) Formulate this as a two-person game and solve it.

1.45.a Answer: The matrix is

You/Stranger	H	T
H	3	-2
T	-2	1

The saddle point is $X^* = \left(\frac{3}{8}, \frac{5}{8}\right) = Y^*$. This is obtained from solving $3x - 2(1-x) = -2x + (1-x)$ and $3y - 2(1-y) = -2y + (1-y)$. The value of this game is $v = -\frac{1}{8}$, so this is definitely a game you should not play.

(b) Suppose the stranger decides to play the strategy $\bar{Y} = (\frac{1}{3}, \frac{2}{3})$. Find a best response and the expected payoff.

1.45.b Answer: If the stranger plays the strategy $\bar{Y} = (\frac{1}{3}, \frac{2}{3})$, then your best response strategy is $\bar{X} = (0, 1)$ that means you will call Tails all the time. The reason is because

$$E(X, \bar{Y}) = (x, 1-x)A\,\bar{Y}^T = -\frac{1}{3}x,$$

which is maximized at $x = 0$. This results in an expected payoff to you of zero. $\bar{Y}$ is not a winning strategy for the stranger.

1.46 Suppose that the batter in the baseball game Example 1.8 hasn't done his homework to learn the percentages in the game matrix. So, he uses the strategy $X^* = \left(\frac{1}{3}, \frac{1}{3}, \frac{1}{3}\right)$. What is the pitcher's best response strategy?

1.46 Answer: Since the expected payoff to player I using X^* is $X^*A = (0.28, 0.2933, 0.27)$, the smallest of these is 0.27. That means the best response for player II is to always play column 3. $Y = (0, 0, 1)$.

1.47 In general, if we have two payoff functions $f(x, y)$ for player I and $g(x, y)$ for player II, suppose that both players want to maximize their own payoff functions with the variables that they control. Then $y^* = y^*(x)$ is a best response of player II to x if

$$g(x, y^*(x)) = \max_y g(x, y), \quad y^* \in \arg\max_y g(x, y).$$

and $x^* = x^*(y)$ is a best response of player I to y if

$$f(x^*(y), y) = \max_x f(x, y), \quad x^* \in \arg\max_x f(x, y).$$

---

[1] This question appeared in a column by Marilyn Vos Savant in Parade Magazine on March 31, 2002.

(a) Find the best responses if $f(x, y) = (C - x - y)x$ and $g(x, y) = (D - x - y)y$, where C and D are constants.

1.47.a Answer: We can find the first derivatives and set to zero:

$$\frac{\partial f}{\partial x} = C - 2x - y = 0 \Rightarrow x^*(y) = \frac{C - y}{2},$$

and

$$\frac{\partial g}{\partial y} = D - x - 2y = 0 \Rightarrow y^*(x) = \frac{D - x}{2}.$$

These are the best responses since the second partials $f_{xx} = g_{yy} = -2 < 0$.

(b) Solve the best responses and show that the solution x^*, y^* satisfies $f(x^*, y^*) \geq f(x, y^*)$ for all x, and $g(x^*, y^*) \geq g(x^*, y)$ for all y.

1.47.b Answer: Best responses are $x = \frac{C-y}{2}$, $y = \frac{D-x}{2}$, which can be solved to give $x^* = \frac{(2C-D)}{3}$, $y^* = \frac{(2D-C)}{3}$.
  Next,

$$f(x^*, y^*) = \frac{(D - 2C)^2}{9} \quad \text{and} \quad f(x, y^*) = \frac{(4C - 2D - 3x)x}{3}.$$

The maximum of $f(x, y^*)$ is $\frac{(D-2C)^2}{9}$, achieved at $x = \frac{2C-D}{3}$, which means it is true that $f(x^*, y^*) \geq f(x, y^*)$ for all x.

CHAPTER TWO

Solution Methods for Matrix Games

2.1 Solution of Some Special Games

Problems

2.1 In a simplified analysis of a football game, suppose that the offense can only choose a pass or run, and the defense can choose only to defend a pass or run. Here is the matrix in which the payoffs are the average yards gained:

		Defense	
Offense		Run	Pass
Run		1	8
Pass		10	0

The offense's goal is to maximize the average yards gained per play. Find $v(A)$ and the optimal strategies using the explicit formulas. Check your answers by solving graphically as well.

2.1 Answer: If we use the 2×2 formulas, we get $\det(A) = -80$, $A^* = \begin{bmatrix} 0 & -8 \\ -10 & 1 \end{bmatrix}$, so

$$X^* = \frac{(1\ 1)A^*}{(1\ 1)A^*\begin{bmatrix}1\\1\end{bmatrix}} = \left(\frac{10}{17}, \frac{7}{17}\right), \quad Y^{*T} = \frac{A^*\begin{bmatrix}1\\1\end{bmatrix}}{(1\ 1)A^*\begin{bmatrix}1\\1\end{bmatrix}} = \left(\frac{8}{17}, \frac{9}{17}\right)$$

$$\text{value}(A) = \frac{\det(A)}{(1\ 1)A^*\begin{bmatrix}1\\1\end{bmatrix}} = \frac{80}{17}.$$

Solutions Manual to Accompany Game Theory: An Introduction, Second Edition. E.N. Barron.
© 2013 John Wiley & Sons, Inc. Published 2013 by John Wiley & Sons, Inc.

If we use the graphical method, the lines cross where $x + 10(1 - x) = 8x \Rightarrow x = \frac{10}{17}$. The rest also checks.

2.2 Suppose that an offensive pass against a run defense now gives 12 yards per play on average to the offense (so the 10 in the previous matrix changes to a 12). Believe it or not, the offense should pass less, not more. Verify that and give a game theory (not math) explanation of why this is so.

2.2 Answer: Using the formulas we get with $A = \begin{bmatrix} 1 & 8 \\ 12 & 0 \end{bmatrix}$, $\det(A) = -96$, and $A^* = \begin{bmatrix} 0 & -8 \\ -12 & 1 \end{bmatrix}$,

$$X^* = \left(\frac{12}{19}, \frac{7}{19} \right), \quad Y^* = \left(\frac{8}{19}, \frac{11}{19} \right), \quad v(A) = \frac{96}{19} = 5.0526.$$

Since $\frac{7}{19} = 0.3684 < \frac{7}{17} = 0.4118$, we see that the passing percentage goes down even though the payoff is increased for passing. The game theory explanation of this is that the defense, knowing that the payoff for a pass is higher, will increase the percentage of time it will defend against the pass (i.e., $\frac{11}{19} > \frac{9}{17}$).

2.3 Does the same phenomenon occur for the defense? To answer this, compare the original game in Problem 2.1 to the new game in which the defense reduces the number of yards per run to 6 instead of 8 when defending against the pass. What happens to the optimal strategies?

2.3 Answer: The new matrix is $A = \begin{bmatrix} 1 & 6 \\ 10 & 0 \end{bmatrix}$ and the optimal strategies using the formulas become

$$X^* = \left(\frac{2}{3}, \frac{1}{3} \right), \quad Y^* = \left(\frac{2}{5}, \frac{3}{5} \right), \quad v(A) = 4.$$

The defense will guard against the pass **more**.

2.4 If the matrix of the game has an inverse, the formulas can be simplified to

$$X^* = \frac{(1\ 1)A^{-1}}{(1\ 1)A^{-1}\begin{bmatrix} 1 \\ 1 \end{bmatrix}}, \quad Y^{*T} = \frac{A^{-1}\begin{bmatrix} 1 \\ 1 \end{bmatrix}}{(1\ 1)A^{-1}\begin{bmatrix} 1 \\ 1 \end{bmatrix}}, \quad v(A) = \frac{1}{(1\ 1)A^{-1}\begin{bmatrix} 1 \\ 1 \end{bmatrix}},$$

where $A^{-1} = \frac{1}{\det(A)}\begin{bmatrix} a_{22} & -a_{12} \\ -a_{21} & a_{11} \end{bmatrix}$. This requires that $\det(A) \neq 0$. Construct an example to show that even if A^{-1} does not exist, the original formulas with A^{-1} replaced by A^* still hold but now $v(A) = 0$.

2.4 Answer: Many examples are possible. Here is one example. Consider $A = \begin{bmatrix} 1 & -1 \\ -1 & 1 \end{bmatrix}$.

This matrix has $\det(A) = 0$ so A^{-1} does not exist. But, $A^* = \begin{bmatrix} 1 & 1 \\ 1 & 1 \end{bmatrix}$, and

$$X^* = \frac{(1\ 1)A^*}{(1\ 1)A^* \begin{bmatrix} 1 \\ 1 \end{bmatrix}} = \left(\frac{1}{2}, \frac{1}{2} \right),$$

$$Y^{*T} = \frac{A^* \begin{bmatrix} 1 \\ 1 \end{bmatrix}}{(1\ 1)A^* \begin{bmatrix} 1 \\ 1 \end{bmatrix}} = \left(\frac{1}{2}, \frac{1}{2} \right)^T,$$

$$v(A) = \frac{\det(A)}{(1\ 1)A^* \begin{bmatrix} 1 \\ 1 \end{bmatrix}} = 0.$$

2.5 Show that the formulas in the theorem are exactly what you would get if you solved the 2×2 game graphically (i.e., using Theorem A.8) under the assumption that no pure saddle point exists.

2.5 Answer: Since there is no pure saddle, the optimal strategies are completely mixed. Then Theorem A.8 tells us for $X^* = (x, 1 - x)$,

$$E(X^*, 1) = E(X^*, 2) \Rightarrow xa_{11} + (1 - x)a_{21} = xa_{12} + (1 - x)a_{22},$$

which gives $x = x^* = \frac{a_{22} - a_{21}}{a_{11} - a_{12} - a_{21} + a_{22}}$. Then

$$1 - x^* = 1 - \frac{a_{22} - a_{21}}{a_{11} - a_{12} - a_{21} + a_{22}} = \frac{a_{11} - a_{12}}{a_{11} - a_{12} - a_{21} + a_{22}}.$$

Since $A^* = \begin{bmatrix} a_{22} & -a_{12} \\ -a_{21} & a_{11} \end{bmatrix}$,

$$X^* = \frac{(1\ 1)A^*}{(1\ 1)A^* \begin{bmatrix} 1 \\ 1 \end{bmatrix}} = \frac{(a_{22} - a_{21}, -a_{12} + a_{11})}{a_{11} - a_{12} - a_{21} + a_{22}}$$

the formulas match.

The rest of the steps are similar to find Y^* and $v(A)$.

2.6 Solve the 2×2 games using the formulas and check by solving graphically:

$$\text{(a)} \begin{bmatrix} 4 & -3 \\ -9 & 6 \end{bmatrix}; \quad \text{(b)} \begin{bmatrix} 8 & 99 \\ 29 & 6 \end{bmatrix}; \quad \text{(c)} \begin{bmatrix} -32 & 4 \\ 74 & -27 \end{bmatrix}.$$

2.6 Answer: (a) $X^* = \left(\frac{15}{22}, \frac{7}{22} \right)$, $Y^* = \left(\frac{9}{22}, \frac{13}{22} \right)$, $v = -\frac{3}{22}$.
  (b) $X^* = \left(\frac{23}{114}, \frac{91}{114} \right)$, $Y^* = \left(\frac{31}{38}, \frac{7}{38} \right)$, $v = \frac{941}{38}$.
  (c) $X^* = \left(\frac{101}{137}, \frac{36}{137} \right)$, $Y^* = \left(\frac{31}{137}, \frac{106}{137} \right)$, $v = -\frac{568}{137}$.

2.7 Give an example to show that when optimal pure strategies exist, the formulas in the theorem won't work.

2.7 Answer: One possible example is $A = \begin{bmatrix} 8 & 9 \\ 6 & 2 \end{bmatrix}$, which obviously has a saddle at payoff 8, so that the actual optimal strategies are $X^* = (1, 0)$, $Y^* = (1, 0)$, $v(A) = 8$. However, if we simply apply the formulas, we get the nonoptimal alleged solution $X^* = (\frac{4}{5}, \frac{1}{5})$, $Y^* = (\frac{7}{5}, -\frac{2}{5})$, which is completely incorrect.

2.8 Let $A = (a_{ij})$, i, $j = 1, 2$. Show that if $a_{11} + a_{22} = a_{12} + a_{21}$, then $v^+ = v^-$ or, equivalently, there are optimal pure strategies for both players. This means that if you end up with a zero denominator in the formula for $v(A)$, it turns out that there had to be a saddle in pure strategies for the game and hence the formulas don't apply from the outset.

2.8 Answer: Consider cases. If $a_{12} > a_{22}$, then because $a_{11} + a_{22} = a_{12} + a_{21}$, it must be that $a_{11} > a_{21}$ since $a_{12} - a_{22} = a_{11} - a_{21} > 0$ so there is a saddle point in the first row. The rest of the cases are similar.

2.2 Invertible Matrix Games

Problems

2.9 Consider Example 2.5 and let $p = 0.5$, $a_1 = 9$, $a_2 = 7$, $a_3 = 6$, $a_4 = 1$. Find $v(A)$, X^*, Y^*, solving the game. Find the value of k^* first.

2.9 Answer: If we calculate $f(k) = \frac{k-p}{G_k}$, we get $f(1) = 4.5$, $f(2) = 5.90$, $f(3) = 5.94$ and $f(4) = 2.46$. This gives us $k^* = 3$. Then we calculate $x_i = \frac{1}{a_i G_3}$ and get

$$X^* = \left(\frac{14}{53}, \frac{18}{53}, \frac{21}{53}, 0 \right) = (0.264, 0.339, 0.3962, 0),$$

and $v(A) = \frac{3-p}{G_3} = 5.943$. Finally, to calculate Y^*, we have

$$y_j = \begin{cases} \frac{1}{p} \left(1 - \frac{k^*-p}{a_j G_{k^*}} \right), & \text{if } j = 1, 2, \ldots, k^*, \\ 0, & \text{otherwise,} \end{cases}$$

which results in $Y^* = (0.6792, 0.30188, 0.01886, 0)$. Thus player I will attack target 3 about 40% of the time, while player II will defend target 3 with probability only about 2%.

2.10 Consider the matrix game

$$A = \begin{bmatrix} 3 & 5 & 3 \\ 4 & -3 & 2 \\ 3 & 2 & 3 \end{bmatrix}.$$

Show that there is a saddle in pure strategies at $(1, 3)$ and find the value. Verify that $X^* = (\frac{1}{3}, 0, \frac{2}{3})$, $Y^* = (\frac{1}{2}, 0, \frac{1}{2})$ is also an optimal saddle point. Does A have an inverse? Find it and use the formulas in the theorem to find the optimal strategies and value.

2.10 Answer: Since $v^- = 3 = v^+$ there is a pure saddle and it is at $(1, 3)$ since that element gives both v^- and v^+. The matrix does have an inverse given by

$$A^{-1} = \begin{bmatrix} \frac{13}{18} & \frac{1}{2} & -\frac{19}{18} \\ \frac{1}{3} & 0 & -\frac{1}{3} \\ -\frac{17}{18} & -\frac{1}{2} & \frac{29}{18} \end{bmatrix}.$$

The formulas then give

$$X^* = \left(\frac{1}{3}, 0, \frac{2}{3}\right), \quad Y^* = \left(\frac{1}{2}, 0, \frac{1}{2}\right), \quad v(A) = 3.$$

It is easy to check these are optimal.

2.11 Solve the following games:

(a) $\begin{bmatrix} 2 & 2 & 3 \\ 2 & 2 & 1 \\ 3 & 1 & 4 \end{bmatrix}$; (b) $\begin{bmatrix} 5 & 4 & 2 \\ 1 & 5 & 3 \\ 2 & 3 & 5 \end{bmatrix}$; (c) $\begin{bmatrix} 4 & 2 & -1 \\ -4 & 1 & 4 \\ 0 & -1 & 5 \end{bmatrix}$.

(d) The following matrix A does not have an inverse. Use dominance to solve it.

$$A = \begin{bmatrix} -4 & 2 & -1 \\ -4 & 1 & 4 \\ 0 & -1 & 5 \end{bmatrix}.$$

2.11.a Answer: The matrix has a saddle at row 1, column 2; formulas do not apply.

2.11.b Answer: The inverse of A is

$$A^{-1} = \begin{bmatrix} \frac{8}{35} & -\frac{1}{5} & \frac{1}{35} \\ \frac{1}{70} & \frac{3}{10} & -\frac{13}{70} \\ -\frac{1}{10} & -\frac{1}{10} & \frac{3}{10} \end{bmatrix}.$$

The formulas then give $X^* = (\frac{1}{2}, 0, \frac{1}{2})$, $Y^* = (\frac{1}{5}, \frac{9}{20}, \frac{7}{20})$, $v = \frac{7}{2}$. There is another optimal $Y^* = (\frac{1}{2}, 0, \frac{1}{2})$ that we can find by noting that the second column is dominated by a convex combination of the first and third columns (with $\lambda = \frac{2}{3}$). This optimal strategy for player II is not obtained using the formulas. It is in fact optimal since

$$A. \begin{bmatrix} \frac{1}{2} \\ 0 \\ \frac{1}{2} \end{bmatrix} = \begin{bmatrix} \frac{7}{2} \\ 2 \\ \frac{7}{2} \end{bmatrix} \leq \begin{bmatrix} \frac{7}{2} \\ \frac{7}{2} \\ \frac{7}{2} \end{bmatrix}$$

Since $E(2, Y) = 2 < v(A)$, we know the second component of an optimal strategy for player I must have $x_2 = 0$.

2.11.c Answer: Since

$$A^{-1} = \begin{pmatrix} \frac{8}{35} & -\frac{1}{5} & \frac{1}{35} \\ \frac{1}{70} & \frac{3}{10} & -\frac{13}{70} \\ -\frac{1}{10} & -\frac{1}{10} & \frac{3}{10} \end{pmatrix}$$

the formulas give the optimal strategies

$$X^* = \left(\frac{11}{19}, \frac{5}{19}, \frac{3}{19}\right), \quad Y^* = \left(\frac{3}{19}, \frac{28}{57}, \frac{20}{57}\right), \quad v = \frac{24}{19}.$$

2.11.d Answer: The matrix has the last column dominated by column 1. Then player I has row 2 dominated by column 1. The resulting matrix is $\begin{bmatrix} -4 & 2 \\ 0 & -1 \end{bmatrix}$. This game can be solved with 2×2 formulas or graphically. The solution for the original game is then

$$X^* = \left(\frac{1}{7}, 0, \frac{6}{7}\right), \quad Y^* = \left(\frac{3}{7}, \frac{4}{7}, 0\right), \quad v(A) = -\frac{4}{7}.$$

2.12 To underscore that the formulas can be used only if you end up with legitimate strategies, consider the game with matrix

$$A = \begin{bmatrix} 1 & 5 & 2 \\ 4 & 4 & 4 \\ 6 & 3 & 4 \end{bmatrix}.$$

(a) Does this matrix have a saddle in pure strategies? If so find it.

2.12.a Answer: We calculate $v^- = 4, v^+ = 4$ so there is a pure saddle at row 2, column 3, $X^* = (0, 1, 0), Y^* = (0, 0, 1)$.

(b) Find A^{-1}.

2.12.b Answer: The inverse is

$$A^{-1} = \frac{1}{10} \begin{bmatrix} 2 & -7 & 6 \\ 4 & -4 & 2 \\ -6 & \frac{27}{2} & -8 \end{bmatrix}.$$

(c) Without using the formulas, find an optimal **mixed** strategy for player II with two positive components.

2.12.c Answer: We know that $v(A) = 4$ and since $X^* = (0, 1, 0)$ satisfies $E(X^*, j) = 4, j = 1, 2, 3$, we know X^* is optimal for player I.

If we assume that there is a mixed optimal strategy for player II, $Y^* = (y_1, y_2, 0)$, with $y_j > 0, j = 1, 2$, then by Theorem A.8 we have the system of equations

$$y_1 + 5y_2 = 4,$$
$$y_1 + y_2 = 1,$$

which has solution $y_1 = \frac{1}{4}, y_2 = \frac{3}{4}$. We verify that $Y^* = (\frac{1}{4}, \frac{3}{4}, 0)$ is optimal for player II. To do that, calculate

$$E(i, Y^*) = \begin{bmatrix} 4 \\ 4 \\ \frac{15}{4} \end{bmatrix}.$$

Since $E(i, Y^*) \le 4, i = 1, 2, 3$, Theorem A.8 tells us that Y^* is optimal for player II. In a similar way, it is easy to verify that both $Y^* = (0, \frac{2}{3}, \frac{1}{3})$ and $Y^* = (\frac{1}{3}, \frac{2}{3}, 0)$ are optimal but $Y = (\frac{2}{3}, 0, \frac{1}{3})$ is not optimal.

(d) Use the formula to calculate Y^*. Why isn't this optimal? What's wrong?

2.12.d Answer: The formulas give $X^* = (0, 1, 0)$ and $Y^* = (\frac{2}{5}, \frac{4}{5}, -\frac{1}{5})$, which is not optimal because it isn't even a strategy. Adding a constant to A will not affect the outcome of the formula for Y^*, which means that the formulas in this case do not work because the optimal strategy is not completely mixed.

2.13 Show that value$(A + b) =$ value$(A) + b$ for any constant b, where by $A + b = (a_{ij} + b)$ is meant A plus the matrix with all entries $= b$. Show also that (X, Y) is a saddle for the matrix $A + b$ if and only if it is a saddle for A.

2.13 Answer: Use the definition of value and strategy to see that

$$v(A + b) = \min_Y \max_X X (A + b) Y^{*T}$$

$$= \min_Y \max_X \left(X A Y^{*T} + \sum_{i=1}^{n} \sum_{j=1}^{m} b x_i y_j \right)$$

$$= \min_Y \max_X \left(X A Y^{*T} + b \sum_{i=1}^{n} x_i \sum_{j=1}^{m} y_j \right)$$

$$= \min_Y \max_X \left(X A Y^{*T} \right) + b$$

$$= v(A) + b.$$

We also see that $X (A + b) Y^T = X A Y^T + b$
If (X^*, Y^*) is a saddle for A, then

$$X (A + b) Y^{*T} = X A Y^{*T} + b \le X^* A Y^{*T} + b = X^* (A + b) Y^{*T}$$

and

$$X^* (A + b) Y^T = X^* A Y^T + b \ge X^* A Y^{*T} + b = X^* (A + b) Y^{*T}$$

for any X, Y. Together this says that X^*, Y^* is also a saddle for $A + b$. The converse is similar.

2.14 Derive the formula for $X = \dfrac{J_n A^{-1}}{J_n A^{-1} J_n^T}$, assuming the game matrix has an inverse. Follow the same procedure as that in obtaining the formula for Y.

2.14 Answer: If we assume that the optimal strategy for player II is $Y^* = (y_1, \ldots, y_n), y_j > 0, j = 1, 2, \ldots, n$, then by Theorem A.8 we have for player I's strategy $X \in S_n$,

$$X A = v(A) J_m \Rightarrow X A A^{-1} = X = v(A) J_n A^{-1},$$

which gives $X = v(A) J_n A^{-1}$. Next, multiply both sides on the right by J_n^T to get

$$X J_n^T = 1 = v(A) J_n A^{-1} J_n^T$$

since $X J_n^T = \sum_{i=1}^n x_i = 1$. Assuming $J_n A^{-1} J_n^T \neq 0$, we get

$$v(A) = \frac{1}{J_n A^{-1} J_n^T} \quad \text{and} \quad X = \frac{J_n A^{-1}}{J_n A^{-1} J_n^T}.$$

2.15 A magic square game has a matrix in which each row has a row sum that is the same as each of the column sums. For instance, consider the matrix

$$A = \begin{bmatrix} 11 & 24 & 7 & 20 & 3 \\ 4 & 12 & 25 & 8 & 16 \\ 17 & 5 & 13 & 21 & 9 \\ 10 & 18 & 1 & 14 & 22 \\ 23 & 6 & 19 & 2 & 15 \end{bmatrix}.$$

This is a magic square of order 5 and sum 65. Find the value and optimal strategies of this game and show how to solve any magic square game.

2.15 Answer: By the invertible matrix theorem we have

$$X^* = \frac{J_5 A^{-1}}{J_5 A^{-1} J_5^T} = \left(\frac{1}{5}, \frac{1}{5}, \frac{1}{5}, \frac{1}{5}, \frac{1}{5} \right).$$

Using the formulas we also have $v(A) = 13$ and $Y^* = X^*$. This leads us to suspect that each row and column, in the general case, is played with probability $\frac{1}{n}$.

Now, let S denote the common sum of the rows and columns in an $n \times n$ magic square. For any optimal strategy X for player I and Y for player II, we must have

$$E(i, Y) \leq v(A) \leq E(X, j), \quad i, j = 1, 2, \ldots, n.$$

Since $E(X, j) = \sum_{i=1}^n a_{ij} x_i \geq v$, adding for $j = 1, 2, \ldots, n$, we get

$$\sum_{j=1}^n \sum_{i=1}^n a_{ij} x_i = \sum_{i=1}^n \sum_{j=1}^n a_{ij} x_i = S \geq n v(A).$$

Similarly, since $E(i, Y) = \sum_{j=1}^n a_{ij} y_j \leq v$, adding for $i = 1, 2, \ldots, n$, we get

$$\sum_{i=1}^n \sum_{j=1}^n a_{ij} y_j = \sum_{j=1}^n \sum_{i=1}^n a_{ij} y_j = S \leq n v(A).$$

Putting them together, we have $v(A) = \frac{S}{n}$.

Finally, we need to verify that $X^* = Y^* = (\frac{1}{n}, \ldots, \frac{1}{n})$ is optimal. That follows immediately from the fact that

$$E(X^*, j) = \sum_{i=1}^{n} a_{ij} \frac{1}{n} = \frac{S}{n} = v(A),$$

and similarly $E(i, Y^*) = v(A)$, $i = 1, 2, \ldots, n$. We conclude using Theorem A.8.

2.16 Why is the hide-and-seek game in Example 2.4 called that? Determine what happens in the hide-and-seek game if there is at least one $a_k = 0$.

2.16 Answer: One interpretation is that an object is hidden in one of n boxes. The seeker gets a_i if she searches box i and finds the object there, otherwise she gets 0.

If any $a_k = 0$, then the matrix is no longer invertible and we cannot use the invertible matrix theorem to solve the game. However, we can use dominance to see that $v(A) = 0$ and the seeker should look in any box $i \neq k$, while the hider should put the object in box k.

2.17 Solve the hide-and-seek game in Example 2.4 with matrix $A_{n \times n} = (a_{ij})$, $a_{ii} = \frac{i}{i+1}$, $i = 1, 2, \ldots, n$, and $a_{ij} = 0$, $i \neq j$. Find the general solution and give the exact solution when $n = 5$.

2.17 Answer: Since $\det A = \frac{n!}{(n+1)!} = \frac{1}{n+1} > 0$ the matrix has an inverse. The inverse matrix is

$$B_{n \times n} = A^{-1} = (b_{ij}), \text{ where } b_{ij} = \begin{cases} \frac{i+1}{i}, & i = j, i = 1, 2, \ldots, n; \\ 0, & i \neq j. \end{cases}$$

Then, setting $q = \sum_{i=1}^{n} \frac{i+1}{i}$ we have

$$v(A) = \frac{1}{q} \quad \text{and} \quad X^* = Y^* = \frac{1}{q} \left(2, \frac{3}{2}, \frac{4}{3}, \ldots, \frac{n+1}{n} \right).$$

If $n = 5$ we have $q = \frac{437}{60}$ so

$$v(A) = \frac{60}{437}, \quad X^* = Y^* = \frac{60}{437} \left(2, \frac{3}{2}, \frac{4}{3}, \frac{5}{4}, \frac{6}{5} \right) = (0.27, 0.21, 0.18, 0.17, 0.16).$$

Note that you search with decreasing probabilities as the box number increases.

2.18 For the game with matrix

$$\begin{bmatrix} -1 & 0 & 3 & 3 \\ 1 & 1 & 0 & 2 \\ 2 & -2 & 0 & 1 \\ 2 & 3 & 3 & 0 \end{bmatrix},$$

we determine that the optimal strategy for player II is $Y = (\frac{3}{7}, 0, \frac{1}{7}, \frac{3}{7})$. We are also told that player I has an optimal strategy X which is completely mixed. Given that the value of the game is $\frac{9}{7}$, find X.

2.18 Answer: The inverse matrix is

$$A^{-1} = \frac{1}{108} \begin{bmatrix} -15 & 9 & 27 & 15 \\ -9 & 27 & -27 & 9 \\ 19 & -33 & 9 & 17 \\ 12 & 36 & 0 & -12 \end{bmatrix}.$$

Since we know that $v = \frac{9}{7}$, we know that

$$X = \frac{9}{7} J_n A^{-1} = \left(\frac{1}{12}, \frac{13}{28}, \frac{3}{28}, \frac{29}{84} \right).$$

2.19 A triangular game is of the form

$$A = \begin{bmatrix} a_{11} & a_{12} & \cdots & a_{1n} \\ 0 & a_{12} & \cdots & a_{2n} \\ \vdots & \vdots & \vdots & \vdots \\ 0 & 0 & \cdots & a_{nn} \end{bmatrix}.$$

(a) Find conditions under which this matrix has an inverse.

2.19.a Answer: The matrix has an inverse if $\det(A) \neq 0$. Since $\det(A) = a_{11} a_{22} \cdots a_{nn}$, no diagonal entries may be zero.

(b) Now consider

$$A = \begin{bmatrix} 1 & -3 & -5 & 1 \\ 0 & 4 & 4 & -2 \\ 0 & 0 & 8 & 3 \\ 0 & 0 & 0 & 50 \end{bmatrix}.$$

Solve the game by finding $v(A)$ and the optimal strategies.

2.19.b Answer: The inverse matrix is

$$A^{-1} = \begin{bmatrix} 1 & \frac{3}{4} & \frac{1}{4} & -\frac{1}{200} \\ 0 & \frac{1}{4} & -\frac{1}{8} & \frac{7}{400} \\ 0 & 0 & \frac{1}{8} & -\frac{3}{400} \\ 0 & 0 & 0 & \frac{1}{50} \end{bmatrix}.$$

Then, we have

$$v(A) = \frac{40}{91}, \quad X^* = v J_5 A^{-1} = \frac{1}{91}(40, 40, 10, 1),$$

$$Y^* = v A^{-1} J_5 = \frac{1}{910}(798, 57, 47, 8),$$

2.20 Another method that can be used to solve a game uses calculus to find the interior saddle points. For example, consider

$$A = \begin{bmatrix} 4 & -3 & -2 \\ -3 & 4 & -2 \\ 0 & 0 & 1 \end{bmatrix}.$$

A strategy for each player is of the form $X = (x_1, x_2, 1 - x_1 - x_2)$, $Y = (y_1, y_2, 1 - y_1 - y_2)$, so we consider the function $f(x_1, x_2, y_1, y_2) = XAY^T$. Now solve the system of equations

$$f_{x_1} = f_{x_2} = f_{y_1} = f_{y_2} = 0$$

to get X^* and Y^*. If these are legitimate completely mixed strategies, then you can verify that they are optimal and then find $v(A)$. Carry out these calculations for A and verify that they give optimal strategies.

2.20 Answer: We calculate

$$f(x_1, x_2, y_1, y_2) = 1 - y_1 + x_1 + x_1(7y_1 - 3) - y_2 + x_2(7y_2 - 3).$$

We take derivatives and get the system of equations

$$7y_1 - 3 = 0,$$
$$7y_2 - 3 = 0,$$
$$7x_1 - 1 = 0,$$
$$7x_2 - 1 = 0,$$

which has solution

$$X^* = \left(\frac{1}{7}, \frac{1}{7}, \frac{5}{7}\right), \qquad Y^* = \left(\frac{3}{7}, \frac{3}{7}, \frac{1}{7}\right).$$

Then

$$v(A) = f\left(\frac{1}{7}, \frac{1}{7}, \frac{3}{7}, \frac{3}{7}\right) = \frac{1}{7}.$$

Since $E(X^*, j) = \frac{1}{7} = v$, $j = 1, 2, 3$, and $E(i, Y^*) = \frac{1}{7} = v$, $i = 1, 2, 3$, Theorem A.8 tells us that we have solved the game.

2.21 Consider the Cat versus Rat game (Example 1.3). The game matrix is 16×16 and consists of 1s and 0s, but the matrix can be considerably reduced by eliminating dominated rows and columns. The game reduces to the 3×3 game

Cat/Rat	abcj	aike	hlia
dcbi	1	0	1
djke	0	1	1
ekjd	1	1	0

Now solve the game.

2.21 Answer: This is an invertible matrix with inverse

$$A^{-1} = \frac{1}{2} \begin{bmatrix} 1 & -1 & 1 \\ -1 & 1 & 1 \\ 1 & 1 & -1 \end{bmatrix}.$$

We may use the formulas to get $value(A) = \frac{2}{3}$, $X^* = Y^* = (\frac{1}{3}, \frac{1}{3}, \frac{1}{3})$, and the cat and rat play each of the rows and columns of the reduced game with probability $\frac{1}{3}$. This corresponds to *dcbi*, *djke*, *ekjd* for the cat, each with probability $\frac{1}{3}$ and *abcj*, *aike*, *hlia* each with probability $\frac{1}{3}$ for the rat. Equivalent paths exist for eliminated rows and columns.

2.22 In tennis, two players can choose to hit a ball left, center, or right of where the opposing player is standing. Name the two players I and II and suppose that I hits the ball, while II anticipates where the ball will be hit. Suppose that II can return a ball hit right 90% of the time, a ball hit left 60% of the time, and a ball hit center 70% of the time. If II anticipates incorrectly, she can return the ball only 20% of the time. Score a return as $+1$ and a ball not returned as -1. Find the game matrix and the optimal strategies.

2.22 Answer: The game matrix with player I as the row player is

I/II	Right	Left	Center
Right	-0.8	0.6	0.6
Left	0.6	-0.2	0.6
Center	0.6	0.6	-0.4 .

For example, if player I hits the ball Left, and player II anticipates Left, the expected payoff to player I is $0.6(-1) + 0.4(+1) = -0.2$. If player I hits Right, and II anticipates Right, the payoff is $0.9(-1) + 0.1(+1) = -0.8$. If player I hits Center and II anticipates Center, the expected payoff to I is $0.7(-1) + 0.3(+1) = -0.4$. Finally, if I hits the ball and II anticipates incorrectly, the payoff to I is $0.2(-1) + 0.8(+1) = 0.6$.

The matrix has an inverse given by

$$A^{-1} = \frac{1}{109} \begin{bmatrix} -35 & 75 & 60 \\ 75 & -5 & 105 \\ 60 & 105 & -25 \end{bmatrix}.$$

By the invertible formulas we have $v(A) = 0.2626$, and optimal strategies $X^* = (0.24, 0.42, 0.34) = Y^*$. Player I should aim for the center 42% of the time and II should anticipate a ball to the center 42% of the time (not 100% of the time).

2.3 Symmetric Games

Problems

2.23 Find the matrix for a noisy Hamilton–Burr duel and solve the game.

2.23 Answer: The duelists shoot at 10, 6, or 2 paces with accuracies 0.2, 0.4, 1.0, respectively. The accuracies are the same for players. The game matrix becomes

B/H	10	6	2
10	0	−0.6	−0.6
6	0.6	0	−0.2
2	0.6	0.2	0

For example, if Burr decides to fire at 10 paces, and Hamilton fires at 10 also, then the expected payoff to Burr is zero since the duelists have the same accuracies and payoffs (it's a draw if they fire at the same time). If Burr decides to fire at 10 paces and Hamilton decides to not fire, then the outcome depends only on whether or not Burr kills Hamilton at 10. If Burr misses, he dies. The expected payoff to Burr is

$$Prob(B \text{ kills } H \text{ at } 10)(+1) + Prob(B \text{ misses})(-1)$$
$$= 0.2 - 0.8 = -0.6.$$

Similarly, Burr fires at 2 paces and Hamilton fires at 6 paces, then Hamilton's survival depends on whether or not he missed at 6. The expected payoff to Burr is

$$Prob(H \text{ kills } B \text{ at } 6)(-1) + Prob(H \text{ misses } B \text{ at } 6)(+1)$$
$$= -0.4 + 0.6 = 0.2.$$

Calculating the lower value we see that $v^- = 0$ and the upper value is $v^+ = 0$, both of which are achieved at row 3, column 3. We have a pure saddle point and both players should wait until 2 paces to shoot. Makes perfect sense that a duelist would not risk missing.

2.24 Each player displays either one or two fingers and simultaneously guesses how many fingers the opposing player will show. If both players guess correctly or both incorrectly, the game is a draw. If only one guesses correctly, that player wins an amount equal to the total number of fingers shown by both players. Each pure strategy has two components: the number of fingers to show and the number of fingers to guess. Find the game matrix and the optimal strategies.

2.24 Answer: A strategy is a pair (a, b) where $a = 1, 2$, $b = 1, 2$, and a is the number of fingers to show, while b is the number of fingers to guess. The game matrix with the rules of the problem becomes

I/II	$(1, 1)$	$(1, 2)$	$(2, 1)$	$(2, 2)$
$(1, 1)$	0	2	−3	0
$(1, 2)$	−2	0	0	3
$(2, 1)$	3	0	0	−4
$(2, 2)$	0	−3	4	0

This is a symmetric game with solution $v = 0$ and an infinite number of saddle points. The extreme saddle points are

$$X^* = Y^* = \left(0, \frac{4}{7}, \frac{3}{7}, 0\right) \quad \text{and} \quad X^* = Y^* = \left(0, \frac{3}{5}, \frac{2}{5}, 0\right).$$

To see this, we have

$$XA \geq (v, v, v, v), v = 0 \Rightarrow (1) : -2x_2 + 3x_3 \geq 0 \Rightarrow x_3 \geq \frac{2}{3}x_2$$

$$(2) : 2x_1 - 3x_4 \geq 0 \Rightarrow x_1 \geq \frac{3}{2}x_4$$

$$(3) : -3x_1 + 4x_4 \geq 0 \Rightarrow x_4 \geq \frac{3}{4}x_1$$

$$(4) : 3x_2 - 4x_3 \geq 0 \Rightarrow x_2 \geq \frac{4}{3}x_3.$$

Now, by (2) and (3),

$$x_1 \geq \frac{3}{2}x_4 \geq \frac{3}{2}\frac{3}{4}x_1 = \frac{9}{8}x_1 \Rightarrow x_1 = 0,$$

and

$$x_4 \geq \frac{3}{4}x_1 \geq \frac{3}{4}\frac{3}{2}x_4 = \frac{9}{8}x_4 \Rightarrow x_4 = 0.$$

Thus $x_2 + x_3 = 1$ and by (1) and (4),

$$x_3 \geq \frac{2}{3}x_2 = \frac{2}{3}(1 - x_3) \Rightarrow x_3 \geq \frac{2}{5}$$

and

$$x_2 \geq \frac{4}{3}x_3 = \frac{4}{3}(1 - x_2) \Rightarrow x_2 \geq \frac{4}{7}.$$

Combining,

$$1 = x_2 + x_3 \Rightarrow \frac{3}{7} \geq x_3 \geq \frac{2}{5}.$$

Similarly,

$$1 = x_2 + x_3 \Rightarrow \frac{3}{5} \geq x_2 \geq \frac{4}{7}.$$

We conclude that $X^* = (0, x, 1 - x, 0)$, $\frac{3}{5} \geq x \geq \frac{4}{7}$, is optimal for player I. By symmetry, it is also optimal for player II.

2.25 This exercise shows that symmetric games are more general than they seem at first and in fact this is the main reason they are important. Assuming that $A_{n \times m}$ is **any** payoff matrix with $value(A) > 0$, define the matrix B that will be of size $(n + m + 1) \times (n + m + 1)$, by

$$B = \begin{bmatrix} 0 & A & -\vec{1} \\ -A^T & 0 & \vec{1} \\ \vec{1} & -\vec{1} & 0 \end{bmatrix}.$$

The notation $\vec{1}$, for example, in the third row and first column, is the $1 \times n$ matrix consisting of all 1s. B is a skew symmetric matrix and it can be shown that if $P = (p_1, \ldots, p_n, q_1, \ldots, q_m, \gamma)$ is an optimal strategy for matrix B, then, setting

$$b = \sum_{i=1}^{n} p_i = \sum_{j=1}^{m} q_j > 0, \quad x_i = \frac{p_i}{b}, \quad y_j = \frac{q_j}{b},$$

we have $X = (x_1, \ldots, x_n)$, $Y = (y_1, \ldots, y_m)$ as a saddle point for the game with matrix A. In addition, $value(A) = \frac{\gamma}{b}$. The converse is also true. Verify all these points with the matrix

$$A = \begin{bmatrix} 5 & 2 & 6 \\ 1 & \frac{7}{2} & 2 \end{bmatrix}.$$

2.25 Answer: The matrix B is given by

$$B = \begin{bmatrix} 0 & 0 & 5 & 2 & 6 & -1 \\ 0 & 0 & 1 & \frac{7}{2} & 2 & -1 \\ -5 & -1 & 0 & 0 & 0 & 1 \\ -2 & -\frac{7}{2} & 0 & 0 & 0 & 1 \\ -6 & -2 & 0 & 0 & 0 & 1 \\ 1 & 1 & -1 & -1 & -1 & 0 \end{bmatrix}.$$

Now we use the method of symmetry to solve the game with matrix B. Obviously, $v(B) = 0$, and if we let $P = (p_1, \ldots, p_6)$ denote the optimal strategy for player I, we have

$$P\,A \geq (0, 0, 0, 0, 0, 0) \Rightarrow$$

$$-5p_3 - 2p_4 - 6p_5 + p_6 \geq 0$$

$$-p_3 - \frac{7}{2}p_4 - 2p_5 + p_6 \geq 0$$

$$5p_1 + p_2 - p_6 \geq 0$$

$$2p_1 + \frac{7}{2}p_2 - p_6 \geq 0$$

$$6p_1 + 2p_2 - p_6 \geq 0$$

$$-p_1 - p_2 + p_3 + p_4 + p_5 \geq 0$$

$$p_1 + p_2 + p_3 + p_4 + p_5 + p_6 = 1.$$

This system of inequalities has one and only one solution given by

$$P = \left(\frac{5}{53}, \frac{6}{53}, \frac{3}{53}, \frac{8}{53}, 0, \frac{31}{53} \right).$$

As far as the solution of the game with matrix B we have $v(B) = 0$, and P is the optimal strategy for both players I and II.

Now to find the solution of the original game, we have (now with a new meaning for the symbols p_i) $n = 2, m = 3$, and

$$p_1 = \frac{5}{53}, \quad p_2 = \frac{6}{53}, \quad q_1 = \frac{3}{53}, \quad q_2 = \frac{8}{53}, \quad q_3 = 0, \quad \gamma = \frac{31}{53}.$$

Then $b = p_1 + p_2 = \frac{11}{53}, b = q_1 + q_2 + q_3 = \frac{3+8}{53} = \frac{11}{53}$. Now define the strategy for the original game with A,

$$X^* = (x_1, x_2), \quad x_1 = \frac{p_1}{b} = \frac{5}{11}, \quad x_2 = \frac{p_2}{b} = \frac{6}{11}$$

and

$$Y^* = (y_1, y_2, y_3), \quad y_j = \frac{q_j}{b} \Rightarrow y_1 = \frac{3}{11}, \quad y_2 = \frac{8}{11}, \quad y_3 = 0.$$

Finally, $v(A) = \frac{\gamma}{b} = \frac{\frac{31}{53}}{\frac{11}{53}} = \frac{31}{11}$.

The problem is making the statement that this is indeed the solution of our original game. We verify that by calculating using the original matrix A,

$$E(i, Y^*) = \frac{31}{11} = v(A), \; i = 1, 2 \quad \text{and} \quad E(X^*, j) = \frac{31}{11}, j = 1, 2,$$

$$E(X^*, 3) = \frac{42}{11} \geq v(A)$$

By Theorem A.8, we conclude that X^*, Y^* is optimal and $v(A) = \frac{31}{11}$.

2.4 Matrix Games and Linear Programming

Problems

2.26 Use any method to solve the games with the following matrices:

(a) $\begin{bmatrix} 0 & 3 & 3 & 2 & 4 \\ 4 & 4 & 3 & 1 & 4 \end{bmatrix}$; (b) $\begin{bmatrix} 4 & 4 & -4 & -1 \\ -4 & -2 & 4 & 4 \\ 2 & -4 & -1 & -5 \\ -3 & 1 & 0 & -4 \end{bmatrix}$; (c) $\begin{bmatrix} 2 & -5 & 3 & 0 \\ -4 & -5 & -5 & -6 \\ 3 & -4 & -1 & -2 \\ 0 & 4 & 1 & 3 \end{bmatrix}$.

2.26 Answer:
(a) $X^* = (\frac{3}{5}, \frac{2}{5})$, $Y^* = (\frac{1}{5}, 0, 0, \frac{4}{5}, 0)$, $v = \frac{8}{5}$.
(b) $X^* = (\frac{21}{53}, \frac{24}{53}, \frac{8}{53}, 0)$, $Y^* = (\frac{23}{53}, \frac{4}{53}, \frac{26}{53}, 0)$, $v = \frac{4}{53}$;
(c) $X^* = (\frac{9}{55}, 0, \frac{1}{5}, \frac{7}{11})$, $Y^* = (\frac{34}{55}, \frac{2}{11}, \frac{1}{5}, 0)$, $v = \frac{51}{55}$.

2.27 Solve this game using the two methods of setting it up as a linear program:

$$\begin{bmatrix} -2 & 3 & 3 & 4 & 1 \\ 3 & -2 & -5 & 2 & 4 \\ 4 & -5 & -1 & 4 & -1 \\ 2 & -4 & 3 & 4 & -3 \end{bmatrix}.$$

2.27 Answer: For Method 1, the LP problem for each player is

Player I

Min $z_1 = p_1 + p_2 + p_3 + p_4$
$4p_1 + 9p_2 + 10p_3 + 8p_4 \geq 1$
$9p_1 + 4p_2 + p_3 + 2p_4 \geq 1$
$9p_1 + p_2 + 5p_3 + 9p_4 \geq 1$
$10p_1 + 8p_2 + 10p_3 + 10p_4 \geq 1$
$7p_1 + 10p_2 + 5p_3 + 3p_4 \geq 1$
$p_i \geq 0$

Player II

Max $z_2 = q_1 + q_2 + q_3 + q_4 + q_5$
$4q_1 + 9q_2 + 9q_3 + 10q_4 + 7q_5 \leq 1$
$9q_1 + 4q_2 + q_3 + 8q_4 + 10q_5 \leq 1$
$10q_1 + q_2 + 5q_3 + 10q_4 + 5q_5 \leq 1$
$8q_1 + 2q_2 + 9q_3 + 10q_4 + 3q_5 \leq 1$
$q_j \geq 0$

For Method 2, the LP problems are as follows:

Player I

Max v
$-2x_1 + 3x_2 + 4x_3 + 2x_4 \geq v$
$3x_1 - 2x_2 - 5x_3 - 4x_4 \geq v$
$3x_1 - 5x_2 - x_3 + 3x_4 \geq v$
$4x_1 + 2x_2 + 4x_3 + 4x_4 \geq v$
$x_1 + 4x_2 - x_3 - 3x_4 \geq v$
$x_i \geq 0$
$\sum x_i = 1$

Player II

Min v
$-2y_1 + 3y_2 + 3y_3 + 4y_4 + y_5 \leq v$
$3y_1 - 2y_2 - 5y_3 + 2y_4 + 4y_5 \leq v$
$4y_1 - 5y_2 - y_3 + 4y_4 - y_5 \leq v$
$2y_1 - 4y_2 + 3y_3 + 4y_4 - 3y_5 \leq v$
$y_j \geq 0$
$\sum y_j = 1$

Both methods result in the solution

$$X^* = \left(\frac{47}{82}, \frac{10}{41}, \frac{15}{82}, 0 \right), \quad Y^* = \left(\frac{22}{41}, \frac{14}{41}, \frac{5}{41}, 0, 0 \right), \quad v = \frac{13}{41}.$$

In each of the following problems, set up the payoff matrices and solve the games using the linear programming method with both formulations, that is, both with and without transforming variables. You may use Maple/Mathematica or any software you want to solve these linear programs.

2.28 Consider the Submarine versus Bomber game. The board is a 3×3 grid.

1	2	3
4	5	6
7	8	9

3×3 Submarine-bomber game.

A submarine (which occupies two squares) is trying to hide from a plane that can deliver torpedoes. The bomber can fire one torpedo at a square in the grid. If it is occupied by a part of the submarine, the sub is destroyed (score 1 for the bomber). If the bomber fires at an unoccupied square, the sub escapes (score 0 for the bomber). The bomber should be the row player.

(a) Formulate this game as a matrix and solve it. (Hint: there are 12 pure strategies for the sub and 9 for the bomber.)

2.28.a Answer: The matrix is

Bomber/Sub	12	23	36	69	98	87	74	41	25	65	85	45
1	1	0	0	0	0	0	0	1	0	0	0	0
2	1	1	0	0	0	0	0	0	1	0	0	0
3	0	1	1	0	0	0	0	0	0	0	0	0
4	0	0	0	0	0	0	1	1	0	0	0	1
5	0	0	0	0	0	0	0	0	1	1	1	1
6	0	0	1	1	0	0	0	0	0	1	0	0
7	0	0	0	0	0	1	1	0	0	0	0	0
8	0	0	0	0	1	1	0	0	0	0	1	0
9	0	0	0	1	1	0	0	0	0	0	0	0

We use Method 2 Linear Programming to solve this game with Mathematica. The linear programming problems are as follows:

Player I

Max v

$x_1 + x_2 \geq v$
$x_2 + x_3 \geq v$
$x_3 + x_6 \geq v$
$x_6 + x_9 \geq v$
$x_8 + x_9 \geq v$
$x_7 + x_8 \geq v$
$x_4 + x_7 \geq v$
$x_1 + x_4 \geq v$
$x_2 + x_5 \geq v$
$x_5 + x_6 \geq v$
$x_5 + x_8 \geq v$
$x_4 + x_5 \geq v$
$x_i \geq 0$
$\sum x_i = 1$

Player II

Min v

$y_1 + y_8 \leq v$
$y_1 + y_2 + y_9 \leq v$
$y_2 + y_3 \leq v$
$y_7 + y_8 + y_{12} \leq v$
$y_9 + y_{10} + y_{11} + y_{12} \leq v$
$y_3 + y_4 + y_{10} \leq v$
$y_6 + y_7 \leq v$
$y_5 + y_6 + y_{11} \leq v$
$y_4 + y_5 \leq v$
$y_j \geq 0$
$\sum y_j = 1$

This problem has solution $v(A) = \frac{1}{4}$ with optimal strategies

$$X^* = \left(0, \frac{1}{4}, 0, \frac{1}{4}, 0, \frac{1}{4}, 0, \frac{1}{4}, 0\right),$$

and

$$Y^* = \left(0, \frac{1}{4}, 0, \frac{1}{4}, 0, \frac{1}{4}, 0, 0, 0, 0, 0, \frac{1}{4}\right).$$

(b) Using symmetry the game can be reduced to a 3×2 game. Find the reduction and then solve the resulting game.

2.28.b Answer: The sub's strategies can be reduce to (S1) hide in a pair of squares that include the center square (5); (S2) hide in a pair which does not include the center square (5). The bomber's strategies can be reduced to (B1) fire at a corner, (B2) fire at a square in the middle of each side, (B3) fire at the center square. The payoff matrix is then reduced to

Bomber/Sub	(S1)	(S2)
(B1)	0	$\frac{1}{4}$
(B2)	$\frac{1}{4}$	$\frac{1}{4}$
(B3)	1	0

For example, if we play $B1$ against $S2$ this means that the bomber will fire at 1 of 4 corners while the sub will hide along the edges because of the configuration of the grid, that will result in a one in four chance of hitting the sub. That's where the $\frac{1}{4}$ comes from.

This reduced game has a pure saddle point at $(B2, S2)$. Then $v(A) = \frac{1}{4}$; the bomber fires at one of the four middle side squares with equal probability; the sub hides in one of the eight locations that does not include the center square with equal probability. This results in the same strategies without using symmetry but in compact form.

2.29 There are two presidential candidates, John and Dick, who will choose which states they will visit to garner votes. Suppose that there are four states that are in play, but candidate John only has the money to visit three states. If a candidate visits a state, the gain (or loss) in poll numbers is indicated as the payoff in the matrix

John/Dick	State 1	State 2	State 3	State 4
State 1	1	-4	6	-2
State 2	-8	7	2	1
State 3	11	-1	3	-3

Solve the game.

2.29 Answer: We cannot use the invertible matrix theorem, but we can use linear programming. To set this up we have, in Maple,

```
> with(LinearAlgebra):with(simplex):
> A:=Matrix([[1,-4,6,-2],[-8,7,2,1],[11,-1,3,-3]]);
> X:=Vector(3,symbol= x);
> B:=Transpose(X).A;
> cnstx:=seq(B[j] >=v,j=1..4),seq(x[i] >= 0,i=1..3),
                        add(x[i],i=1..3)=1;
> maximize(v,cnstx);
```

This results in $X^* = (0, \frac{14}{23}, \frac{9}{23})$ and $v(A) = -\frac{13}{23}$. It looks like John should not bother with State 1, and spend most of his time in State 2. For Dick, the problem is

```
> Y:=<y[1],y[2],y[3],y[4]>;
> B:=A.Y;
> cnsty:=seq(B[j]<=w,j=1..3),seq(y[j] >=0,j=1..4),
                        add(y[j],j=1..4)=1;
> minimize(w,cnsty);
```

The result is $Y^* = (\frac{4}{23}, 0, 0, \frac{19}{23})$. Dick should skip States 2 and 3 and put everything into States 1 and 4. John should think about raising more money to be competitive in State 4 because otherwise, John suffers a net loss of $-\frac{13}{23}$.

2.30 Assume that we have a silent duel but the choice of a shot may be taken at 10, 8, 6, 4, or 2 paces. The accuracy, or probability of a kill is 0.2, 0.4, 0.6, 0.8, and 1, respectively, at the paces. Set up and solve the game.

2.30 Answer: We indicate how to solve this in Maple. First define the accuracy function:

```
p2 := x->piecewise(x = 0, .2, x = .2, .4, x = .4, .6,
x = .6, .8, x = .8, 1) ;
```

Next define the payoff function:

```
u:=(x,y)->piecewise(x<y,
1*p1(x)+(-1)*(1-p1(x))*p2(y)+(0)*(1-p1(x))*(1-p2(y)),
x>y,(-1)*p2(y)+(1)*(1-p2(y))*p2(x)+0*(1-p2(y))*(1-p1(x)),
x=y,0*p1(x)*p2(x)+(1)*p1(x)*(1-p2(x))+(-1)*(1-p1(x))*p2(x)
+0*(1-p1(x))*(1-p2(x)));
```

Finally, set up the matrix:

```
with(LinearAlgebra):
A:=Matrix([[u1(0,0),u1(0,.2),u1(0,.4),u1(0,.6),u1(0,.8)],
        [u1(.2,0),u1(.2,.2),u1(.2,.4),u1(.2,.6),u1(.2,.8)],
        [u1(.4,0),u1(.4,.2),u1(.4,.4),u1(.4,.6),u1(.4,.8)],
        [u1(.6,0),u1(.6,.2),u1(.6,.4),u1(.6,.6),u1(.6,.8)],
        [u1(.8,0),u1(.8,.2),u1(.8,.4),u1(.8,.6),u1(.8,.8)]]);
```

The game matrix is

$$\begin{bmatrix} 0.0 & -0.12 & -0.28 & -0.44 & -0.6 \\ 0.12 & 0.0 & 0.04 & -0.08 & -0.2 \\ 0.28 & -0.04 & 0.0 & 0.28 & 0.2 \\ 0.44 & 0.08 & -0.28 & 0.0 & 0.6 \\ 0.6 & 0.2 & -0.2 & -0.6 & 0 \end{bmatrix}$$

The game is symmetric and has solution by linear programming given by

$$X^* = \left(0, \frac{5}{11}, \frac{5}{11}, 0, \frac{1}{11}\right) = Y^*, \quad v = 0.$$

2.31 Consider the Cat versus Rat game in Example 1.3. Suppose each player moves exactly three segments. The payoff to the cat is 1 if they meet and 0 otherwise. Find the matrix and solve the game.

2.31 Answer: Each player has 8 possible strategies but many of them are dominated. The reduced matrix is

Cat/Rat	1	2	3
A	1	1	0
B	0	1	1

Referring to the Figure 1.1 in the book, we have the strategies for Rat: $1 = hgf$, $2 = hlk$, $3 = abc$, and for Cat, $A = djk$, $B = ekj$.

The solution is the cat should take small loops and the Rat should stay along the edges of the maze. $X^* = (\frac{1}{2}, \frac{1}{2})$, $Y^* = (\frac{1}{2}, 0, \frac{1}{2})$ and $v = \frac{1}{2}$.

2.32 Drug runners can use three possible methods for running drugs through Florida: small plane, main highway, or backroads. The cops know this, but they can only patrol one of these methods at a time. The street value of the drugs is $100,000 if the drugs reach New York using the main highway. If they use the backroads, they have to use smaller-capacity cars so the street value drops to $80,000. If they fly, the street value is $150,000. If they get stopped, the drugs and the vehicles are impounded, they get fined, and they go to prison. This represents a loss to the drug kingpins of $90,000, by highway, $70,000 by backroads, and $130,000 if by small plane. On the main highway, they have a 40% chance of getting caught if the highway is being patrolled, a 30% chance on the backroads, and a 60% chance if they fly a small plane (all assuming that the cops are patrolling that method). Set this up as a zero sum game assuming that the cops want to minimize the drug kingpins gains, and solve the game to find the best strategies the cops and drug lords should use.

2.32 Answer: The pure strategies are labeled plane (*P*), highway (*H*), roads (*R*), for each player. The drug runner chooses one of those to try to get to New York, and the cops choose one of those to patrol. The game matrix in which the drug runner is the row player becomes

$$A = \begin{bmatrix} -18 & 150 & 150 \\ 100 & 24 & 100 \\ 80 & 80 & 35 \end{bmatrix}.$$

For example, if drug runner plays H and cops patrol H, the drug runner's expected payoff is $(-90)(0.4) + (100)(0.6) = 24$. The saddle point is

$$X^* = (0.144, 0.3183, 0.5376) \quad \text{and} \quad Y^* = (0.4628, 0.3651, 0.1721).$$

The drug runners should use the backroads more than half the time, but the cops should patrol the backroads only about 17% of the time.

2.33 LUC is about to play UIC for the state tennis championship. LUC has two players: A and B, and UIC has three players: X, Y, Z. The following facts are known about the relative abilities of the players: X always beats B; Y always beats A; A always beats Z. In any other match each player has a probability $\frac{1}{2}$ of winning. Before the matches, the coaches must decide on who plays who. Assume that each coach wants to maximize the expected number of matches won (these are singles matches and there are two games so each coach must pick who will play game 1 and who will play game 2. The players do not have to be different in each game).

Set this up as a matrix game. Find the value of the game and the optimal strategies.

2.33 Answer: LUC is the row player and UIC the column player. The strategies for LUC are AA, AB, BA, BB. UIC has strategies XX, XY, ..., indicated in the table

LUC/UIC	XX	XY	YX	XZ	ZX	YY	YZ	ZY	ZZ
AA	0	−1	−1	1	1	−2	0	0	2
AB	−1	0	−2	0	0	−1	−1	1	1
BA	−1	−2	0	0	0	−1	1	−1	1
BB	−2	−1	−1	−1	−1	0	0	0	0

The numbers in the table are obtained from the rules of the game. For example, if LUC plays AB and UIC plays XY, then the expected payoff to LUC on each match is zero and hence on both matches it is zero.

Solving this game with Maple/Mathematica/Gambit, we get the saddle point

$$X^* = \left(\frac{1}{2}, 0, 0, \frac{1}{2}\right), \quad Y^* = \left(\frac{1}{2}, 0, 0, 0, 0, \frac{1}{2}, 0, 0, 0\right)$$

so LUC should play AA and BB with equal probability, and UIC should play XX, YY with equal probability. Z is going to sit out. The value of the game is $v = -1$, and it looks like LUC loses the tournament.

2.34 Player II chooses a number $j \in \{1, 2, \ldots, n\}$, $n \geq 2$, while I tries to guess what it is by guessing an $i \in \{1, 2, \ldots, n\}$. If the guess is correct (i.e., $i = j$) then II pays

+1 to player I. If $i > j$ player II pays player I the amount b^{i-j} where $0 < b < 1$ is fixed. If $i < j$, player I wins nothing.

(a) Find the game matrix.

2.34.a Answer: According to the description of the problem, the matrix is

$$A = \begin{bmatrix} 1 & 0 & 0 & 0 & \cdots & 0 \\ b & 1 & 0 & 0 & \cdots & 0 \\ b^2 & b & 1 & 0 & \cdots & 0 \\ \vdots & \vdots & \vdots & \vdots & \vdots & \vdots \\ b^{n-1} & b^{n-2} & b^{n-3} & b^{n-4} & \cdots & 1 \end{bmatrix}$$

(b) Solve the game.

2.34.b Answer:

Assuming that each row is played with positive probability we may set up the system of equations $X A = v J_n$ to get

$$\begin{aligned} x_1 + b x_2 + \cdots + b^{n-1} x_n &= v \\ x_2 + b x_3 + \cdots + b^{n-2} x_n &= v \\ x_3 + \cdots + b^{n-3} x_n &= v \\ &\vdots \quad \vdots \\ x_{n-1} + b x_n &= v \\ x_n &= v. \end{aligned}$$

Using reverse substitution, we get

$$\begin{aligned} & \Rightarrow x_n = v \\ x_{n-1} + bv = v & \Rightarrow x_{n-1} = v(1-b) \\ x_{n-2} + b x_{n-1} + b^2 x_n = v & \Rightarrow x_{n-2} = v(1-b) \\ \vdots & \\ x_1 + b x_2 + \cdots b^{n-1} x_n = v & \Rightarrow x_1 = v(1-b). \end{aligned}$$

We have $X = v(1-b, 1-b, \ldots, 1-b, 1)$. To find v, all the components must add to 1 and

$$\sum_{i=1}^{n} x_i = (n-1)v(1-b) + v = 1 \Rightarrow v = \frac{1}{1 + (n-1)(1-b)}.$$

The computation for the optimal Y is similar. We get

$$Y = v(1, 1-b, 1-b, \ldots, 1-b), \quad v = \frac{1}{1 + (n-1)(1-b)}.$$

2.35 Consider the asymmetric silent duel in which the two players may shoot at paces $(10, 6, 2)$ with accuracies $(0.2, 0.4, 1.0)$ for player I, but, since II is a better shot, $(0.6, 0.8, 1.0)$ for player II. Given that the payoffs are $+1$ to a survivor and -1 otherwise (but 0 if a draw), set up the matrix game and solve it.

2.35 Answer: The game matrix is

$$\begin{bmatrix} -0.40 & -0.44 & -0.6 \\ -0.28 & -0.40 & -0.2 \\ -0.2 & -0.6 & 0 \end{bmatrix}.$$

To use Maple to get these matrices:

```
Accuracy functions:
>p1:=x->piecewise(x=0,.2,x=.2,.4,x=.4,.4,x=.6,.4,x=.8,1);
>p2:=x->piecewise(x=0,0.6,x=.2,.8,x=.4,.8,x=.6,.8,x=.8,1);
Payoff function
>u1:=(x,y)->
    piecewise(x<y,1*p1(x)+
                  (-1)*(1-p1(x))*p2(y)+
                  (0)*(1-p1(x))*(1-p2(y)),
        x>y,(-1)*p2(y)+(1)*(1-p2(y))*p2(x)+0*(1-p2(y))*(1-p1(x)),
        x=y,0*p1(x)*p2(x)+(1)*p1(x)*(1-p2(x))+(-1)*(1-p1(x))*p2(x)
              +0*(1-p1(x))*(1-p2(x)));
>with(LinearAlgebra):
>A:=Matrix([[u1(0,0),u1(0,.4),u1(0,.8)],
            [u1(.4,0),u1(.4,.4),u1(.4,.8)],
            [u1(.8,0),u1(.8,.4),u1(.8,.8)]]);
```

The solution of this game is

$$X^* = (0, 1, 0), \quad Y^* = (0, 1, 0), \quad v(A) = -0.4.$$

2.36 Suppose that there are four towns connected by a highway as in the following diagram:

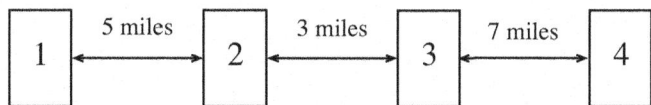

Assume that 15% of the total populations of the four towns are nearest to town 1, 30% nearest to town 2, 20% nearest to town 3, and 35% nearest to town 4. There are two superstores, say, I and II, thinking about building a store to serve these four towns. If both stores are in the same town or in two different towns but with the same distance to a town, then I will get a 65% market share of business. Each store gets 90% of the business of the town in which they put the store if they are in two different towns. If store I is closer to a town than II is, then I gets 90% of the business of that town. If store I is farther than store II from a town, store I still gets 40% of that town's business, except for the town II is in. Find the payoff matrix and solve the game.

2.36 Answer: To find the game matrix with player I as the row player, if I locates the store in town 1, and II locates the store in town 4, for example, then the payoff to I, in terms of market share, is

$$(0.9)(15) + (0.9)(30) + (0.4)(20) + (0.9)(35) = 52\%.$$

Similarly, if store I is in town 1 and store II is in town 2, then the expected payoff to I is

$$(0.9)(15) + (0.1)(30) + (0.4)(20) + (0.4)(35) = 38.5\%.$$

The game matrix becomes

$$A = \begin{bmatrix} 65 & 38.5 & 41.5 & 52 \\ 78 & 65 & 56.5 & 62 \\ 78 & 58.5 & 65 & 62 \\ 63 & 48.5 & 51.5 & 65 \end{bmatrix}.$$

The saddle point is $X^* = (0, 0.43, 0.57, 0)$ and $Y^* = (0, 0.57, 0.43, 0)$.

2.37 Two farmers are having a fight over a disputed 6-yard-wide strip of land between their farms. Both farmers think that the strip is theirs. A lawsuit is filed and the judge orders them to submit a confidential settlement offer to settle the case fairly. The judge has decided to accept the settlement offer that concedes the most land to the other farmer. In case both farmers make no concession or they concede equal amounts, the judge will favor farmer II and award her all 6 yards. Formulate this as a constant sum game assuming that both farmers' pure strategies must be the yards that they concede: 0, 1, 2, 3, 4, 5, 6. Solve the game. What if the judge awards 3 yards if equal concessions?

2.37 Answer: The matrix is

	0	1	2	3	4	5	6
0	0	1	2	3	4	5	6
1	5	0	2	3	4	5	6
2	4	4	0	3	4	5	6
3	3	3	3	0	4	5	6
4	2	2	2	2	0	5	6
5	1	1	1	1	1	0	6
6	0	0	0	0	0	0	0

Then, $X^* = (0.43, 0.086.0.152, 0.326, 0, 0, 0)$, $Y^* = (0.043, 0.217, 0.413, 0.326, 0, 0, 0)$, $v = 2.02174$. This says 43% of the time farmer I should concede nothing, while farmer II should concede 2 about 41% of the time. Since the value to farmer I is about 2, farmer II will get about 4 yards. The rules definitely favor farmer II.

In the second case in which the judge awards 3 yards to each farmer if they make equal concessions, each farmer should concede 2 yards and the value of the game is 3. Observe that the only difference between this game's matrix and the previous game is that the matrix has 3 on the main diagonal. This game, however, has a pure saddle point.

2.38 Two football teams, B and P, meet for the Superbowl. Each offense can play run right (RR), run left (RL), short pass (SP), deep pass (DP), or screen pass (ScP). Each defense can choose to blitz (BL), defend a short pass (DSP), or defend a long pass (DLP), or defend a run (DR). Suppose that team B does a statistical analysis and determines the following payoffs when they are on defense:

B/P	RR	RL	SP	DP	ScP
BL	-5	-7	-7	5	4
DSP	-6	-5	8	6	3
DLP	-2	-3	-8	6	-5
DR	3	3	-5	-15	-7

A negative payoff represents the number of yards gained by the offense, so a positive number is the number of yards lost by the offense on that play of the game. Solve this game and find the optimal mix of plays by the defense and offense. (**Caution:** This is a matrix in which you might want to add a constant to ensure $v(A) > 0$. Then subtract the constant at the end to get the real value. You do not need to do that with the strategies.)

2.38 Answer: The saddle point is

$$X^* = \left(\frac{23}{226}, \frac{165}{452}, \frac{73}{452}, \frac{42}{113} \right),$$

$$= (0.102, 0.365, 0.162, 0.372)$$

$$Y^* = \left(\frac{169}{452}, \frac{55}{226}, 0, \frac{61}{452}, \frac{28}{113} \right)$$

$$= (0.374, 0.243, 0.0, 0.135, 0.245).$$

The value is $v = -\frac{431}{226} = -1.907$. According to this, team P should never play a short pass but team B should defend against SP about 16% of the time. Also, on average, P will lose about 2 yards per play. They should pack it in.

2.39 Let $a > 0$. Use the graphical method to solve the game in which player II has an infinite number of strategies with matrix

$$\begin{bmatrix} a & 2a & \frac{1}{2} & 2a & \frac{1}{4} & 2a & \frac{1}{6} & \cdots \\ a & 1 & 2a & \frac{1}{3} & 2a & \frac{1}{5} & 2a & \cdots \end{bmatrix}.$$

Pick a value for a and solve the game with a finite number of columns to see what's going on.

2.39 Answer: The solution is $X = (\frac{1}{2}, \frac{1}{2})$, $Y = (1, 0, 0...)$, $v = a$. If you graph the lines for player I versus the columns ending at column n, you will see that the lines that determine the optimal strategy for I will correspond to column 1 and the column with $(2a, 1/n)$, or $(1/n, 2a)$. This gives $x^* = \frac{a}{(2a - 1/n)}$. Letting $n \to \infty$ gives $x^* = \frac{1}{2}$. For player II you can see from the graph that the last column and the first column are the only ones being used, which leads to the matrix for player II for a fixed n of $\begin{bmatrix} a & 2a \\ a & \frac{1}{n} \end{bmatrix}$. For all large enough n so that $2a > 1/n$, we have a saddle point at row 1 column 1, independent of n, which means $Y^* = (1, 0, 0, \ldots)$.

For example, if we take $a = \frac{3}{2}$, and truncate it at column 5, we get the matrix

$$\begin{bmatrix} \frac{3}{2} & 3 & \frac{1}{2} & 3 & \frac{1}{4} \\ \frac{3}{2} & 1 & 3 & \frac{1}{3} & 3 \end{bmatrix}$$

and the graph is shown.

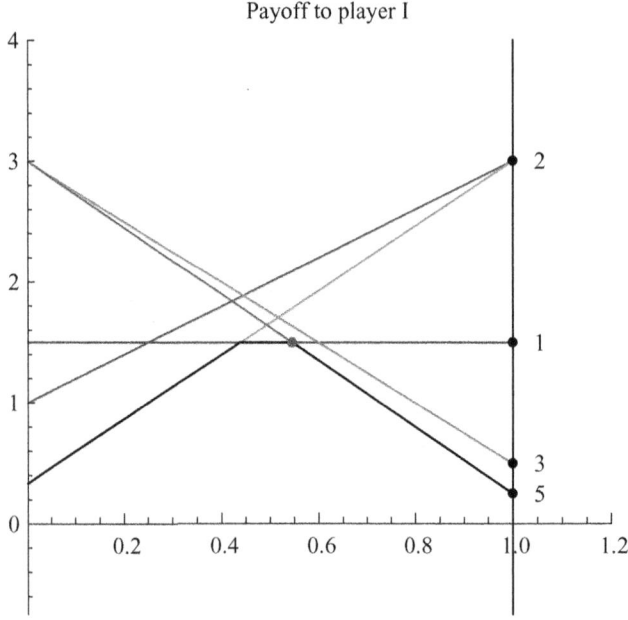

Payoff to player I

The solution of this game is

$$v(A) = \frac{3}{2}, \quad X^* = (0.545107, 0.454893), \quad Y^* = (1, 0, 0, 0, 0).$$

2.40 A Latin square game is a square game in which the matrix A is a Latin square. A Latin square of size n has every integer from 1 to n in each row and column.

(a) Solve the game of size 5

$$A = \begin{bmatrix} 1 & 2 & 3 & 4 & 5 \\ 2 & 4 & 1 & 5 & 3 \\ 3 & 5 & 4 & 2 & 1 \\ 4 & 1 & 5 & 3 & 2 \\ 5 & 3 & 2 & 1 & 4 \end{bmatrix}.$$

2.40.a Answer: If we use Method 2 to solve this by linear programming, we have the problems

Player I	Player II
Max v	Min v

$$x_1 + 2x_2 + 3x_3 + 4x_4 + 5x_5 \geq v \qquad y_1 + 2y_2 + 3y_3 + 4y_4 + 5y_5 \leq v$$
$$2x_1 + 4x_2 + 5x_3 + x_4 + 3x_5 \geq v \qquad 2y_1 + 4y_2 + y_3 + 5y_4 + 3y_5 \leq v$$
$$3x_1 + x_2 + 4x_3 + 5x_4 + 2x_5 \geq v \qquad 3y_1 + 5y_2 + 4y_3 + 2y_4 + y_5 \leq v$$
$$4x_1 + 5x_2 + 2x_3 + 3x_4 + x_5 \geq v \qquad 4y_1 + y_2 + 5y_3 + 3y_4 + 2y_5 \leq v$$
$$5x_1 + 3x_2 + x_3 + 2x_4 + 4x_5 \geq v \qquad 5y_1 + 3y_2 + 2y_3 + y_4 + 4y_5 \leq v$$
$$x_i \geq 0 \qquad\qquad\qquad\qquad y_j \geq 0$$
$$\sum x_i = 1 \qquad\qquad\qquad\qquad \sum y_j = 1.$$

If we add the inequalities for player I, we see that

$$15(x_1 + x_2 + x_3 + x_4 + x_5) \geq 5v \Rightarrow 15 \geq 5v \Rightarrow 3 \geq v.$$

Now add the inequalities for player II and get $v \geq 3$, and we conclude that $v(A) = 3$.

We could use software to solve the programs, but instead we will make the judicious guess that $X^* = Y^* = (\frac{1}{5}, \frac{1}{5}, \frac{1}{5}, \frac{1}{5}, \frac{1}{5})$, since all rows and columns contain the same numbers in the same proportions. Now we only need to verify.

First, a direct computation shows $E(X^*, Y^*) = 3$.

Second, $E(i, Y^*) = E(X^*, j) = 3$ for all rows i and columns j. By the conditions to be a saddle point, this is enough to verify that (X^*, Y^*) is a saddle point.

(b) Prove that a Latin square game of size n has $v(A) = \frac{(n+1)}{2}$. You may need the fact that $1 + 2 + \cdots + n = \frac{n(n+1)}{2}$.

2.40.b Answer: Based on the previous part, it is a reasonable guess that $X^* = (\frac{1}{n}, \ldots, \frac{1}{n}) = Y^*$ is a saddle point. If that is the case, then

$$v = E(X^*, Y^*) = X^* A Y^{*T} = X^* \begin{bmatrix} \frac{n(n+1)}{2} & \frac{1}{n} \\ & \vdots \\ \frac{n(n+1)}{2} & \frac{1}{n} \end{bmatrix} = \frac{1}{n} \cdot n \cdot \frac{n(n+1)}{2} \frac{1}{n} = \frac{(n+1)}{2}.$$

Now we may verify all of this by checking $E(i, Y^*) = E(X^*, j) = \frac{(n+1)}{2} = v$.

2.41 Consider the game with matrix $A = \begin{bmatrix} -3 & -3 & 2 \\ -1 & 3 & -2 \\ 3 & -1 & -2 \\ 2 & 2 & 3 \end{bmatrix}$.

(a) Show that

$$X^* = \left(\frac{3}{8}\left(1 + \frac{\lambda}{3}\right), \quad \frac{5}{16}(1 - \lambda), \quad \frac{5}{16}(1 - \lambda), \frac{\lambda}{2} \right), \quad Y^* = \left(\frac{1}{4}, \frac{1}{4}, \frac{1}{2} \right)$$

is not a saddle point of the game if $0 \leq \lambda \leq 1$.

2.41.a Answer: First, we have to make sure that X^* is a legitimate strategy. The components add up to 1 and each component is ≥ 0 as long as $0 \leq \lambda \leq 1$ so it is legitimate.

Next, if we calculate $E(X^*, Y^*) = X^* A Y^{*T} = \frac{1}{2}(3\lambda - 1)$ for any λ. Thus, we must have $v(A) = \frac{1}{2}(3\lambda - 1)$ and $v(A) = -\frac{1}{2}$ if $\lambda = 0$.

Now we have to check that (X^*, Y^*) is a saddle point and $v(A)$ is really the value. For that, we check

$$X^* A = \left[-\tfrac{1}{2}, -\tfrac{1}{2}, 3\lambda - \tfrac{1}{2} \right] \geq \left[v, v, v \right].$$

The last inequalities are **not true** (unless $\lambda = 0$) since

$$-\frac{1}{2} \leq -\frac{1}{2} + \frac{3}{2}\lambda = \frac{1}{2}(3\lambda - 1) = v.$$

Finally, we check

$$A\, Y^{*T} = \begin{bmatrix} -\tfrac{1}{2} \\ -\tfrac{1}{2} \\ -\tfrac{1}{2} \\ \tfrac{5}{2} \end{bmatrix} \leq \begin{bmatrix} v \\ v \\ v \\ v \end{bmatrix}.$$

The last inequalities are not all true because $-\frac{1}{2} \leq -\frac{1}{2} + \frac{3}{2}\lambda = v$, but

$$\frac{5}{2} \leq v = \frac{1}{2}(3\lambda - 1) \Rightarrow \lambda \geq 2,$$

and that contradicts $0 \leq \lambda \leq 1$. Thus everything falls apart.

(b) Solve the game.

2.41.b Answer: First note that the first row can be dropped because of strict dominance. Then we are left with the matrix $\begin{bmatrix} -1 & 3 & -2 \\ 3 & -1 & -2 \\ 2 & 2 & 3 \end{bmatrix}$.

The solution of the game is $X^* = (0, 0, 0, 1)$, $Y^* = (\frac{1}{4}, \frac{3}{4}, 0)$, $v(A) = 2$.

2.42 Two players, Curly and Shemp, are betting on the toss of a fair coin. Shemp tosses the coin, hiding the result from Curly. Shemp looks at the coin. If the coin is heads, Shemp says that he has heads and demands $1 from Curly. If the coin is tails, then Shemp may tell the truth and pay Curly $1, or he may lie and say that he got a head and demands $1 from Curly. Curly can challenge Shemp whenever Shemp demands $1 to see whether Shemp is telling the truth, but it will cost him $2 if it turns out that Shemp was telling the truth. If Curly challenges the call and it turns out that Shemp was lying, then Shemp must pay Curly $2. Find the matrix and solve the game.

2.42 Answer: Shemp has the two strategies to (1) tell the truth or (2) call H no matter what. Curly also has two strategies: (1) believe a call of tails or heads, or (2) believe a call of tails and challenge a call of heads. The game matrix to Shemp as the row player is

Shemp/Curly	(1)	(2)
(1)	0	$\frac{1}{2}$
(2)	1	0

For example, if Shemp tells the truth and Curly believes tails but challenges heads, the expected payoff to Shemp is

$$(-1)\frac{1}{2} + (2)\frac{1}{2} = \frac{1}{2}.$$

Similarly, if Shemp always calls heads and Curly believes the call no matter what, Shemp's expected payoff is

$$(-1)\frac{1}{2} + (+1)\frac{1}{2} = 0.$$

The remaining entries are similar.

Since $1 - x = \frac{1}{2}x \Rightarrow x = \frac{2}{3}$, and $\frac{1}{2}(1 - y) = y \Rightarrow y = \frac{1}{3}$, Shemp should call the actual toss two-thirds of the time and lie one-third of the time. Curly should believe the call one-third of the time and believe tails called but challenge heads called the rest of the time.

2.43 Wiley Coyote is waiting to nab Speedy who must emerge from a tunnel with three exits, A, B and C. B and C are relatively close together, but far away from A. Wiley can lie in wait for Speedy near an exit, but then he will catch Speedy only if Speedy uses this exit. But Wiley has two other options. He can wait between B and C, but then if Speedy comes out of A he escapes while if he comes out of B or C, Wiley can only get him with probability p. Wiley's last option is to wait at a spot overlooking all three exits, but then he catches Speedy with probability q no matter which exit Speedy uses.

(a) Find the matrix for this game for arbitrary p, q.

2.43.a Answer: The game matrix is

Wiley/Speedy	A	B	C
A	1	0	0
B	0	1	0
C	0	0	1
B-C	0	p	p
A-B-C	q	q	q

This reflects the rules that if Wiley waits at the right exit, Speedy will be caught. If Wiley waits between B and C, Speedy is caught with probability p if he exits from B or C, but he gets away if he comes out of A. If Wiley overlooks all 3 exits he has equal probability q of catching Speedy no matter what Speedy chooses.

(b) Show that using convex dominance, Wiley's options of waiting directly at an exit dominate one or the other or both of his other two options if $p < \frac{1}{2}$ or if $q < \frac{1}{3}$.

2.43.b Answer: If any convex combination of the first 3 rows dominates row 4, we can drop row 4. Dominance requires then that we must have numbers $\lambda_1, \lambda_2, \lambda_3$, such that $\lambda_1 > 0, \lambda_2 > p, \lambda_3 > p$ where $\lambda_1 + \lambda_2 + \lambda_3 = 1, \lambda_i \geq 0$. Adding gives $1 = \lambda_1 + \lambda_2 + \lambda_3 \geq 2p \Rightarrow p < \frac{1}{2}$ for strict domination. Similarly, for row 5, strict domination would imply $\lambda_1 + \lambda_2 + \lambda_3 > 3q \Rightarrow q < \frac{1}{3}$. Thus, if $p < \frac{1}{2}$ or $q < \frac{1}{3}$, Wiley should pick an exit and wait there.

(c) Let $p = 3/4, q = 2/5$. Solve the game.

2.43.c Answer: The linear programming setup for this problem is

$$
\left(
\begin{array}{l}
\text{Player I} \\
\text{Max } v \\
x_1 + \frac{2x_5}{5} \geq v \\
x_2 + \frac{3x_4}{4} + \frac{2x_5}{5} \geq v \\
x_3 + \frac{3x_4}{4} + \frac{2x_5}{5} \geq v \\
x_i \geq 0 \\
\sum x_i = 1
\end{array}
\right)
\qquad
\left(
\begin{array}{l}
\text{Player II} \\
\text{Min } v \\
y_1 \leq v \\
y_2 \leq v \\
y_3 \leq v \\
\frac{3y_2}{4} + \frac{3y_3}{4} \leq v \\
\frac{2y_1}{5} + \frac{2y_2}{5} + \frac{2y_3}{5} \leq v \\
y_j \geq 0 \\
\sum y_j = 1
\end{array}
\right)
$$

The solution of the game by linear programming is $X^* = (\frac{3}{7}, 0, 0, \frac{4}{7}, 0)$, $Y^* = (\frac{3}{7}, \frac{3}{7}, \frac{1}{7})$, and $v(A) = \frac{3}{7}$. Since B and C are symmetric, another possible $Y^* = (\frac{3}{7}, \frac{1}{7}, \frac{3}{7})$. A little less than half the time Wiley should wait outside A, and the rest of the time he should overlook all three exits. Speedy should exit from A with probability $\frac{3}{7}$ and either B or C with probability $\frac{4}{7}$.

2.44 Left and Right bet either \$1 or \$2. If the total amount wagered is even, then Left takes the entire pot; if it is odd, then Right takes the pot.

(a) Set up the game matrix and analyze this game.

2.44.a Answer: The game matrix is

L/R	1	2
1	1	−1
2	−2	2

Remember that the numbers represent the payoffs ($=$ winnings) for Left and do not include the money that Left puts into the pot.

Since $x - 2(1 - x) = -x + 2(1 - x) \Rightarrow x = \frac{2}{3}$, and $y - (1 - y) = -2 + 2(1 - y) \Rightarrow y = \frac{1}{2}$, Left should bet \$1 with probability $\frac{2}{3}$ and Right should bet \$1 with probability $\frac{1}{2}$. We have $v = 0$, $X^* = (\frac{2}{3}, \frac{1}{3})$, $Y^* = (\frac{1}{2}, \frac{1}{2})$.

(b) Suppose now that the amount bet is not necessarily 1 or 2: Left and Right choose their bet from a set of positive integers. Suppose the sets of options are the following:
1. Left and Right can bet any number in $L = R = \{1, 2, 3, 4, 5, 6\}$;
2. Left and Right can bet any number in $L = R = \{2, 4, 6, 8, 9, 13\}$.
Analyze each of these cases.

2.44.b Answer: In the first case, the game matrix is

L/R	1	2	3	4	5	6
1	1	−1	3	−1	5	−1
2	−2	2	−2	4	−2	6
3	1	−3	3	−3	5	−3
4	−4	2	−4	4	−4	6
5	1	−5	3	−5	5	−5
6	−6	2	−6	4	−6	6

Start with eliminating dominated rows and columns. For example, row 1 dominates rows 3 and 5; column 2 dominates columns 4 and 6; then row 2 dominates rows 4 and 6, and column 2 dominates columns 3 and 5. The matrix reduces to the 2×2 case $A = \begin{bmatrix} 1 & -1 \\ -2 & 2 \end{bmatrix}$, which has the same solution as the first part. Left should bet 1 two-thirds of the time; Right should bet 1 half the time, and it is a fair game.

For the second case, the matrix is

L/R	2	4	6	8	9	13
2	2	4	6	8	−2	−2
4	2	4	6	8	−4	−4
6	2	4	6	8	−6	−6
8	2	4	6	8	−8	−8
9	−9	−9	−9	−9	9	13
13	−13	−13	−13	−13	9	13

By dominance this reduces to the game

L/R	2	9
2	2	−2
9	−9	9

This game has solution $v = 0$, $X^* = (\frac{9}{11}, \frac{2}{11})$, $Y^* = (\frac{7}{22}, \frac{15}{22})$. Left should bet 2 with probability $\frac{9}{11}$ and bet 9 with probability $\frac{2}{11}$. Right should bet the same amounts but with probabilities $\frac{7}{22}$ and $\frac{15}{22}$, respectively.

(c) Suppose Left may bet any amount in $L = \{1, 2, 31, 32\}$ and Right any amount in $R = \{2, 5, 16, 17\}$. The payoffs are as follows:

1. If Left $\in \{1, 2\}$ and Right $\in \{2, 5\}$ and Left+Right even $\Rightarrow$ Right pays Left the sum of the amounts each bet.

2. If Left $\in \{1, 2\}$ and Right $\in \{2, 5\}$ and Left+Right odd $\Rightarrow$ Left pays Right the sum of the amounts each bet.

3. If Left $\in \{31, 32\}$ **or** Right $\in \{16, 17\}$, and Left+Right odd $\Rightarrow$ Right pays Left the sum of the amounts each bet.

4. If Left $\in \{31, 32\}$ **or** Right $\in \{16, 17\}$, and Left+Right even $\Rightarrow$ Left pays Right the sum of the amounts each bet.

2.44.c Answer: Left and Right each have four strategies. The game matrix is

L/R	2	5	16	17
1	-3	6	17	-18
2	4	-7	-18	19
31	33	-36	47	-48
32	-34	37	-48	49

This game is completely mixed and the matrix has an inverse so we could use the invertible matrix game formulas, or we could use the linear programming method. We get the solution

$$v(A) = -\frac{1}{152}, \quad X^* = (50, 48, 181, 153)\frac{1}{456}, \quad Y^* = (10, 10, 67, 65)\frac{1}{152}.$$

In decimal terms,

$$v(A) = -0.0066, X^* = (0.109, 0.105, 0.397, 0.336),$$
$$Y^* = (0.066, 0.066, 0.441, 0.428)$$

The game is not quite fair to Left, and both players prefer to play the bigger numbers.

2.45 The evil Don Barzini has imprisoned Tessio and three of his underlings in the Corleone family (total value 4) somewhere in Brooklyn, and Clemenza and his two assistants somewhere in Queens (total value 3). Sonny sets out to rescue either Tessio or Clemenza and their associates; Barzini sets out to prevent the rescue but he doesn't know where Sonny is headed. If they set out to the same place, Barzini has an even chance of arriving first, in which case Sonny returns without rescuing anybody. If Sonny arrives first he rescues the group. The payoff to Sonny is determined as the number of Corleone family members rescued. Describe this scenario as a matrix game and determine the optimal strategies for each player.

2.45 Answer: Sonny has two strategies: (1) Brooklyn and (2) Queens. Barzini also has two strategies: (1) Brooklyn and (2) Queens. If Sonny heads for Brooklyn, Barzini also heads for Brooklyn. Sonny's expected payoff is

$$0\left(\frac{1}{2}\right) + 4\left(\frac{1}{2}\right) = 2.$$

Sonny/Barzini	(1)	(2)
(1)	2	4
(2)	3	$\frac{3}{2}$

This game is easily solved using $2x + 3(1 - x) = 4x + \frac{3}{2}(1 - x) \Rightarrow x = \frac{3}{7}$, and $2y + 4(1 - y) = 3y + \frac{3}{2}(1 - y)$ that gives $y = \frac{5}{7}$. The optimal strategies are as follows:

$$X^* = \left(\frac{3}{7}, \frac{4}{7}\right), \quad Y^* = \left(\frac{5}{7}, \frac{2}{7}\right), \quad v = \frac{18}{7}.$$

Sonny should head for Queens with probability $\frac{4}{7}$, and Barzini should head for Brooklyn with probability $\frac{5}{7}$. On average, Sonny rescues between 2 and 3 capos.

2.46 You own two companies Uno and Due. At the end of a fiscal year, Uno should have a tax bill of \$3 million and Due should have a tax bill of \$1 million. By cooking the books you can file tax returns making it look as though you should not have to pay any tax at all for either Uno or Due, or both. Due to limited resources, the IRS can only audit one company. If a company with cooked books is audited, the fraud is revealed and all unpaid taxes will be levied plus a penalty of $p\%$ of the tax due.

(a) Set this up as a matrix game. Your payoffs will depend on p.

2.46.a Answer:

Your strategies are $H^U H^D$, $H^U C^D$, $C^U H^D$, $C^U C^D$, where $H = $ honest, $C = $ cheat and the superscript refers to the company. The IRS has two strategies: investigate Uno and investigate Due. Here is the matrix:

IRS/You	$H^U H^D$	$H^U C^D$	$C^U H^D$	$C^U C^D$
U	4	3	$4+3p$	$3+3p$
D	4	$4+p$	1	$1+p$

The IRS is the row player because the IRS wants to maximize the tax receipts, while you want to minimize them. For example, if the IRS investigates Uno and you play $H^U H^D$, you simply pay your total bill of 4 million. If, however, you play $C^U H^D$, then you are caught and you have to pay the 4 million you owe as well as a penalty of $3p$, a percentage of the 3 million owed on Uno. The rest of the entries are similar.

(b) In the case the fine is $p = 50\%$, what strategy should you adopt to minimize your expected tax bill?

2.46.b Answer: If $p = \frac{1}{2}$ the matrix is

IRS/You	$H^U H^D$	$H^U C^D$	$C^U H^D$	$C^U C^D$
U	4	3	$\frac{11}{2}$	$\frac{8}{2}$
D	4	$\frac{11}{2}$	1	$\frac{3}{2}$

The solution of this game is $v(A) = \frac{7}{2}$, $X^* = (\frac{2}{3}, \frac{1}{3})$, and $Y^* = (0, \frac{2}{3}, 0, \frac{1}{3})$. The IRS should audit Uno $\frac{2}{3}$ of the time (because it is worth more to the IRS) and you should play $H^U C^D$ $\frac{2}{3}$ of the time and cheat on both of them $\frac{1}{3}$ of the time.

(c) For what values of p is complete honesty optimal?

2.46.c Answer: We want $H^U H^D$ to be an optimal pure strategy and we want to know what p should be for that to occur. If it is an optimal strategy, then it must be the case that $v(A) = 4$.

Consider the graphical method for solving the game. The first step is to graph all the lines

$$z = E(X, 1) = 4x + 4(1 - x),$$
$$z = E(X, 2) = 3x + (4 + p)(1 - x),$$
$$z = E(X, 3) = (4 + 3p)x + (1 - x),$$
$$z = E(X, 4) = (3 + 3p)x + (1 + p)(1 - x).$$

Since we want the value of the game to be 4, if all the columns are used with positive probability in an optimal strategy for You, then in all four equations $z = 4$. Solve for x in terms of p to get the four possible solutions:

$$0 \le x_1 \le 1, \quad x_2 = \frac{p}{1+p}, \quad x_3 = \frac{1}{1+p}, \quad x_4 = \frac{1}{2}\frac{3-p}{1+p}$$

The first x_1 is any number in $[0, 1]$. Clearly $x_2, x_3 \in [0, 1]$ for any $0 \le p \le 1$. But $x_4 \in [0, 1]$ requires that $3 - p \le 2 + 2p \Rightarrow p \ge 1$.

If the value of the game must be 4, and the optimal strategy for You must be $Y^* = (1, 0, 0, 0)$, then the two lines involved in finding Y^* must include the line corresponding to column 1. This will be true if $p \ge 1$, but is not true for $p < 1$. We conclude that complete honesty will be optimal if the fine is a minimum of 100% of the tax owed.

The figure shows what happens when $p = 1.3$. In this case, $X^* = (0.565, 0.435)$, $Y^* = (1, 0, 0, 0)$, $v = 4$.

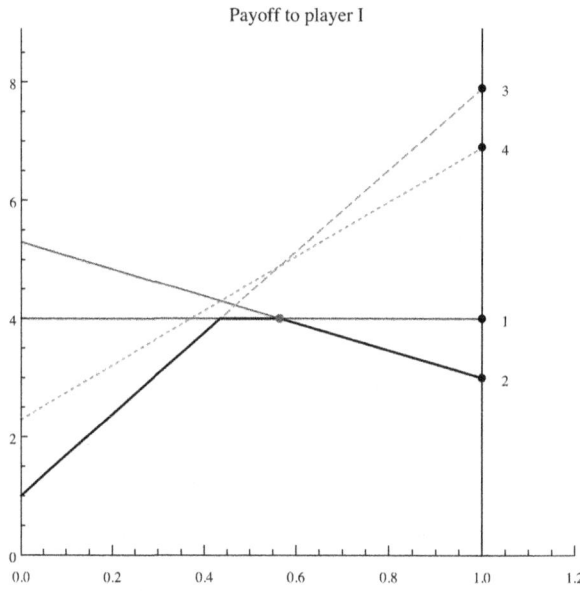

Payoff to player I

2.5 Appendix: Linear Programming and the Simplex Method

Problems

2.47 Use the simplex method to solve the games with the following matrices. Check your answers using Maple and verify that the strategies obtained are indeed optimal.

$$\text{(a)}\begin{bmatrix} 0 & 3 & 3 & 2 & 4 \\ 4 & 4 & 3 & 1 & 4 \end{bmatrix}; \quad \text{(b)}\begin{bmatrix} 4 & 4 & -4 & -1 \\ -4 & -2 & 4 & 4 \\ 2 & -4 & -1 & -5 \\ -3 & 1 & 0 & -4 \end{bmatrix}; \quad \text{(c)}\begin{bmatrix} 2 & -5 & 3 & 0 \\ -4 & -5 & -5 & -6 \\ 3 & -4 & -1 & -2 \\ 0 & 4 & 1 & 3 \end{bmatrix}.$$

2.47 Answer: (a) $X^* = (\frac{3}{5}, \frac{2}{5})$, $Y^* = (\frac{1}{5}, 0, 0, \frac{4}{5}, 0)$

(b) $X^* = (\frac{21}{53}, \frac{24}{53}, \frac{8}{53}, 0)$, $Y^* = (\frac{23}{53}, \frac{4}{53}, \frac{26}{53}, 0)$, $v = \frac{4}{53}$.

(c) $X^* = (\frac{9}{55}, 0, \frac{1}{5}, \frac{7}{11})$, $Y^* = (\frac{34}{55}, \frac{2}{11}, \frac{1}{5}, 0)$, $v = \frac{51}{55}$.

2.48 Solve this problem using Maple in two ways:

$$\begin{bmatrix} -2 & 3 & 3 & 4 & 1 \\ 3 & -2 & -5 & 2 & 4 \\ 4 & -5 & -1 & 4 & -1 \\ 2 & -4 & 3 & 4 & -3 \end{bmatrix}.$$

2.48 Answer: The solution is

$$X^* = \left(\frac{47}{82}, \frac{10}{41}, \frac{15}{82}, 0\right), \quad Y^* = \left(\frac{22}{41}, \frac{14}{41}, \frac{5}{41}, 0, 0\right), \quad v = \frac{13}{41}.$$

The linear programs by both methods is as follows:

Method 1:

Player I

Min $z_1 = p_1 + p_2 + p_3 + p_4$

$4p_1 + 9p_2 + 10p_3 + 8p_4 \geq 1$

$9p_1 + 4p_2 + p_3 + 2p_4 \geq 1$

$9p_1 + p_2 + 5p_3 + 9p_4 \geq 1$

$10p_1 + 8p_2 + 10p_3 + 10p_4 \geq 1$

$7p_1 + 10p_2 + 5p_3 + 3p_4 \geq 1$

$p_i \geq 0$

Player II

Max $z_2 = q_1 + q_2 + q_3 + q_4 + q_5$

$4q_1 + 9q_2 + 9q_3 + 10q_4 + 7q_5 \leq 1$

$9q_1 + 4q_2 + q_3 + 8q_4 + 10q_5 \leq 1$

$10q_1 + q_2 + 5q_3 + 10q_4 + 5q_5 \leq 1$

$8q_1 + 2q_2 + 9q_3 + 10q_4 + 3q_5 \leq 1$

$q_j \geq 0.$

Method 2:

Player I

Max v

$-2x_1 + 3x_2 + 4x_3 + 2x_4 \geq v$

$3x_1 - 2x_2 - 5x_3 - 4x_4 \geq v$

$3x_1 - 5x_2 - x_3 + 3x_4 \geq v$

$4x_1 + 2x_2 + 4x_3 + 4x_4 \geq v$

$x_1 + 4x_2 - x_3 - 3x_4 \geq v$

$x_i \geq 0$

$\sum x_i = 1$

Player II

Min v

$-2y_1 + 3y_2 + 3y_3 + 4y_4 + y_5 \leq v$

$3y_1 - 2y_2 - 5y_3 + 2y_4 + 4y_5 \leq v$

$4y_1 - 5y_2 - y_3 + 4y_4 - y_5 \leq v$

$2y_1 - 4y_2 + 3y_3 + 4y_4 - 3y_5 \leq v$

$y_j \geq 0$

$\sum y_j = 1.$

Both methods give the same solution.

2.6 Review Problems

Problems

Give the precise definition, complete the statement, or answer True or False. If False, give the correct statement.

2.49 Complete the statement: In order for (X^*, Y^*) to be a saddle point and v to be the value, it is necessary and sufficient that $E(X^*, j)$ _____, for all _____ and $E(i, Y^*)$ _____ for all _____.

2.49 Answer: In order for (X^*, Y^*) to be a saddle point and v to be the value, it is necessary and sufficient that $E(X^*, j) \geq v$, for all $j = 1, 2, \ldots, m$ and $E(i, Y^*) \leq v$ for all $i = 1, 2, \ldots, n$.

2.50 State the von Neumann Minimax theorem for two-person zero sum games with matrix A.

2.50 Answer: The von Neumann minimax theorem says that for any $n \times m$ matrix A, we have

$$\min_{Y \in S_m} \max_{X \in S_n} X A Y^T = \max_{X \in S_n} \min_{Y \in S_m} X A Y^T.$$

2.51 Suppose (X^*, Y^*) is a saddle point, $v(A)$ is the value of the game and $x_k > 0$ for the kth component of X^*. What is $E(k, Y^*)$? What is the most you can say if $x_k = 0$ for $E(k, Y^*)$?

2.51 Answer: $E(k, Y^*) = v(A)$. If $x_k = 0$, $E(k, Y^*) \leq v(A)$.

2.52 Suppose A is a game matrix and Y^0 is a given strategy for player II. Define what it means for X^* to be a best response strategy to Y^0 for player I.

2.52 Answer:

$$X^* A Y^{0T} = \max_{X \in S_n} X A Y^{0T}.$$

2.53 Suppose $A_{n \times n}$ has an inverse A^{-1}. What three assumptions do you need to be able to use the formulas

$$v := 1/J_n A^{-1} J_n^T, \quad X^* = v J_n A^{-1}, \quad Y^* = v A^{-1} J_n^T$$

to conclude that $v = v(A)$ and (X^*, Y^*) are optimal mixed strategies?

2.53 Answer:
1. A has an inverse.
2. $J_n A^{-1} J_n^T \neq 0$
3. $v(A) \neq 0$

2.54 $v(A)$ is the value and (X^*, Y^*) is a saddle point of the game with matrix A if and only if . . . .

2.54 Answer:

$$E(X, Y^*) \leq E(X^*, Y^*) \leq E(X^*, Y^*)$$

for every mixed strategy X for player I and Y for player II.

2.55 $v(A)$ is the value and (i^*, j^*) is a pure saddle point of the game with matrix A if and only if . . . .

2.55 Answer:

$$a_{ij^*} \leq a_{i^*j^*} \leq a_{i^*j}$$

for $i = 1, 2, \ldots, n$, $j = 1, 2, \ldots, m$.

2.56 For any game $v^+ \leq v(A) \leq v^-$.

2.56 Answer: True.

2.57 For any matrix game $E(X, Y) = XAY^T$ and the value satisfies
$v(A) = \max_X \min_Y E(X, Y) = \min_Y \max_X E(X, Y)$.

2.57 Answer: True.

2.58 If (X^*, Y^*) is a saddle point for a matrix game, then X^* is a best response strategy to Y^* for player I, but the reverse is not necessarily true.

2.58 Answer: False. The reverse is true.

2.59 If Y is an optimal strategy for player II in a zero sum game and $E(i, Y) < \text{value}(A)$ for some row i, then for any optimal strategy X for player I, we must have . . ..

2.59 Answer: $x_i = 0$.

2.60 The graphical method for solving a $2 \times m$ game won't work if . . ..

2.60 Answer: . . . there is a pure saddle point.

2.61 (X^*, Y^*) is a saddle point if and only if $\min_{j=1,2,\ldots,m} E(X^*, j) = \max_{i=1,2,\ldots,n} E(i, Y^*)$.

2.61 Answer: True.

2.62 If (Y^0, X^0) is a saddle point for the game A, then $E(Y^0, X) \geq E(Y^0, X^0) \geq E(Y, X^0)$, for all strategies X for player II and Y for player I.

2.62 Answer: True.

2.63 If $X^* = (x_1^*, \ldots, x_n^*)$ is an optimal strategy for player I and $x_1^* = 0$, then we may drop row 1 when we look for the optimal strategy for player II.

2.63 Answer: False. Only if you know $x_1 = 0$ for *every* optimal strategy for player I.

2.64 For a game matrix $A_{n \times m}$ and given strategies $(X^* = (x_1, \ldots, x_n)$, $Y^* = (y_1, \ldots, y_m)$, then $E(X^*, Y^*) = \sum_{i=1}^{n} x_i E(i, Y^*) = \sum_{j=1}^{m} y_j E(X^*, j)$.

2.64 Answer: True.

2.65 If $i^* = 2$, $j^* = 3$, is a pure saddle point in the game with matrix $A_{3 \times 4}$, then using mixed strategies, $X^* = $_____, $Y^* = $_____ is a mixed saddle point for A.

2.65 Answer: $X^* = (0, 1, 0)$, $Y^* = (0, 0, 1, 0)$.

2.66 For a 2×2 game with matrix $A = \begin{pmatrix} a_{11} & a_{12} \\ a_{21} & a_{22} \end{pmatrix}$, the value of the game is $\dfrac{\det(A)}{J_2 A^* J_2^T}$, where $A^* = \begin{pmatrix} \underline{\quad} & \underline{\quad} \\ \underline{\quad} & \underline{\quad} \end{pmatrix}$.

2.66 Answer: This is True assuming that $J_2 A^* J_2^T \neq 0$, and $A^* = \begin{bmatrix} a_{22} & -a_{12} \\ -a_{21} & a_{11} \end{bmatrix}$.

2.67 The invertible matrix theorem says that if A^{-1} exists, then $v(A) = \frac{1}{J_n A^{-1} J_n^T}$ is the value of the game.

2.67 Answer: False. There's more to it than that.

2.68 If $f(x, y) = (x, 1 - x) \begin{pmatrix} a_{11} & a_{12} \\ a_{21} & a_{22} \end{pmatrix} \begin{pmatrix} y \\ 1 - y \end{pmatrix}$, then a mixed saddle point for the game $X^* = (x^*, 1 - x^*)$, $Y^* = (y^*, 1 - y^*)$, $0 < x^*, y^* < 1$, satisfies

$$\frac{\partial f}{\partial x}(x^*, y^*) = \frac{\partial f}{\partial y}(x^*, y^*) = 0.$$

2.68 Answer: True.

2.69 If A is a skew symmetric game with $v(A) > 0$, then there is always a saddle point in which $X^* = -Y^*$.

2.69 Answer: True! This is a point of logic. Since any skew symmetric game must have $v(A) = 0$, the sentence has a false premise. A false premise always implies anything, even something that is false on it's own (since $-Y^*$ can't even be a strategy).

2.70 If A is skew symmetric then for any strategy X for player I, $E(X, X) = 0$.

2.70 Answer: True.

CHAPTER THREE

Two-Person Nonzero Sum Games

3.1 The Basics

Problems

3.1 Show that (X^*, Y^*) is a saddle point of the game with matrix A if and only if (X^*, Y^*) is a Nash equilibrium of the bimatrix game $(A, -A)$.

3.1 Answer: Use the definitions. For example, if X^*, Y^* is a Nash equilibrium, then $X^* A Y^{*T} \geq X A Y^{*T}$, $\forall X$, and $X^*(-A)Y^{*T} \geq X^*(-A)Y^T$, $\forall Y$. Then

$$X^* A Y^T \geq X^* A Y^{*T} \geq X A Y^{*T}, \quad \forall\, X \in S_n, Y \in S_m.$$

3.2 Suppose that a married couple, both of whom have just finished medical school, now have choices regarding their residencies. One of the new doctors has three choices of programs, while the other has two choices. They value their prospects numerically on the basis of the program itself, the city, staying together, and other factors, and arrive at the bimatrix

$$\begin{bmatrix} (5.2, 5.0) & (4.4, 4.4) & (4.4, 4.1) \\ (4.2, 4.2) & (4.6, 4.9) & (3.9, 4.3) \end{bmatrix}.$$

(a) Find all the pure Nash equilibria. Which one should be played?

3.2.a Answer: There are two pure Nash equilibria at $X = (0, 1)$, $Y = (0, 1, 0)$, and $X = (1, 0)$, $Y = (1, 0, 0)$.

$$\begin{bmatrix} \boxed{(5.2, 5.0)} & (4.4, 4.4) & (4.4, 4.1) \\ (4.2, 4.2) & \boxed{(4.6, 4.9)} & (3.9, 4.3) \end{bmatrix}.$$

Note that column 3 is dominated and may be dropped. There is also a mixed Nash at $X = (\frac{7}{13}, \frac{6}{13})$, $Y = (\frac{1}{6}, \frac{5}{6}, 0)$, $E_{\mathrm{I}} = \frac{68}{15}$, $E_{\mathrm{II}} = \frac{301}{65}$. We will see how to calculate mixed Nash equilibria in later sections.

Solutions Manual to Accompany Game Theory: An Introduction, Second Edition. E.N. Barron.
© 2013 John Wiley & Sons, Inc. Published 2013 by John Wiley & Sons, Inc.

Here is a table summarizing the results.

x_1	x_2	y_1	y_2	y_3	$E_I(X^*, Y^*)$	$E_{II}(X^*, Y^*)$
0.538462	0.461538	0.166667	0.833333	0	4.53333	4.63077
0	1.0	0	1.0	0	4.6	4.9
1.0	0	1.0	0	0	5.2	5.

It certainly seems that the last Nash equilibrium will be the one used since that one gives both players the largest payoffs.

(b) Find the safety levels for each player.

3.2.b Answer: The safety levels use the concepts and techniques of zero sum games. For player I, we have

$$A = \begin{bmatrix} 5.2 & 4.4 & 4.4 \\ 4.2 & 4.6 & 3.9 \end{bmatrix}.$$

We easily compute $v(A) = 4.4$, $X^* = (1, 0)$, $Y^* = (0, 0, 1)$. $X^* = (1, 0)$ **is the maxmin strategy for player I.** For player II, we need to deal with B^T, which is given by

$$B^T = \begin{bmatrix} 5.0 & 4.2 \\ 4.4 & 4.9 \\ 4.1 & 4.3 \end{bmatrix}.$$

This game has solution $v(B^T) = 4.63$, $X^* = (\frac{5}{13}, \frac{8}{13}, 0)$, $Y^* = (\frac{7}{13}, \frac{6}{13})$. By definition $X^* = (\frac{5}{13}, \frac{8}{13}, 0)$ is the **maxmin strategy for player II.**

3.3 Consider the bimatrix game that models the game of chicken:

I/II	Turn	Straight
Turn	(19, 19)	(−42, 68)
Straight	(68, −42)	(−45, −45)

Two cars are headed toward each other at a high rate of speed. Each player has two options: turn off, or continue straight ahead. This game is a macho game for reputation, but leads to mutual destruction if both play straight ahead.

(a) There are two pure Nash equilibria. Find them.

3.3.a Answer: The two pure Nash equilibria are at (*Turn,Straight*) and (*Straight,Turn*).

(b) Verify by using the definition of mixed Nash equilibrium that the mixed strategy pair $X^* = (\frac{3}{52}, \frac{49}{52})$, $Y^* = (\frac{3}{52}, \frac{49}{52})$ is a Nash equilibrium, and find the expected payoffs to each player.

3.3.b Answer: To verify that $X^* = (\frac{3}{52}, \frac{49}{52})$, $Y^* = (\frac{3}{52}, \frac{49}{52})$ is a Nash equilibrium, we need to show that

$$E_I(X^*, Y^*) \geq E_I(X, Y^*) \quad \text{and} \quad E_{II}(X^*, Y^*) \geq E_{II}(X^*, Y),$$

for all strategies X, Y. Calculating the payoffs, we get

$$E_{\mathrm{I}}(X, Y^*) = (x, 1-x)\begin{bmatrix} 19 & -42 \\ 68 & -45 \end{bmatrix}\begin{bmatrix} 3 \\ 49 \end{bmatrix}\frac{1}{52} = -\frac{2001}{52}$$

for any $X = (x, 1-x)$, $0 \le x \le 1$. Similarly,

$$E_{\mathrm{II}}(X^*, Y) = \frac{1}{52}(3, 49)\begin{bmatrix} 19 & 68 \\ -42 & -45 \end{bmatrix}\begin{bmatrix} y \\ 1-y \end{bmatrix} = -\frac{2001}{52}.$$

Since $E_{\mathrm{I}}(X^*, Y^*) = E_{\mathrm{II}}(X^*, Y^*) = -\frac{2001}{52}$, we have shown (X^*, Y^*) is a Nash equilibrium.

(c) Find the safety levels for each player.

3.3.c Answer: For player I, $A = \begin{bmatrix} 19 & -42 \\ 68 & -45 \end{bmatrix}$. Then $v(A) = -42$ and that is the safety level for I. The maxmin strategy for player I is $X = (1, 0)$. For player II, $B^T = \begin{bmatrix} 19 & -42 \\ 68 & -45 \end{bmatrix}$ so the safety level for II is also $v(B^T) = -42$. The maxmin strategy for player II is $X^{B^T} = (1, 0)$.

3.4 We may eliminate a row or a column by dominance. If $a_{ij} \ge a_{i'j}$ for *every* column j, and we have strict inequality for *some* column j, then we may eliminate row i'. If player I drops row i', then the entire pair of numbers in that row are dropped. Similarly, if $b_{ij} \ge b_{ij'}$ for every row i, and we have strict inequality for some row i, then we may drop column j', and all the pairs of payoffs in that column. If the inequalities are all strict, this is called **strict dominance**, otherwise it is **weak dominance**.

(a) By this method, solve the game

$$\begin{bmatrix} (3, -1) & (2, 1) \\ (-1, 7) & (1, 3) \\ (4, -3) & (-2, 9) \end{bmatrix}.$$

3.4.a Answer: First, row 2 is dominated for player I, so get rid of it. Second, column 1 is dominated for player II, so get rid of it. Finally, the payoff at $(2, 1)$ is a solution; that is, $X^* = (1, 0, 0)$, $Y^* = (0, 1)$, $E_{\mathrm{I}} = 2$, $E_{\mathrm{II}} = 1$.

(b) Find the safety levels for each player.

3.4.b Answer: Safety for I is 2, and safety for II is 1.

3.5 When we have weak dominance in a game, the order of removal of the dominated row or column makes a difference. Consider the matrix

$$\begin{bmatrix} (10, 0) & (5, 1) & (4, -2) \\ (10, 1) & (5, 0) & (1, -1) \end{bmatrix}.$$

(a) Suppose player I moves first. Find the result.

3.5.a Answer: Weak dominance by row 1 and then strict dominance by column 2 gives the Nash equilibrium at $(5, 1)$.

(b) What happens if player II moves first?

3.5.b Answer: Player II drops the third column by strict dominance. This gives the matrix

$$\begin{bmatrix} (10, 0) & (5, 1) \\ (10, 1) & (5, 0) \end{bmatrix}.$$

Then we have two pure Nash equilibria at $(10, 1)$ and $(5, 1)$.

3.6 Suppose two merchants have to choose a location along the straight road. They may choose any point in $\{1, 2, \ldots, n\}$. Assume there is exactly one customer at each of these points and a customer will always go to the nearest merchant. If the two merchants are equidistant to a customer then they share that customer, that is, $\frac{1}{2}$ the customer goes to each store. For example, if $n = 11$ and if player I chooses location 3 and player II chooses location 8, then the payoff to I is 5 and the payoff to II is 6.

(a) Suppose $n = 5$. Find the bimatrix and find the Nash equilibrium.

3.6.a Answer: The game matrix is

I/II	1	2	3	4	5
1	$\frac{5}{2}, \frac{5}{2}$	$1, 4$	$\frac{3}{2}, \frac{7}{2}$	$2, 3$	$\frac{5}{2}, \frac{5}{2}$
2	$4, 1$	$\frac{5}{2}, \frac{5}{2}$	$2, 3$	$\frac{5}{2}, \frac{5}{2}$	$3, 2$
3	$\frac{7}{2}, \frac{3}{2}$	$3, 2$	$\boxed{\frac{5}{2}, \frac{5}{2}}$	$3, 2$	$\frac{7}{2}, \frac{3}{2}$
4	$3, 2$	$\frac{5}{2}, \frac{5}{2}$	$2, 3$	$\frac{5}{2}, \frac{5}{2}$	$4, 1$
5	$\frac{5}{2}, \frac{5}{2}$	$2, 3$	$\frac{3}{2}, \frac{7}{2}$	$1, 4$	$\frac{5}{2}, \frac{5}{2}$

Using dominance, we may eliminate every row but row 3 and every column but column 3. The unique Nash equilibrium is at $(3, 3)$ with expected payoff $\frac{5}{2}$ to each merchant.

(b) What do you think is the Nash equilibrium in general if $n = 2k + 1$, that is, n is an odd integer.

3.6.b Answer: The Nash equilibrium consists of both players choosing location k with payoff $\frac{2k+1}{2}$ to each player.

3.7 Two airlines serve the route ORD to LAX. Naturally, they are in competition for passengers who make their decision based on airfares alone. Lower fares attract more passengers and increases the load factor (the number of bodies in seats). Suppose the bimatrix is given as follows where each airline can choose to set the fare at Low = \$250 or High = \$700.

	Low	High
Low	$-50, -10$	$175, -20$
High	$-100, 200$	$100, 100$

The numbers are in millions. Find the pure Nash equilibria.

3.7 Answer: There is one pure Nash at Low, Low. Both airlines should set their fares low.

3.8 Consider the game with bimatrix

	A	B	C
a	$1, 1$	$3, x$	$2, 0$
b	$2x, 3$	$2, 2$	$3, 1$
c	$2, 1$	$1, x$	$x^2, 4$

(a) Find x so that the game has no pure Nash equilibria.

3.8.a Answer: Place a bar over the largest number in each row and column:

	A	B	C
a	$1, \bar{1}$	$\bar{3}, x$	$2, 0$
b	$2x, \bar{3}$	$\bar{2}, 2$	$\bar{3}, 1$
c	$\bar{2}, 1$	$1, x$	$x^2, \bar{4}$

No pure Nash equilibria will exist if $2x < 2, x < 1, x^2 < 3$ which means we need $x < 1$.

(b) Find x so that the game has exactly two pure Nash equilibria.

3.8.b Answer: In order for the game to have exactly two Nash equilibria we need $2x > 2$ and $1 < x^2 < 3$, i.e., $1 < x < \sqrt{3}$. Then there will be two Nash equilibria at $(2x, 3)$ and $(3, x)$.

(c) Find x so that the game has exactly three pure Nash equilibria. Is there an x so that there are four pure Nash equilibria?

3.8.c Answer: Starting with the matrix

	A	B	C
a	$1, \bar{1}$	$3, x$	$2, 0$
b	$2x, \bar{3}$	$\bar{2}, 2$	$\bar{3}, 1$
c	$\bar{2}, 1$	$1, x$	$x^2, \bar{4}$

we see that there will be exactly three pure Nash equilibria if we have $x^2 > 3, x > 1, 2x > 2$. This requires $x > \sqrt{3}$.

In order to get four Nash, we would also need $(1, x)$ to be a Nash equilibrium for $x > \sqrt{3}$. However, no value of x can affect player I's payoff, which is what we need to get a fourth Nash.

3.2 2 × 2 Bimatrix Games, Best Response, Equality of Payoffs

Problems

3.9 Suppose we have a two-person matrix game that results in the following best response functions. Find the Nash equilibria if they exist.

(a)

$$BR_1(y) = \begin{cases} 0, & \text{if } y \in [0, \tfrac{1}{4}); \\ [0, 1], & \text{if } y = \tfrac{1}{4}; \\ 1, & \text{if } y \in (\tfrac{1}{4}, 1]. \end{cases} \qquad BR_2(x) = \begin{cases} 1, & \text{if } x \in [0, \tfrac{1}{2}); \\ [0, 1], & \text{if } x = \tfrac{1}{2}; \\ 0, & \text{if } x \in (\tfrac{1}{2}, 1]. \end{cases}$$

3.9.a Answer: Graph $x = BR_1(y)$ and $y = BR_2(x)$ on the same set of axes. Where the graphs intersect is the Nash equilibrium (they could intersect at more than one point). The unique Nash equilibrium is $X^* = (\tfrac{1}{2}, \tfrac{1}{2})$, $Y^* = (\tfrac{1}{4}, \tfrac{3}{4})$.

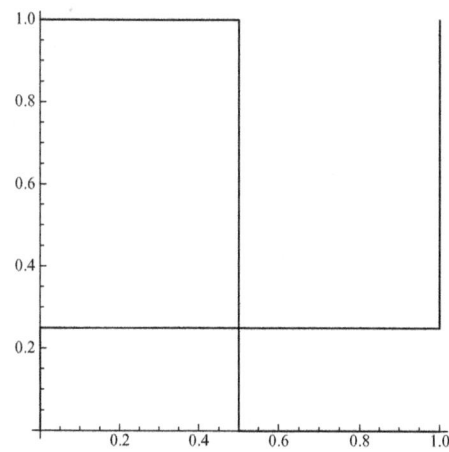

(b)

$$BR_1(y) = \{x \in [0, 1] \mid 1 - 3y \leq x \leq 1 - 2y\},$$
$$BR_2(x) = \{y \in [0, 1] \mid x \leq y \leq \tfrac{2}{3}x + \tfrac{1}{3}\}.$$

3.9.b Answer: Here is a plot of the four lines bounding the best response sets.

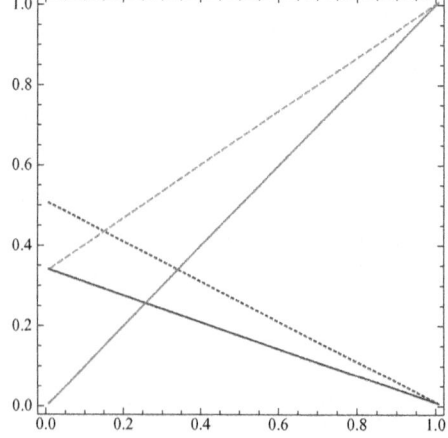

The intersection of the two sets constitutes all the Nash equilibria. Mathematica characterizes the intersection as

$$\left(y = \frac{1}{4}, x = \frac{1}{4} \right) \bigcup \left(\frac{1}{4} < y \le \frac{1}{3}, 1 - 3y \le x \le y \right)$$

$$\bigcup \left(\frac{1}{3} < y < \frac{3}{7}, \frac{(-1 + 3y)}{2} \le x \le 1 - 2y \right)$$

$$\bigcup \left(y = \frac{3}{7}, x = \frac{1}{7} \right),$$

You can see from the figure that there are lots of Nash equilibria. For example $x^* = \frac{1}{4}, y^* = \frac{1}{4}$, is one.

3.10 Complete the verification that the inequalities in Proposition 3.2.3 are sufficient for a Nash equilibrium in a 2 × 2 game.

3.10 Answer: We have to show that the two inequalities

$$(3)\ E_{\text{II}}(X^*, 1) \le E_{\text{II}}(X^*, Y^*)$$
$$(4)\ E_{\text{II}}(X^*, 2) \le E_{\text{II}}(X^*, Y^*)$$

implies that $E_{\text{II}}(X^*, Y) \le E_{\text{II}}(X^*, Y^*)$ for any strategy $Y = (y, 1 - y) \in S_2$. Multiply (3) by y and (4) by $1 - y$ and add to get

$$y E_{\text{II}}(X^*, 1) + (1 - y) E_{\text{II}}(X^*, 2) = E_{\text{II}}(X^*, Y) \le\ E_{\text{II}}(X^*, Y^*),$$

which is what we needed to show.

3.11 Apply the method of this section to analyze the modified study–party game:

I/II	Study	Party
Study	(2, 2)	(3, 1)
Party	(1, 3)	(4, 4)

Find all Nash equilibria and graph the rational reaction sets.

3.11 Answer: We calculate the best response functions

$$x = BR_1(y) = \arg\max_{0 \le x \le 1} (x, 1 - x) \begin{bmatrix} 2 & 3 \\ 1 & 4 \end{bmatrix} \begin{bmatrix} y \\ 1 - y \end{bmatrix} = \begin{cases} 0 & 0 \le y < \frac{1}{2} \\ [0, 1] & y = \frac{1}{2} \\ 1 & \frac{1}{2} < y \le 1 \end{cases}$$

and

$$y = BR_2(x) = \arg\max_{0 \le y \le 1} (x, 1 - x) \begin{bmatrix} 2 & 1 \\ 3 & 4 \end{bmatrix} \begin{bmatrix} y \\ 1 - y \end{bmatrix} = \begin{cases} 0 & 0 \le x < \frac{1}{2} \\ [0, 1] & x = \frac{1}{2} \\ 1 & \frac{1}{2} < x \le 1 \end{cases}$$

Graphing these two functions shows they intersect at $(0, 0)$, $(\frac{1}{2}, \frac{1}{2})$, $(1, 1)$.

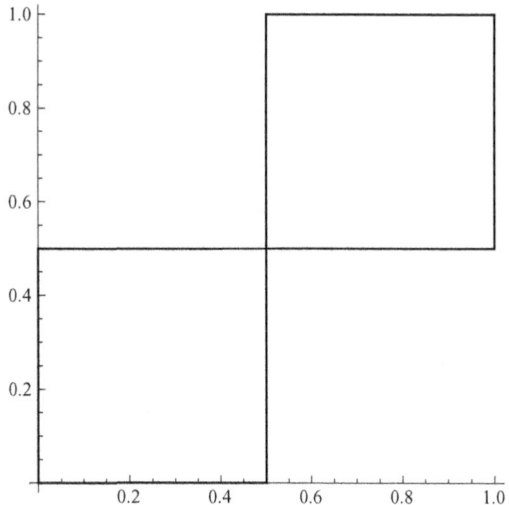

The Nash equilibria are $X_1 = (\frac{1}{2}, \frac{1}{2})$, $Y_1 = (\frac{1}{2}, \frac{1}{2})$, $X_2 = (0, 1) = Y_2$, $X_3 = (1, 0) = Y_3$.

3.12 Consider the Stop–Go game. Two drivers meet at an intersection at the same time. They have the options of stopping and waiting for the other driver to continue, or going.

Here is the payoff matrix in which the player who stops while the other goes loses a bit less than if they both stopped.

I/II	Stop	Go
Stop	1, 1	$1 - \varepsilon, 2$
Go	2, $1 - \varepsilon$	0, 0

Assume that $0 < \varepsilon < 1$. Find all Nash equilibria.

3.12 Answer: There are two pure Nash equilibria at (Stop, Go) and (Go, Stop). The mixed Nash is $X^* = Y^* = (\frac{1-\varepsilon}{2-\varepsilon}, \frac{1}{2-\varepsilon})$. To see where that comes from, assume that both rows and columns are played with positive probabilities. Using equality of payoffs, we solve the equations

$$(x, 1 - x) \begin{bmatrix} 1 & 2 \\ 1 - \varepsilon & 0 \end{bmatrix} = (v_{\text{II}}, v_{\text{II}}) \Rightarrow x + (1 - \varepsilon)(1 - x) = 2x \Rightarrow x = \frac{1 - \varepsilon}{2 - \varepsilon},$$

and

$$\begin{bmatrix} 1 & 1 - \varepsilon \\ 2 & 0 \end{bmatrix} \begin{bmatrix} y \\ 1 - y \end{bmatrix} = \begin{bmatrix} v_{\text{I}} \\ v_{\text{I}} \end{bmatrix} \Rightarrow y + (1 - \varepsilon)(1 - y) = 2y \Rightarrow y = \frac{1 - \varepsilon}{2 - \varepsilon}.$$

The payoffs for the mixed Nash is $v_{\text{I}} = v_{\text{II}} = \frac{2-2\varepsilon}{2-\varepsilon}$. Observe that the probability both players Go is $\left(\frac{1}{2-\varepsilon}\right)^2$. It is a dangerous intersection.

3.13 Determine all Nash equilibria and graph the rational reaction sets for the game.

$$\begin{bmatrix} (-10, 5) & (2, -2) \\ (1, -1) & (-1, 1) \end{bmatrix}.$$

3.13 Answer: To find the best response functions calculate

$$x = BR_1(y) = \arg\max_{0 \le x \le 1} (x, 1-x) \begin{bmatrix} -10 & 2 \\ 1 & -1 \end{bmatrix} \begin{bmatrix} y \\ 1-y \end{bmatrix} = \begin{cases} 0 & \frac{3}{14} < y \le 1 \\ [0, 1] & y = \frac{3}{14} \\ 1 & 0 \le y < \frac{3}{14} \end{cases}$$

and

$$y = BR_2(x) = \arg\max_{0 \le y \le 1} (x, 1-x) \begin{bmatrix} 5 & -2 \\ -1 & 1 \end{bmatrix} \begin{bmatrix} y \\ 1-y \end{bmatrix} = \begin{cases} 0 & 0 \le x < \frac{2}{9} \\ [0, 1] & x = \frac{2}{9} \\ 1 & \frac{2}{9} < x \le 1. \end{cases}$$

Here is the figure:

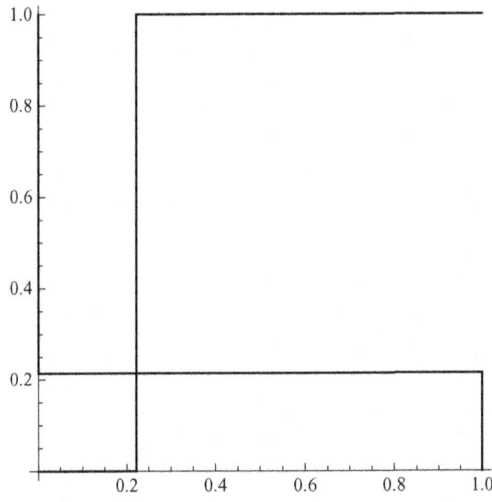

The only Nash is $X = (\frac{2}{9}, \frac{7}{9})$, $Y = (\frac{3}{14}, \frac{11}{14})$, with payoffs $-\frac{4}{7}, \frac{1}{3}$. The rational reaction sets intersect at only this Nash.

3.14 There are two companies each with exactly one job opening. Suppose firm 1 offers the pay p_1 and firm 2 offers pay p_2 with $p_1 < p_2 < 2p_1$. There are two prospective applicants each of whom can apply to only one of the two firms. They make simultaneous and independent decisions. If exactly one applicant applies to a company, that applicant gets the job. If both apply to the same company, the firm flips a fair coin to decide who is hired and the other is unemployed (payoff zero).

(a) Find the game matrix.

3.14.a Answer: The pure strategies are applying to one of the two firms. The matrix is

I/II	Apply firm1	Apply firm2
Apply firm1	$\frac{1}{2}p_1, \frac{1}{2}p_1$	p_1, p_2
Apply firm2	p_2, p_1	$\frac{1}{2}p_2, \frac{1}{2}p_2$

(b) Find all Nash equilibria.

3.14.b Answer: Since $p_1 < p_2 < 2p_1$, there are two pure Nash equilibria in which one player applies to firm 1 and the other applies to firm 2. Of course, since they make their decision simultaneously and independently, they can't really arrange this ahead of time.

There is a mixed Nash equilibrium at $X^* = Y^* = (\frac{2p_1-p_2}{p_1+p_2}, \frac{2p_2-p_1}{p_1+p_2})$, which is obtained by equality of payoffs:

$$\frac{1}{2}p_1y_1 + p_1y_2 = p_2y_1 + \frac{1}{2}p_2y_2$$

$$y_1 + y_2 = 1,$$

which has solution $y_2 = \frac{2p_2-p_1}{p_1+p_2}$. Under the mixed Nash equilibrium, each player is more likely to apply to firm 2, in the proportion of the firm's pay to the total. The expected payoff to each player is $\frac{3}{2}\frac{p_1p_2}{p_1+p_2}$.

3.15 This problem looks at the general 2×2 game to compute a mixed Nash equilibrium:

$$\begin{bmatrix} (A, a) & (B, b) \\ (C, c) & (D, d) \end{bmatrix}.$$

(a) Assume $a - b + d - c \neq 0, d \neq c$. Use equality of payoffs to find $X^* = (x, 1 - x)$.

3.15.a Answer: Let $X^* = (x, 1 - x)$. By Equality of Payoffs, assuming both columns are played with positive probability, we get the equations

$$E_{\text{II}}(X^*, 1) = E_{\text{II}}(X^*, 2) \Rightarrow ax + c(1 - x) = bx + d(1 - x)$$

$$\Rightarrow x = \frac{d - c}{a - b + d - c}.$$

Then $X^* = (\frac{d-c}{a-b+d-c}, \frac{a-b}{a-b+d-c})$. In order for this to work X^* must be a legitimate strategy with positive components.

(b) What do you have to assume about A, B, C, D to find $Y^* = (y, 1 - y), 0 < y < 1$? Now find y.

3.15.b Answer: Let $Y^* = (y, 1 - y)$. Assuming both rows are played with positive probability,

$$E_{\text{I}}(1, Y^*) = E_{\text{I}}(2, Y^*) \Rightarrow Ay + B(1 - y) = Cy + D(1 - y) \Rightarrow y = \frac{D - B}{A - C + D - B}.$$

This requires that $A - C + D - B \neq 0$ and $D \neq B$. Then $Y^* = (\frac{D-B}{A-C+D-B}, \frac{A-C}{A-C+D_B})$ and the components must be positive.

3.16 This game is the nonzero sum version of Pascal's wager (see Example 1.21).

You/God	God Reveals	God Hidden
Believe	α, A	$-\beta, B$
Don't Believe	$-\gamma, -\Gamma$	$0, -\Delta$

Since God choosing to not exist is paradoxical, we change God's strategies to Reveals, or is Hidden. Your payoffs are explained as in Example 1.21. If you choose Believe, God obtains the positive payoffs $A > 0$, $B > 0$. If you choose to Not Believe, and God chooses Reveal, then you receive $-\gamma$ but God also receives $-\Gamma$. If God chooses to remain Hidden, then He receives $-\Delta$ if you choose to Not Believe. Assume $A, B, \Gamma, \Delta > 0$ and $\alpha, \beta, \gamma > 0$.

(a) Determine when there are only pure Nash equilibria and find them under those conditions.

3.16.a Answer: If $A > B$ then there is a pure Nash at (α, A); if $-\Delta > -\Gamma$ then there is a pure Nash at $(0, -\Delta)$.

(b) Find conditions under which a single mixed Nash equilibrium exists and determine what it is.

3.16.b Answer: Using the equality of payoff theorem or the previous problem, we have

$$X^* = \left(\frac{\Gamma - \Delta}{A + \Gamma - B - \Delta}, \frac{A - B}{A + \Gamma - B - \Delta} \right) \; Y^* = \left(\frac{\beta}{\alpha + \beta + \gamma}, \frac{\alpha + \gamma}{\alpha + \beta + \gamma} \right).$$

This is a mixed strategy if $0 < \frac{\Gamma - \Delta}{A + \Gamma - B - \Delta} < 1$ and $0 < \frac{\beta}{\alpha + \beta + \gamma} < 1$.

(c) Find player I's best response to the strategy $Y^0 = (\frac{1}{2}, \frac{1}{2})$.

3.16.c Answer: We have $E_I(X, Y^0) = X A Y^{0T} = f(x)$, where $f(x) = x \frac{\alpha - \beta + \gamma}{2} - \frac{\gamma}{2}$. Player I finds his best response by maximizing $f(x)$ over $x \in [0, 1]$. We easily find

$$x^* = \begin{cases} 1, & \text{if } \frac{\alpha - \beta + \gamma}{2} - \frac{\gamma}{2} > 0; \\ 0, & \text{if } \frac{\alpha - \beta + \gamma}{2} - \frac{\gamma}{2} < 0; \\ [0, 1], & \text{if } \frac{\alpha - \beta + \gamma}{2} - \frac{\gamma}{2} = 0. \end{cases}$$

If $x^* = 1$, $E_I(X^*, Y^0) = \frac{\alpha - \beta}{2}$. If $x^* < 1$, $E_I(X^*, Y^0) = \frac{-\gamma}{2}$. As long as $\alpha > \beta$, player I should believe all the time.

3.17 Verify by checking against pure strategies that the mixed strategies $X^* = (\frac{3}{4}, 0, \frac{1}{4})$ and $Y^* = (0, \frac{1}{3}, \frac{2}{3})$ is a Nash equilibrium for the game with matrix

$$\begin{bmatrix} (x, 2) & (3, 3) & (1, 1) \\ (y, y) & (0, z) & (2, w) \\ (a, 4) & (5, 1) & (0, 7) \end{bmatrix},$$

where x, y, z, w, and a are arbitrary.

3.17 Answer: We need to check that $E_I(X^*, Y^*) \geq E_I(i, Y^*)$, $i = 1, 2, 3$, and $E_{II}(X^*, Y^*) \geq E_{II}(X^*, j)$, $j = 1, 2, 3$. Calculating, we get

$$E_I(X^*, Y^*) = X^* A Y^{*T} = \frac{5}{3}, \quad E_I(i, Y^*) = \left(\frac{5}{3}(i = 1), \frac{4}{3}(i = 2), \frac{5}{3}(i = 3)\right).$$

Each of these numbers is $\leq \frac{5}{3}$. Next,

$$E_{II}(X^*, Y^*) = X^* B Y^{*T} = \frac{5}{2}, \quad E_{II}(X^*, j) = \left(\frac{5}{2}(j = 1), \frac{5}{2}(j = 2), \frac{5}{2}(j = 3)\right).$$

Each of these are actually equal to $\frac{5}{2}$. We conclude that X^*, Y^* is indeed a Nash equilibrium.

3.18 Two radio stations (WSUP and WHAP) have to choose formats for their broadcasts. There are three possible formats: Rhythm and Blues (RB), Elevator Music (EM), or all talk (AT). The audiences for the three formats are 50%, 30%, and 20%, respectively. If they choose the same formats they will split the audience for that format equally, while if they choose different formats, each will get the total audience for that format.

(a) Model this as a nonzero sum game.

3.18.a Answer: The payoff matrix is

WSUP/WHAP	RB	EM	AT
RB	25, 25	50, 30	50, 20
EM	30, 50	15, 15	30, 20
AT	20, 50	20, 30	10, 10

The numbers in the matrix represent the percent of the market obtained by the radio station.

(b) Find all the Nash equilibria.

3.18.b Answer: There are two pure Nash equilibria:

WSUP/WHAP	RB	EM	AT
RB	25, 25	50, 30	50, 20
EM	30, 50	15, 15	30, 20
AT	20, 50	20, 30	10, 10

In addition, the third row and column may be dropped by dominance. This leaves us with finding a mixed Nash for a 2×2 game.

Using equality of payoffs, we have

$$25y + 50(1 - y) = 30y + 15(1 - y) \Rightarrow y = \frac{7}{8},$$

and, by symmetry, $x = \frac{7}{8}$. Therefore, a mixed Nash is $X^* = (\frac{7}{8}, \frac{1}{8}, 0) = Y^*$, with payoffs $E_{WSUP} = E_{WHAP} = 28.125$. Note that both players would do better with the pure Nash's. Later, we will see that the mixed Nash is evolutionary stable.

3.19 In a modified story of the prodigal son, a man had two sons, the prodigal and the one who stayed home. The man gave the prodigal son his share of the estate, which he squandered, and told the son who stayed home that all that he (the father) has is his (the son's). When the man died, the prodigal son again wanted his share of the estate. They each tell the judge (it ends up in court) a share amount they would be willing to take, either $\frac{1}{4}, \frac{1}{2}$, or $\frac{3}{4}$. Call the shares for each player $I_i, II_i, i = 1, 2, 3$. If $I_i + II_j > 1$, all the money goes to the game theory society. If $I_i + II_j \leq 1$, then each gets the share they asked for and the rest goes to an antismoking group.

(a) Find the game matrix and find all pure Nash equilibria.

3.19.a Answer: The pure strategies for each son is the share of the estate to claim. The matrix is

I/II	0.25	0.5	0.75
0.25	(0.25, 0.25)	(0.25, 0.50)	(0.25, 0.75)
0.5	(0.50, 0.25)	(0.50, 0.50)	(0, 0)
0.75	(0.75, 0.25)	(0, 0)	(0, 0)

There are three pure Nash equilibria: (1) row 1, column 3, (2) row 2, column 2, and (3) row 3, column 1.

(b) Find at least two distinct mixed Nash equilibria using the equality of payoffs Theorem C.4?

3.19.b Answer:

The table contains the pure and mixed Nash points as well as the payoffs.

X	Y	E_I	E_{II}
$(\frac{1}{3}, \frac{1}{6}, \frac{1}{2})$	$(\frac{1}{3}, \frac{1}{6}, \frac{1}{2})$	$\frac{1}{4}$	$\frac{1}{4}$
$(\frac{2}{3}, \frac{1}{3}, 0)$	$(0, \frac{1}{2}, \frac{1}{2})$	$\frac{1}{4}$	$\frac{1}{2}$
$(\frac{1}{3}, 0, \frac{2}{3})$	$(\frac{1}{3}, 0, \frac{2}{3})$	$\frac{1}{4}$	$\frac{1}{4}$
$(0, \frac{1}{2}, \frac{1}{2})$	$(\frac{2}{3}, \frac{1}{3}, 0)$	$\frac{1}{2}$	$\frac{1}{4}$
$(1, 0, 0)$	$(0, 0, 1)$	$\frac{1}{4}$	$\frac{3}{4}$
$(0, 1, 0)$	$(0, 1, 0)$	$\frac{1}{2}$	$\frac{1}{2}$
$(0, 0, 1)$	$(1, 0, 0)$	$\frac{3}{4}$	$\frac{1}{4}$

To use the equality of payoffs theorem to find these you must consider all possibilities in which at least two rows (or columns) are played with positive probability. For example, suppose that columns 2 and 3 are used by player II with positive probability. Equality of payoffs then tells us

$$E_{II}(X^*, 2) = E_{II}(X^*, 3) \Rightarrow 0.5x_1 + 0.5x_2 = 0.75x_1 \Rightarrow x_2 = 0.5x_1.$$

Hence, $X = (x_1, 0.5x_1, 1 - 1.5x_1)$ is a solution as long as $0 < x_1 < \frac{2}{3}$.

If $x_1 = 0$, we get $X = (0, 0, 1)$, which is part of a pure Nash but is not obtained from equality of payoffs. If $x_1 = \frac{2}{3}$ then $X = (\frac{2}{3}, \frac{1}{3}, 0)$. By equality of payoffs we then obtain

$$E_I(1, Y^*) = E_I(2, Y^*) \Rightarrow 0.25y_1 + 0.25y_2 + 0.25y_3 = 0.5y_1 + 0.5y_2 \Rightarrow y_1 + y_2 = 0.5.$$

Then, $Y = (y_1, \frac{1}{2} - y_1, \frac{1}{2})$ is the solution if $0 \leq y_1 < \frac{1}{2}$. Observe that $y_1 = \frac{1}{2}$ is not consistent with our original assumption that $y_2 > 0, y_3 > 0$, but $y_1 = 0$ is. In that case, $Y = (0, \frac{1}{2}, \frac{1}{2})$ and we have the Nash equilibrium $X = (\frac{2}{3}, \frac{1}{3}, 0)$, $Y = (0, \frac{1}{2}, \frac{1}{2})$. Any of the remaining Nash equilibria in the tables are obtained similarly.

If we assume $Y = (y_1, y_2, y_3)$ is part of a mixed Nash with $y_j > 0$, $j = 1, 2, 3$, then equality of payoffs gives the equations

$$0.25x_1 + 0.25x_2 + 0.25x_3 = v_{\mathrm{I}},$$
$$0.5x_1 + 0.5x_2 = v_{\mathrm{I}},$$
$$0.75x_1 = v_{\mathrm{I}},$$
$$x_1 + x_2 + x_3 = 1.$$

The first and fourth equation give $v_{\mathrm{I}} = 0.25$, Then the third equation gives $x_1 = \frac{1}{3}$, then $x_2 = \frac{1}{6}$, and finally $x_3 = \frac{1}{2}$. Since $X = (\frac{1}{3}, \frac{1}{6}, \frac{1}{2})$, we then use equality of payoffs to find Y. The equations are

$$0.25y_1 + 0.25y_2 + 0.25y_3 = v_{\mathrm{II}},$$
$$0.5y_1 + 0.5y_2 = v_{\mathrm{II}},$$
$$0.75y_1 = v_{\mathrm{II}},$$
$$y_1 + y_2 + y_3 = 1.$$

This has the same solution as before $Y = X$, $v_{\mathrm{I}} = v_{\mathrm{II}} = 0.25$, and this is consistent with our original assumption that all components of Y are positive.

3.20 Willard is a salesman with an expense account for travel. He can steal (S) from the account by claiming false expenses or be honest (H) and accurately claim the expenses incurred. Willard's boss is Fillmore. Fillmore can check (C) into the expenses claimed or trust (T) that Willard is honest. If Willard cheats on his expenses he benefits by the amount b assuming he isn't caught by Fillmore. If Fillmore doesn't check then Willard gets away with cheating. If Willard is caught cheating he incurs costs p, which may include getting fired and paying back the amount he stole. Since Willard is a clever thief, we let $0 < \alpha < 1$ be the probability that Willard is caught if Fillmore investigates.

If Fillmore investigates he incurs cost c no matter what. Finally, let λ be Fillmore's loss if Willard cheats on his expenses but is not caught. Assume all of $b, p, \alpha, \lambda > 0$.

(a) Find the game bimatrix.

3.20.a Answer: Each player has two strategies. The game matrix is

Willard/Fillmore	Trust	Check
Honest	$0, 0$	$0, -c$
Steal	$b, -\lambda$	$\alpha(-p) + (1 - \alpha)b, -\lambda(1 - \alpha) - c$

The only entry that needs explanation is (Steal, Check). If Willard steals and Fillmore checks, we have the expected payoff to Willard

payoff if caught $\times$ *Prob*(caught) + payoff if not caught $\times$ *Prob*(not caught)
$= (-p)\alpha + (b)(1 - \alpha)$

The expected payoff to Fillmore is

> payoff to Fillmore if not caught × *Prob*(not caught) − cost of checking
> $= (-\lambda)(1 - \alpha) - c.$

(b) Determine conditions on the parameters so that there is one mixed Nash equilibrium and find it.

3.20.b Answer: If we assume that both rows and columns are played with positive probability, equality of payoffs gives the solution

$$X^* = \left(1 - \frac{c}{\alpha\lambda}, \frac{c}{\alpha\lambda}\right) \quad \text{and} \quad Y^* = \left(1 - \frac{b}{\alpha(b + p)}, \frac{b}{\alpha(b + p)}\right).$$

The expected payoffs are

$$E_\text{I}(X^*, Y^*) = 0, \quad \text{and} \quad E_\text{II}(X^*, Y^*) = -\frac{c}{\alpha}.$$

Now, in order for X^*, Y^* to be a completely mixed strategies with both components positive, we need

$$c < \alpha\lambda \quad \text{and} \quad b < \frac{\alpha p}{1 - \alpha}.$$

If $c > \alpha\lambda$ then $-\lambda > -\lambda(1 - \alpha) - c$ and we would have a pure Nash equilibrium at Steal, Trust. If $b > \frac{\alpha p}{1-\alpha}$, and $c < \alpha\lambda$, then there is a pure Nash equilibrium at Steal, Check.

3.21 Use the equality of payoffs Theorem C.4 to solve the welfare game. In the welfare game the state, or government, wants to aid a pauper if he is looking for work but not otherwise. The pauper looks for work only if he cannot depend on welfare, but he may not be able to find a job even if he looks. The game matrix is

G/P	Look for work	Be a bum
Welfare	(3, 2)	(−1, 3)
No welfare	(−1, 1)	(0, 0)

Find all Nash equilibria and graph the rational reaction sets.

3.21 Answer: We find the best response functions. For the government,

$$x = BR_2(y) = \arg\max_{0 \le x \le 1} (x, 1 - x)\begin{bmatrix} 3 & -1 \\ -1 & 0 \end{bmatrix}\begin{bmatrix} y \\ 1 - y \end{bmatrix}$$

$$= \begin{cases} 0, & \text{if } 0 \le y < \frac{1}{5}; \\ [0, 1], & \text{if } y = \frac{1}{5}; \\ 1, & \text{if } \frac{1}{5} < y \le 1. \end{cases}$$

Similarly,

$$y = BR_1(x) = \arg\max_{0 \le y \le 1} (x, 1-x) \begin{bmatrix} 2 & 3 \\ 1 & 0 \end{bmatrix} \begin{bmatrix} y \\ 1-y \end{bmatrix} = \begin{cases} 1, & \text{if } 0 \le x < \frac{1}{2}; \\ [0,1], & \text{if } x = \frac{1}{2}; \\ 0, & \text{if } \frac{1}{2} < x \le 1. \end{cases}$$

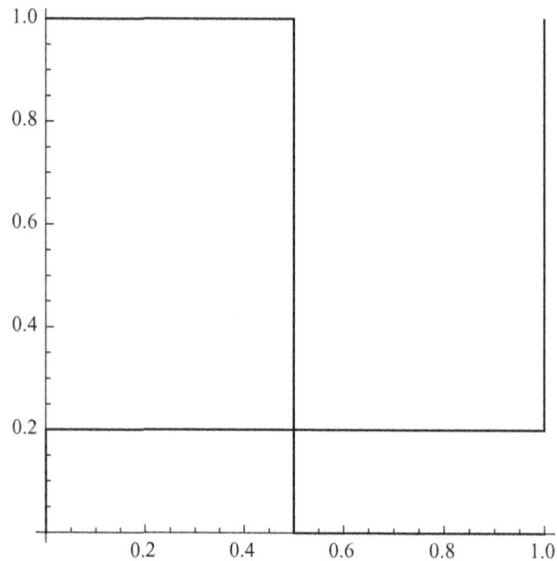

The Nash equilibrium is $X = (\frac{1}{2}, \frac{1}{2})$, $Y = (\frac{1}{5}, \frac{4}{5})$ with payoffs $E_I = -\frac{1}{5}$, $E_{II} = \frac{3}{2}$. The government should aid half the paupers, but 80% of paupers should be bums (or the government predicts that 80% of welfare recipients will not look for work). The rational reaction sets intersect only at the mixed Nash point.

3.3 Interior Mixed Nash Points by Calculus

Problems

3.22 Write down the equations

$$\left. \begin{aligned} \sum_{j=1}^{m} y_j [a_{kj} - a_{nj}] = 0, && k = 1, 2, \ldots, n-1; \\ \sum_{i=1}^{n} x_i [b_{is} - b_{im}] = 0, && s = 1, 2, \ldots, m-1; \\ x_n = 1 - \sum_{i=1}^{n-1} x_i, \quad y_m = 1 - \sum_{j=1}^{m-1} y_j \end{aligned} \right\}$$

for the game.

$$\begin{bmatrix} (2,0) & (3,2) & (4,1) \\ (0,2) & (4,0) & (3,1) \end{bmatrix}.$$

Try to solve the equations. What, if anything, goes wrong?

3.22 Answer: The equations become with $n = 2$, $m = 3$,

$$k = 1, \quad y_1(a_{11} - a_{21}) + y_2(a_{12} - a_{22}) + y_3(a_{13} - a_{23}) = 0,$$
$$s = 1, \quad x_1(b_{11} - b_{13}) + x_2(b_{21} - b_{23}) = 0,$$

which reduces to

$$2y_1 - y_2 + y_3 = 0 \quad \text{and} \quad y_1 + y_2 + y_3 = 1,$$
$$-x_1 + x_2 = 0 \quad \text{and} \quad x_1 + x_2 = 1.$$

We have for $X = (x_1, x_2)$ the solution $X^* = (\frac{1}{2}, \frac{1}{2})$. The player II part of the Nash equilibrium has two possibilities, $Y^* = (\frac{1}{3}, \frac{2}{3}, 0)$, or $Y^* = (0, \frac{1}{2}, \frac{1}{2})$, but you can't get this from the equations $2y_1 - y_2 + y_3 = 0$ and $y_1 + y_2 + y_3 = 1$ because it is an underdetermined system. It reduces to $y_1 = 2y_2 - 1$, and you can see that $y_1 = \frac{1}{3}, y_2 = \frac{2}{3}$ is one possible solution and so is $y_1 = 0, y_2 = \frac{1}{2}$. The equations require that all components be > 0.

3.23 The game matrix in the welfare problem is

G/P	Look for work	Be a bum
Welfare	$(3, 2)$	$(-1, 3)$
No welfare	$(-1, 1)$	$(0, 0)$

Write the system of equations for an interior Nash and solve them.

3.23 Answer: The system is $0 = 4y_1 - y_2$, $y_1 + y_2 = 1$, which gives $y_1 = \frac{1}{5}, y_2 = \frac{4}{5}$.

3.24 Consider the game with bimatrix

$$\begin{bmatrix} (5,4) & (3,6) \\ (6,3) & (1,1) \end{bmatrix}.$$

(a) Find the safety levels and maxmin strategies for each player.

3.24.a Answer: For player I, $A = \begin{bmatrix} 5 & 3 \\ 6 & 1 \end{bmatrix}$ and there is a saddle at row 1, column 2. The safety level for player I is $v(A) = 3$ and the maxmin strategy is $X^A = (1, 0)$.

For player II, $B = \begin{bmatrix} 4 & 6 \\ 3 & 1 \end{bmatrix}$, and $B^T = \begin{bmatrix} 4 & 3 \\ 6 & 1 \end{bmatrix}$. The safety level for player II is $v(B^T) = 3$, and the maxmin strategy for player II is $X^{B^T} = (1, 0)$.

(b) Find all Nash equilibria using the equality of payoffs theorem for the mixed strategy.

3.24.b Answer: There are no pure Nash equilibria. The mixed Nash is obtained from

$$4x + 3(1 - x) = 6x + 1 - x \Rightarrow x = \tfrac{1}{2} \quad \text{and}$$
$$5y + 3(1 - y) = 6y + 1 - y \Rightarrow y = \tfrac{2}{3}.$$

The mixed Nash is $X = (\tfrac{1}{2}, \tfrac{1}{2})$, $Y = (\tfrac{2}{3}, \tfrac{1}{3})$. In addition, the payoff to player I is $\tfrac{7}{2}$ and to player II it is $\tfrac{13}{3}$.

(c) Verify that the mixed Nash equilibrium is individually rational.

3.24.c Answer: Two strategies are individually rational if $E_I(X^*, Y^*) \geq v(A)$ and $E_{II}(X^*, Y^*) \geq v(B^T)$, that is, they get at least their safety levels. In our case, $E_I(X^*, Y^*) = \tfrac{7}{2} \geq v(A) = 3$ and $E_{II}(X^*, Y^*) = \tfrac{13}{3} \geq v(B^T) = 3$.

(d) Verify that $X^* = (\tfrac{1}{4}, \tfrac{3}{4})$, $Y^* = (\tfrac{5}{8}, \tfrac{3}{8})$ is not a Nash equilibrium.

3.24.d Answer: We have to check if $E_I(X^*, Y^*) \geq E_I(i, Y^*)$, $i = 1, 2$ and $E_{II}(X^*, Y^*) \geq E_{II}(X^*, j)$, $j = 1, 2)$. If any of these inequalities fail, then (X^*, Y^*) is not a Nash equilibrium. We have

$$E_I(X^*, Y^*) = 4.15625 \quad \text{and} \quad E_I(i, Y^*) = \begin{cases} 4.25 & i = 1, \\ 4.125 & i = 2. \end{cases}$$

The Nash condition fails for X^*. Let's check for Y^*:

$$E_{II}(X^*, Y^*) = 2.875 \quad \text{and} \quad E_{II}(X^*, j) = \begin{cases} 3.25 & j = 1, \\ 2.25 & j = 2. \end{cases}$$

The Nash condition fails for Y^*.

3.25 In this problem, we consider the analog of the invertible matrix Theorem B.2 for zero sum games. Consider the nonzero sum game $(A_{n \times n}, B_{n \times n})$ and suppose that A^{-1} and B^{-1} exist.

(a) Show that if $J_n A^{-1} J_n^T \neq 0$, $J_n B^{-1} J_n^T \neq 0$ and

$$X^* = \frac{J_n B^{-1}}{J_n B^{-1} J_n^T}, \qquad Y^{*T} = \frac{A^{-1} J_n^T}{J_n A^{-1} J_n^T}$$

are strategies, then (X^*, Y^*) is a Nash equilibrium, and

$$v_I = \frac{1}{J_n A^{-1} J_n^T} = E_I(X^*, Y^*), \qquad v_{II} = \frac{1}{J_n B^{-1} J_n^T} = E_{II}(X^*, Y^*).$$

3.25.a Answer: Assuming the formulas actually give strategies we will show that the strategies are a Nash equilibrium. We have

$$X^* A Y^{*T} = \frac{J_n B^{-1}}{J_n B^{-1} J_n^T} A \frac{A^{-1} J_n^T}{J_n A^{-1} J_n^T} = \frac{J_n B^{-1} J_n^T}{J_n B^{-1} J_n^T} \cdot \frac{1}{J_n A^{-1} J_n^T} = \frac{1}{J_n A^{-1} J_n^T}.$$

and for any strategy $X \in S_n$,

$$XAY^{*T} = XA\frac{A^{-1}J_n^T}{J_nA^{-1}J_n^T} = \frac{XJ_n^T}{J_nA^{-1}J_n^T} = \frac{1}{J_nA^{-1}J_n^T} = v_1.$$

Similarly,

$$X^*BY^{*T} = \frac{J_nB^{-1}}{J_nB^{-1}J_n^T}B\frac{A^{-1}J_n^T}{J_nA^{-1}J_n^T} = \frac{1}{J_nB^{-1}J_n^T}\frac{J_nA^{-1}J_n^T}{J_nA^{-1}J_n^T} = \frac{1}{J_nB^{-1}J_n^T}.$$

and for any $Y \in S_n$,

$$X^*BY^T = \frac{J_nB^{-1}}{J_nB^{-1}J_n^T}BY^T = \frac{J_nY^T}{J_nB^{-1}J_n^T} = \frac{1}{J_nB^{-1}J_n^T} = v_{II}.$$

(b) Use the previous result to solve the game

$$\begin{bmatrix} (0,9) & (8,2) & (7,3) \\ (4,7) & (2,8) & (9,1) \\ (10,-1) & (5,5) & (1,9) \end{bmatrix}.$$

3.25.b Answer: We have

$$A^{-1} = \begin{bmatrix} -\frac{1}{16} & \frac{27}{688} & \frac{29}{344} \\ \frac{1}{8} & -\frac{35}{344} & \frac{7}{172} \\ 0 & \frac{5}{43} & -\frac{2}{43} \end{bmatrix}, \qquad B^{-1} = \frac{1}{604}\begin{bmatrix} 67 & -3 & -22 \\ -64 & 84 & 12 \\ 43 & -47 & 58 \end{bmatrix},$$

$$X^* = \frac{J_nB^{-1}}{J_nB^{-1}J_n^T} = \left(\frac{23}{64}, \frac{17}{64}, \frac{3}{8}\right),$$

and

$$Y^{*T} = \frac{A^{-1}J_n^T}{J_nA^{-1}J_n^T} = \left(\frac{21}{67}, \frac{22}{67}, \frac{24}{67}\right).$$

These are both legitimate strategies. Finally, $v_I = \frac{344}{67}$, $v_{II} = \frac{151}{32}$.

(c) Suppose the game is 2×2 with matrices $A_{2\times2}$, $B_{2\times2}$. What is the analog of the formulas for a saddle point when the matrices do not have an inverse? Use the formulas you obtain to solve the games:

$$(1) \begin{bmatrix} (1,3) & (-1,-2) \\ (-2,-1) & (4,0) \end{bmatrix} \quad \text{and} \quad (2) \begin{bmatrix} (1,3) & (-2,-2) \\ (-2,-1) & (4,0) \end{bmatrix}$$

3.25.c Answer: Set $A^* = \begin{bmatrix} a_{22} & -a_{12} \\ -a_{21} & a_{11} \end{bmatrix}$ and $B^* = \begin{bmatrix} b_{22} & -b_{12} \\ -b_{21} & b_{11} \end{bmatrix}$. Then, from the formulas derived in Problem 3.15 we have, assuming both rows and columns are played with positive probability,

$$X^* = \left(\frac{b_{22} - b_{21}}{b_{11} - b_{12} + b_{22} - b_{21}}, \frac{b_{11} - b_{12}}{b_{11} - b_{12} + b_{22} - b_{21}} \right) = \frac{J_n B^*}{J_n B^* J_n^T},$$

and

$$Y^{*T} = \left(\frac{a_{22} - a_{12}}{a_{11} - a_{21} + a_{22} - a_{12}}, \frac{a_{11} - a_{21}}{a_{11} - a_{21} + a_{22} - a_{12}} \right) = \frac{A^* J_n^T}{J_n A^* J_n^T}.$$

(X^*, Y^*) is a Nash equilibrium if they are strategies, and then

$$v_{\mathrm{I}} = \frac{\det(A)}{J_n A^* J_n^T} = E_{\mathrm{I}}(X^*, Y^*), \qquad v_{\mathrm{II}} = \frac{\det(B)}{J_n B^* J_n^T} = E_{\mathrm{II}}(X^*, Y^*).$$

Of course, if A or B does not have an inverse, then $\det(A) = 0$ or $\det(B) = 0$.

For the game in (1), $A^* = \begin{bmatrix} 4 & 1 \\ 2 & 1 \end{bmatrix}$ and $B^* = \begin{bmatrix} 0 & 2 \\ 1 & 3 \end{bmatrix}$. We get

$$X^* = \frac{J_n B^*}{J_n B^* J_n^T} = \left(\frac{1}{6}, \frac{5}{6} \right), \qquad Y^{*T} = \frac{A^* J_n^T}{J_n A^* J_n^T} = \left(\frac{5}{8}, \frac{3}{8} \right).$$

and

$$v_{\mathrm{I}} = \frac{\det(A)}{J_n A^* J_n^T} = \frac{1}{4}, \qquad v_{\mathrm{II}} = \frac{\det(B)}{J_n B^* J_n^T} = -\frac{1}{3}.$$

There are also two pure Nash equilibria $X_1^* = (0, 1)$, $Y_1^* = (0, 1)$ and $X_2^* = (1, 0)$, $Y_2^* = (1, 0)$.

For the game in (2), $A^* = \begin{bmatrix} 4 & 2 \\ 2 & 1 \end{bmatrix}$ and $B^* = \begin{bmatrix} 0 & 2 \\ 1 & 3 \end{bmatrix}$. We get

$$X^* = \frac{J_n B^*}{J_n B^* J_n^T} = \left(\frac{1}{6}, \frac{5}{6} \right), \qquad Y^{*T} = \frac{A^* J_n^T}{J_n A^* J_n^T} = \left(\frac{2}{3}, \frac{1}{3} \right).$$

and

$$v_{\mathrm{I}} = \frac{\det(A)}{J_n A^* J_n^T} = 0, \qquad v_{\mathrm{II}} = \frac{\det(B)}{J_n B^* J_n^T} = -\frac{1}{3}.$$

In this part of the problem A does not have an inverse and $\det(A) = 0$. There are also two pure Nash equilibria $X_1^* = (0, 1)$, $Y_1^* = (0, 1)$ and $X_2^* = (1, 0)$, $Y_2^* = (1, 0)$.

3.26 Chicken Game. In this version of the game, the matrix is

I/II	Straight	Swerve
Straight	$-a, -a$	$2, 0$
Swerve	$0, 2$	$1, 1$

Find the mixed Nash equilibrium and show that as $a > 0$ increases, so does the expected payoff to each player under the mixed Nash.

3.26 Answer: The mixed Nash is found from equality of payoffs or calculus. We get $X^* = Y^* = (\frac{1}{1+a}, \frac{a}{1+a})$, and the expected payoff to each player is $E_I = E_{II} = \frac{a}{1+a}$. As a increases, so does the expected payoff to each player because $f(a) = \frac{a}{1+a}$ has $f'(a) = \frac{1}{(1+a)^2} > 0$. As a increases, the probability of using row 1 and column 1 decreases.

3.27 Find all possible Nash equilibria for the game and the rational reaction sets:

$$\begin{bmatrix} (a, a) & (0, 0) \\ (0, 0) & (b, b) \end{bmatrix}.$$

Consider all cases $(a > 0, b > 0)$, $(a > 0, b < 0)$, and so on.

3.27 Answer: If $ab < 0$, there is exactly one strict Nash equilibrium: (i) $a > 0, b < 0 \Rightarrow$ the Nash point is $X^* = Y^* = (1, 0)$; (ii) $a < 0, b > 0 \Rightarrow$ the Nash point is $X^* = Y^* = (0, 1)$. The rational reaction sets are simply the lines along the boundary of the square $[0, 1] \times [0, 1]$: either $x = 0, y = 0$ or $x = 1, y = 1$.

If $a > 0, b > 0$ there are three Nash equilibria $X_1 = Y_1 = (1, 0)$, $X_2 = Y_2 = (0, 1)$, and the mixed Nash $X_3 = Y_3 = (\frac{b}{a+b}, \frac{a}{a+b})$. The mixed Nash is obtained from equality of payoffs:

$$xa = (1 - x)b \Rightarrow x = \frac{b}{a + b} \quad \text{and} \quad ya = (1 - y)b \Rightarrow y = \frac{b}{a + b}.$$

If $a < 0, b < 0$, there are three Nash equilibria $X_1 = (1, 0)$, $Y_1 = (0, 1)$, $X_2 = (0, 1)$, $Y_2 = (1, 0)$, and the mixed Nash $X_3 = Y_3 = (\frac{b}{a+b}, \frac{a}{a+b})$. The rational reaction sets are piecewise linear lines as in previous figures which intersect at the Nash equilibria.

3.28 Show that a 2×2 symmetric game

$$A = \begin{bmatrix} a_{11} & a_{12} \\ a_{21} & a_{22} \end{bmatrix}, \qquad B = A^T$$

has exactly the same set of Nash equilibria as does the symmetric game with matrix

$$A' = \begin{bmatrix} a_{11} - a & a_{12} - b \\ a_{21} - a & a_{22} - b, \end{bmatrix}, \qquad B = A'^T$$

for any a, b.

3.28 Answer: The pure Nash equilibria are clearly equivalent. For the interior mixed Nash, the calculus method shows that the partial derivatives in the appropriate variables of the payoff functions lead to equations for the Nash equilibrium independent of a, b.

You may also calculate directly that

$$E'(X, Y) = XA'Y^T = XAY^T - (a \ b)Y^T = E(X, Y) - (a \ b)Y^T.$$

Therefore, $E'(X^*, Y^*) \geq E'(X, Y^*)$ for all X, if and only if $E(X^*, Y^*) \geq E(X, Y^*)$, for all X.

Similarly,

$$E'(X, Y) = E(X, Y) - X\begin{bmatrix} a \\ b \end{bmatrix}$$

and $E'(X^*, Y^*) \geq E'(X^*, Y)$ for all Y if and only if $E(X^*, Y^*) \geq E(X^*, Y)$ for all Y.

3.29 Consider the game with matrix

$$\begin{bmatrix} (2, 0) & (3, 1) & (4, -1) \\ (0, 2) & (4, 0) & (3, 1) \end{bmatrix}.$$

(a) Find the best response sets for each player.

3.29.a Answer: First we define $f(x, y_1, y_2) = XAY^T$, where $X = (x, 1 - x)$, $Y = (y_1, y_2, 1 - y_1 - y_2)$. We calculate

$$f(x, y_1, y_2) = 3 - 3y_1 + y_2 + x(1 + y_1 - 2y_2)$$

and

$$x = BR_1(y_1, y_2) = \begin{cases} 0, & \frac{1}{2} \leq y_2 \leq 1, \text{ and } 0 \leq y_1 < -1 + 2y_2; \\ [0, 1], & \frac{1}{2} \leq y_2 \leq 1 \text{ and } y_1 = -1 + 2y_2; \\ 1, & 0 < 1 + y_1 - 2y_2. \end{cases}$$

Similarly,

$$g(x, y_1, y_2) = XBY^T = 1 + y_1 - y_2 + x(3y_2 - 2)$$

and the maximum of g is achieved at (y_1, y_2), where

$$y_1(x) = \begin{cases} 0, & \frac{2}{3} < x \leq 1 \\ [0, 1], & x = \frac{2}{3} \\ 1, & 0 \leq x < \frac{2}{3} \end{cases} \quad \text{and} \quad y_2(x) = \begin{cases} 0, & 0 \leq x < \frac{2}{3}; \\ [0, 1], & x = \frac{2}{3}; \\ 1, & \frac{2}{3} < x \leq 1. \end{cases}$$

The best response set, $BR_2(x)$, for player II is the set of all pairs $(y_1(x), y_2(x))$. This is obtained by hand or using the simple Mathematica commands:

Maximize $[\{g[x, y_1, y_2], y_1 \geq 0, y_2 \geq 0, y_1 + y_2 \leq 1\}, \{y_1, y_2\}]$.

(b) Find the Nash equilibria and verify that they are in the intersection of the best response sets.

3.29.b Answer: This is a little tricky because we don't know which columns are used by player II with positive probability. However, if we notice that column 3 is dominated by a convex combination of columns 1 and 2, we may drop column 3. Then the game is a 2×2 game and

$$2(1 - x) = x \Rightarrow x = \frac{2}{3} \Rightarrow X = \left(\frac{2}{3}, \frac{1}{3}\right) \quad \text{and} \quad v_{\text{II}} = \frac{2}{3}.$$

Then, for player II

$$2y + 3(1 - y) = 4(1 - y) \Rightarrow y = \frac{1}{3} \quad \text{and} \quad v_{\text{I}} = \frac{8}{3}.$$

The Nash equilibrium is $X^* = \left(\frac{2}{3}, \frac{1}{3}\right)$, $Y^* = \left(\frac{1}{3}, \frac{2}{3}, 0\right)$. Then $y_1 = \frac{1}{3}, y_2 = \frac{2}{3}, x = \frac{2}{3}$, and by definition $\frac{2}{3} \in BR_1\left(\frac{1}{3}, \frac{2}{3}\right)$. Also, $\left(\frac{1}{3}, \frac{2}{3}\right) \in BR_2\left(\frac{2}{3}\right)$.

3.30 A game called the **battle of the sexes** is a game between a husband and wife trying to decide about cooperation. On a given evening, the husband wants to see wrestling at the stadium, while the wife wants to attend a concert at orchestra hall. Neither the husband nor the wife wants to go to what the other has chosen, but neither do they want to go alone to their preferred choice. They view this as a two-person nonzero sum game with matrix

H/W	Wr	Co
Wr	$(2, 1)$	$(-1, -1)$
Co	$(-1, -1)$	$(1, 2)$

If they decide to cooperate and both go to wrestling, the husband receives 2 and the wife receives 1, because the husband gets what he wants and the wife partially gets what she wants. The rest are explained similarly. This is a model of compromise and cooperation. Use the method of this section to find all Nash equilibria. Graph the rational reaction sets.

3.30 Answer: There are three Nash: $X_1 = (0, 1) = Y_1$, $X_2 = (1, 0) = Y_2$, which represents one player always giving in to the other so the one who wins gets 2 and the other gets 1, and $X_3 = \left(\frac{3}{5}, \frac{2}{5}\right)$, $Y_3 = \left(\frac{2}{5}, \frac{3}{5}\right)$ in which they each get payoff $\frac{1}{5}$. Note that the mixed Nash leads to a lower payoff to *both players*.

The best response functions are

$$x = BR_1(y) = \begin{cases} 0, & 0 \le y < \frac{2}{5}; \\ [0, 1]; & y = \frac{2}{5}; \\ 1, & \frac{2}{5} < y \le 1; \end{cases}$$

$$y = BR_2(x) = \begin{cases} 0, & 0 \le x < \frac{3}{5}; \\ [0, 1], & x = \frac{3}{5}; \\ 1, & \frac{3}{5} < x \le 1. \end{cases}$$

The graph becomes

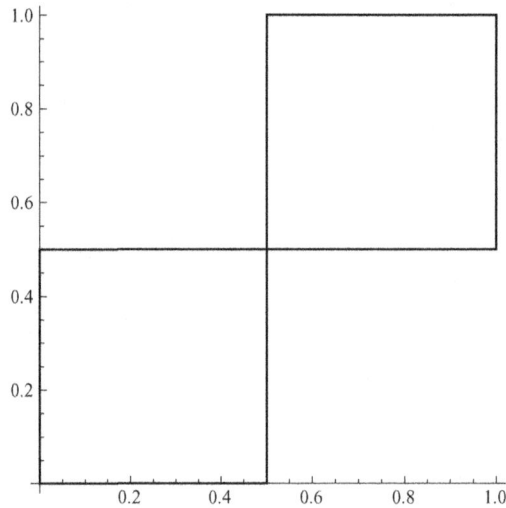

3.31 Hawk–Dove Game. Two companies both want to take over a sales territory. They have the choices of defending the territory and fighting if necessary, or act as if willing to fight but if the opponent fights (F), then backing off (Bo). They look at this as a two-person nonzero sum game with matrix

I/II	F	Bo
F	$(-1, -1)$	$(2, 0)$
Bo	$(0, 2)$	$(0, 0)$

Solve this game and graph the rational reaction sets.

3.31 Answer: There are three Nash equilibria: $X_1 = (1, 0)$, $Y_1 = (0, 1)$, $X_2 = (0, 1)$, $Y_2 = (1, 0)$, $X_3 = (\frac{2}{3}, \frac{1}{3}) = Y_3$. The last Nash gives payoffs $(0, 0)$. The best response functions are

$$x = BR_1(y) = \begin{cases} 0 & \frac{2}{3} < y \le 1; \\ [0, 1] & y = \frac{2}{3}; \\ 1 & 0 \le y < \frac{2}{3}; \end{cases} \quad \text{and} \quad y = BR_2(x) = \begin{cases} 0 & \frac{2}{3} < x \le 1; \\ [0, 1] & x = \frac{2}{3}; \\ 1 & 0 \le x < \frac{2}{3}. \end{cases}$$

The Mathematica commands to solve this and all the 2×2 games is the following:

```
A = {{-1, 2}, {0, 0}}
f[x_, y_] = {x, 1 - x}.A.{y, 1 - y}
Maximize[{f[x, y], 0 <= x <= 1}, {x}]
B = {{-1, 0}, {2, 0}}
g[x_, y_] = {x, 1 - x}.B.{y, 1 - y}
Maximize[{g[x, y], 0 <= y <= 1}, {y}]
Show[{Graphics[Line[{{0, 1}, {0, 2/3}, {1, 2/3}, {1, 0}}]],
  Graphics[Line[{{0, 1}, {2/3, 1}, {2/3, 0}, {1, 0}}]]}, Axes -> True]
```

The last command produces the graph:

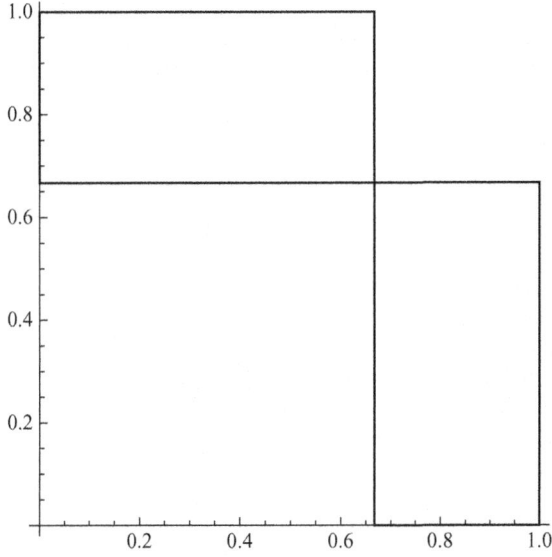

Hawk-Dove best response sets.

3.32 Stag–Hare Game. Two hunters are pursuing a stag. Each hunter has the choice of going after the stag (S), which will be caught if they both go after it and it will be shared equally, or peeling off and going after a rabbit (R). Only one hunter is needed to catch the rabbit and it will not be shared. Look at this as a nonzero sum two-person game with matrix

I/II	S	R
S	$(2, 2)$	$(0, 1)$
R	$(1, 0)$	$(1, 1)$

This assumes that they each prefer stag meat to rabbit meat, and they will each catch a rabbit if they decide to peel off. Solve this game and graph the rational reaction sets.

3.32 Answer: The pure Nash $X = Y = (1, 0)$ in which they both go after the stag gives a payoff of 2 to each player. The mixed Nash $X = Y = (\frac{1}{2}, \frac{1}{2})$ and the pure Nash $X = Y = (0, 1)$, in which they go after the rabbit either all the time or half the time, gives a payoff of 1 to each player. Observe that $X = Y = (1, 0)$ in which they both go after a stag is *Pareto-optimal* in that a player cannot increase her payoff without simultaneously decreasing the other player's payoff. That is not true for $X = Y = (0, 1)$. On the other hand, both players going for rabbit is a Nash equilibrium which is *risk dominant* in that any player who decides to go after a rabbit will definitely, without risk, get a rabbit. That is not true of $X = Y = (1, 0)$ in which if one player decides to go for rabbit the other player is left holding the bag, that is, she gets nothing. The terms *Pareto-optimal* and *risk dominant* are explained more precisely later in this chapter.

3.33 We are given the following game matrix:

$$\begin{bmatrix} 0,0 & 50,40 & 40,50 \\ 40,50 & 0,0 & 50,40 \\ 50,40 & 40,50 & 0,0 \end{bmatrix}.$$

(a) There is a unique mixed Nash equilibrium for this game. Find it.

3.33.a Answer: First note that there is no pure Nash equilibrium. We could use Problem 3.25 to solve this problem or do it directly by equality of payoffs. The equations to solve are

$$50y_2 + 40y_3 = v,$$
$$40y_1 + 50y_3 = v,$$
$$50y_1 + 40y_2 = v,$$
$$y_1 + y_2 + y_3 = 1.$$

Adding the first three equations and using the fourth gives us $v = 30$. Then solving the first three equations gives $Y^* = (\frac{1}{3}, \frac{1}{3}, \frac{1}{3})$. Since the game is symmetric, $X^* = Y^*$. The expected payoff to each player is 30.

(b) Suppose either player deviates from using her Nash equilibrium. Find the best response of the other player and show that it results in a higher payoff.

3.33.b Answer: If player II uses $Y = (y_1, y_2, y_3)$ then the expected payoff to player I is

$$E_1(X, Y) = X A Y^T = x_1(50y_2 + 40y_3) + x_2(50y_3 + 40y_1) + x_3(50y_1 + 40y_2).$$

The best response for player I will be to choose $x_i = 1$ corresponding to the largest coefficient in E_1. The only way to get a mixed strategy as a best response for player I is if the coefficients are all equal, and then they must all be 30, as we have seen. Thus, at least one coefficient must be greater than 30. Assume $50y_2 + 40y_3 > 30$. Then $x_1 = 1, x_2 = x_3 = 0$ and player I's expected payoff will be $50y_2 + 40y_3 > 30$. Thus, no matter what, if player II does not use the mixed Nash, then player I's expected payoff using the best response will be more than 30. This also applies for player II. Thus, if any player deviates from the Nash equilibrium, the best response gives a larger expected payoff to the other player.

3.4 Nonlinear Programming Method for Nonzero Sum Two-Person Games

Problems

3.34 Consider the following bimatrix for a version of the game of chicken (see Problem 3.3):

I/II	Avoid	Straight
Avoid	(1, 1)	(−1, 2)
Straight	(2, −1)	(−3, −3)

(a) What is the explicit objective function for use in the Lemke–Howson algorithm?

3.34.a Answer: The objective function is explicitly set up with $X = (x, 1 - x)$, $Y = (y, 1 - y)$,

$$f(x, y, p, q) = X A Y^T + X B Y^T - p - q = 7x + 7y - 6xy - 6 - p - q.$$

(b) What are the explicit constraints?

3.34.b Answer: The constraints for this problems are

$$A Y^T \leq p J_2^T \Rightarrow \begin{cases} 2y - 1 \leq p \\ 5y - 3 \leq p \end{cases}$$

and

$$X B \leq q J_2 \Rightarrow \begin{cases} 2x - 1 \leq q \\ 5x - 3 \leq q \end{cases}$$

and $0 \leq x \leq 1, 0 \leq y \leq 1$.

(c) Solve the game.

3.34.c Answer: We have three Nash equilibria:

$$\begin{aligned} X_1 &= (1, 0) & Y_1 &= (0, 1) & E_I &= -1, E_{II} = 2, \\ X_2 &= (0, 1) & Y_2 &= (1, 0) & E_I &= 2, E_{II} = -1, \\ X_3 &= (\tfrac{2}{3}, \tfrac{1}{3}) & Y_3 &= (\tfrac{2}{3}, \tfrac{1}{3}) & E_I &= \tfrac{1}{3}, E_{II} = \tfrac{1}{3}. \end{aligned}$$

The Nash equilibria are found here using Mathematica and adjusting the starting point. For example, to find the first point in the table we use

```
FindMaximum[{f[x,y,p,q],0<=x<=1,0<=y<=1, 2 y-1<=p,5y-3<=p,
2 x-1<=q,5x-3<=q},{{x,1},{y,0},{p,2},{q,1}}]
```

The starting point is specified in the last set of curly braces. We get the solution that the maximum of f is zero (as it should be) and $x = 1, y = 0, p = -1, q = 2$.

For the mixed Nash, change the starting point:

```
FindMaximum[{f[x,y,p,q],0<=x<=1,0<=y<=1, 2 y-1<=p,5y-3<=p,
2 x-1<=q,5x-3<=q},{{x,0.5},{y,0.5},{p,2},{q,1}}]
```

We changed the starting point arbitrarily. Then Mathematica gives us the maximum of f is zero, and $x = 0.66, y = 0.66, p = 0.33, q = 0.33$, which is the mixed Nash in the table.

3.35 Suppose you are told that the following is the nonlinear program for solving a game with matrices (A, B):

$$\begin{aligned} \text{obj} = &(90 - 40x)y + (60x + 20)(1 - y) \\ &+ (40x + 10)y + (80 - 60x)(1 - y) - p - q, \end{aligned}$$

$$\text{cnst1} = 80 - 30y \leq p,$$
$$\text{cnst2} = 20 + 70y \leq p,$$
$$\text{cnst3} = 10 + 40x \leq q,$$
$$\text{cnst4} = 80 - 60x \leq q,$$
$$0 \leq x \leq 1, 0 \leq y \leq 1.$$

Find the associated matrices A and B and then solve the problem to find the Nash equilibrium.

3.35 Answer: Since this is a 2×2 game, all we need to do is match up the coefficients in the constraints $AY^T \leq pJ_2^T$ and $XB \leq qJ_2$. $A = \begin{bmatrix} 50 & 80 \\ 90 & 20 \end{bmatrix}$, $B = \begin{bmatrix} 50 & 20 \\ 10 & 80 \end{bmatrix}$. Then use Maple/Mathematica to get $X^* = (0.7, 0.3)$, $Y^* = (0.6, 0.4)$, $p = 62, q = 38$.

3.36 Suppose that the wife in a battle of the sexes game has an additional strategy she can use to try to get the husband to go along with her to the concert, instead of wrestling. Call it strategy Z. The payoff matrix then becomes

H/W	Wr	Co	Z
Wr	(2, 1)	(0, 0)	(−1, 1)
Co	(0, 0)	(1, 2)	(1, 3)

Find all Nash equilibria.

3.36 Answer: Use Maple/Mathematica and adjust the initial point:

$$X_1 = (1, 0) \quad Y_1 = (1, 0, 0) \quad E_{\mathrm{I}} = 2, E_{\mathrm{II}} = 1$$
$$X_2 = (1, 0) \quad Y_2 = (\tfrac{1}{2}, 0, \tfrac{1}{2}) \quad E_{\mathrm{I}} = \tfrac{1}{2}, E_{\mathrm{II}} = 1$$
$$X_3 = (0, 1) \quad Y_3 = (0, 0, 1) \quad E_{\mathrm{I}} = 1, E_{\mathrm{II}} = 3.$$

3.37 Since every two-person **zero sum** game can be formulated as a bimatrix game, show how to modify the Lemke–Howson algorithm to be able to calculate saddle points of zero sum two-person games. Then use that to find the value and saddle point for the game with matrix

$$A = \begin{bmatrix} 2 & -1 & 2 \\ -1 & 2 & -3 \\ 4 & -3 & 5 \\ -3 & \tfrac{1}{2} & -9 \end{bmatrix}.$$

Check your answer by using the linear programming method to solve this game.

3.37 Answer: Take $B = -A$. The algorithm is then

$$\max_{X, Y, p, q} \quad XAY^T - XAY^T - p - q = -p - q$$
subject to
$$AY^T \leq pJ_n^T$$
$$-A^T X^T \leq qJ_m^T \quad \text{(equivalently } XA \geq -qJ_m)$$
$$x_i \geq 0, y_j \geq 0, \quad XJ_n = 1 = YJ_m^T,$$

where $J_k = (1\,1\,1\,\cdots\,1)$ is the $1 \times k$ row vector consisting of all 1's. In addition, $p^* = E_I(X^*, Y^*) = X^*AY^{*T}$, and $q^* = E_{II}(X^*, Y^*) = -X^*AY^{*T} = -p^*$. You can see that this problem is the same as Method 2 of linear programming for solving a game. Thus Lemke–Howson reduces to Method 2.

The Nash equilibrium found using Lemke–Howson is $X^* = (\frac{5}{8}, \frac{3}{8}, 0)$, $Y^* = (0, \frac{5}{8}, \frac{3}{8})$, and the value of the game is $v(A) = \frac{1}{8}$.

3.38 Use nonlinear programming to find all Nash equilibria for the game and expected payoffs for the game with bimatrix

$$\begin{bmatrix} (1, 2) & (0, 0) & (2, 0) \\ (0, 0) & (2, 3) & (0, 0) \\ (2, 0) & (0, 0) & (-1, 6) \end{bmatrix}.$$

3.38 Answer: $X_1 = \left(\frac{1}{2}, \frac{1}{3}, \frac{1}{6}\right)$, $Y_1 = \left(\frac{6}{13}, \frac{5}{13}, \frac{2}{13}\right)$, $E_I = \frac{10}{13}$, $E_{II} = 1$. $X_2 = \left(\frac{3}{4}, 0, \frac{1}{4}\right) = Y_2$ with payoffs $E_I = \frac{5}{4}$, $E_{II} = \frac{3}{2}$. $X_3 = Y_3 = (0, 1, 0)$.

3.39 Consider the game with bimatrix

$$\begin{bmatrix} (-3, -4) & (2, -1) & (0, 6) & (1, 1) \\ (2, 0) & (2, 2) & (-3, 0) & (1, -2) \\ (2, -3) & (-5, 1) & (-1, -1) & (1, -3) \\ (-4, 3) & (2, -5) & (1, 2) & (-3, 1) \end{bmatrix}$$

Find as many Nash equilibria as you can by adjusting the starting point in the Maple or Mathematica commands for Lemke–Howson.

3.39 Answer: The Nash equilibria are as follows:

X^*	Y^*
$(0, 0, 0.333333, 0.666667)$	$(0.25, 0, 0.75, 0)$
$(0, 0.66667, 0.111108, 0.222221)$	$(0.341463, 0.146341, 0.512195, 0)$
$(0.020202, 0.777778, 0, 0.20202)$	$(0, 1, 0, 0)$
$(0.0896435, 0, 0.00464034, 0.905716)$	$(0.25, 0, 0.75, 0)$

The corresponding respective payoffs are as follows:

E_I	E_{II}
-0.25	1
-0.560975	0.3333
2	0.525253
-0.25	2.34465

3.40 Consider the gun duel between Pierre and Bill. Modify the payoff functions so that it becomes a noisy duel with $a_1 = -2, b_1 = -1, c_1 = 2, d_1 = -1, e_1 = 1, f_1 = 2, g_1 = -2, h_1 = 1, k_1 = -1, \ell_1 = 2$, for player I, and $a_2 = -1, b_2 = 1, c_2 = 1, d_2 = 1, e_2 = -1, f_2 = 1, g_2 = 0, h_2 = -1, k_2 = 1, \ell_2 = 1$, for player II. Then solve the game and obtain at least one mixed Nash equilibrium.

3.40 Answer: First we give the payoff functions. For player I, we have

$$u_1(x, y) = \begin{cases} -2 \cdot p_1(x) + (-1)\,(1 - p_1(x)) + 2\,(1 - p_1(x))(1 - p_2(y)), & \text{if } x < y; \\ (-1)\,p_2(y) + (1)\,(1 - p_2(y)) + 2\,(1 - p_2(y))\,(1 - p_1(x)), & \text{if } y < x; \\ (-2)\,p_1(x)p_2(x) + (1)\,p_1(x)(1 - p_2(x)) \\ \quad +(-1)\,(1 - p_1(x))p_2(x) + (2)(1 - p_1(x))(1 - p_2(x)), & \text{if } x = y; \end{cases}$$

where the accuracy functions are given by

$$p_1(x) = p_2(x) = \begin{cases} 0.2, & \text{if } x = 0; \\ 0.6, & \text{if } x = 0.4; \\ 0.8, & \text{if } x = 0.6; \\ 1, & \text{if } x = 0.8. \end{cases}$$

For player II, the payoff is

$$u_2(x, y) = \begin{cases} -1 \cdot p_1(x) + (1)\,(1 - p_1(x)) + (1)\,(1 - p_1(x))(1 - p_2(y)), & \text{if } x < y; \\ (1)\,p_2(y) + (-1)\,(1 - p_2(y)) + (1)\,(1 - p_2(y))\,(1 - p_1(x)), & \text{if } y < x; \\ (0)\,p_1(x)p_2(x) + (-1)\,p_1(x)(1 - p_2(x)) \\ \quad +(1)\,(1 - p_1(x))p_2(x) + (1)(1 - p_1(x))(1 - p_2(x)), & \text{if } x = y. \end{cases}$$

The matrices are as follows:

$$A = \begin{bmatrix} u_1(0, 0) & u_1(0, .4) & u_1(0, .6) & u_1(0, .8) \\ u_1(.4, 0) & u_1(.4, .4) & u_1(.4, .6) & u_1(.4, .8) \\ u_1(.6, 0) & u_1(.6, .4) & u_1(.6, .6) & u_1(.6, .8) \\ u_1(.8, 0) & u_1(.8, .4) & u_1(.8, .6) & u_1(.8, .8) \end{bmatrix} = \begin{bmatrix} 1.20 & -0.56 & -0.88 & -1.2 \\ 1.24 & -0.40 & -1.44 & -1.6 \\ 0.92 & -0.04 & -1.20 & -1.8 \\ 0.6 & -0.2 & -0.6 & -2 \end{bmatrix},$$

$$B = \begin{bmatrix} u_2(0, 0) & u_2(0, .4) & u_2(0, .6) & u_2(0, .8) \\ u_2(.4, 0) & u_2(.4, .4) & u_2(.4, .6) & u_2(.4, .8) \\ u_2(.6, 0) & u_2(.6, .4) & u_2(.6, .6) & u_2(.6, .8) \\ u_2(.8, 0) & u_2(.8, .4) & u_2(.8, .6) & u_2(.8, .8) \end{bmatrix} = \begin{bmatrix} 0.64 & 0.92 & 0.76 & 0.6 \\ -0.28 & 0.16 & -0.12 & -0.2 \\ -0.44 & 0.28 & 0.04 & -0.6 \\ -0.6 & 0.2 & 0.6 & 0 \end{bmatrix}.$$

One Nash equilibrium is $X = (0.71, 0, 0, 0.29)$, $Y = (0, 0, 0.74, 0.26)$. So Pierre fires at 10 paces about 75% of the time and waits until 2 paces about 25% of the time. On the other hand, Bill waits until 4 paces before he takes a shot but 1 out of 4 times waits until 2 paces.

3.5 Correlated Equilibria

Problems

3.41 Let A, B be 2×2 games and $P_{2 \times 2} = (p_{ij})$ a probability distribution. Show that P is a correlated equilibrium if and only if

$$p_{11}(a_{11} - a_{21}) \geq p_{12}(a_{22} - a_{12}),$$
$$p_{22}(a_{22} - a_{12}) \geq p_{21}(a_{11} - a_{21}),$$
$$p_{11}(b_{11} - b_{12}) \geq p_{21}(b_{22} - b_{21}),$$
$$p_{22}(b_{22} - b_{21}) \geq p_{12}(b_{11} - b_{12}).$$

The payoffs in terms of expected social welfare are as follows:

$$u_A = p_{11}a_{11} + p_{22}a_{22} + p_{21}a_{21} + p_{12}a_{12},$$
$$u_B = p_{11}b_{11} + p_{22}b_{22} + p_{21}b_{21} + p_{12}b_{12}.$$

3.41 Answer: This is just rearranging the terms in the inequalities giving the definition of correlated equilibrium:

$$a_{11}p_{11} + a_{12}p_{12} \geq a_{21}p_{11} + a_{22}p_{12} \Leftrightarrow p_{11}(a_{11} - a_{21}) \geq p_{12}(a_{22} - a_{12}),$$
$$a_{21}p_{21} + a_{22}p_{22} \geq a_{11}p_{21} + a_{12}p_{22} \Leftrightarrow p_{22}(a_{22} - a_{12}) \geq p_{21}(a_{11} - a_{21}),$$
$$b_{11}p_{11} + b_{21}p_{21} \geq b_{12}p_{11} + b_{22}p_{21} \Leftrightarrow p_{11}(b_{11} - b_{12}) \geq p_{21}(b_{22} - b_{21}),$$
$$b_{12}p_{12} + b_{22}p_{22} \geq b_{11}p_{12} + b_{21}p_{22} \Leftrightarrow p_{22}(b_{22} - b_{21}) \geq p_{12}(b_{11} - b_{12}).$$

Using P we calculate the expected payoff for each player:

$$E_P(\mathrm{I}) = \sum_{i=1}^{2} \text{Payoff to I if play row } i\, Prob(\text{I plays row } i)$$

$$= \sum_{i=1}^{2} \left(\sum_{j=1}^{2} a_{ij} Prob(\text{I plays row } i \text{ and II plays col } j) \right)$$

$$= \sum_{i=1}^{2} \left(\sum_{j=1}^{2} a_{ij} p_{ij} \right)$$

$$= p_{11}a_{11} + p_{22}a_{22} + p_{21}a_{21} + p_{12}a_{12}.$$

The payoff to player II is similar—just change the matrix.

3.42 Verify that if (X^*, Y^*) is a Nash equilibrium for the game (A, B), then $P = (X^*)^T Y^* = (x_i y_j)$ is a correlated equilibrium.

3.42 Answer: Suppose first that (i^*, j^*) is a pure Nash equilibrium. Then $P = (p_{ij})$ defined by $p_{i^* j^*} = 1$ and $p_{ij} = 0$ if $i \neq i^*, j \neq j^*$ is a correlated equilibrium. Indeed, as a Nash equilibrium we know

$$a_{i^* j^*} \geq a_{ij^*} \quad \text{and} \quad b_{i^* j^*} \geq b_{i^* j}, \quad \forall\, i = 1, 2, \ldots, n, j = 1, 2, \ldots, m.$$

The condition for P to be a correlated equilibrium is

$$\sum_{j=1}^{m} a_{ij} p_{ij} = a_{i^* j^*} \geq \sum_{j=1}^{m} a_{qj} p_{ij} = a_{qj^*}$$

for all rows $i = 1, 2, \ldots, n, q = 1, 2, \ldots, n$, and

$$\sum_{i=1}^{n} b_{ij} p_{ij} = b_{i^* j^*} \geq \sum_{i=1}^{n} b_{ir} p_{ij} = b_{i^* r}$$

for all columns $j = 1, 2, \ldots, m, r = 1, 2, \ldots, m$. But those conditions hold if i^*, j^* is a Nash equilibrium. Thus, P is a correlated equilibrium.

Now suppose $X^* = (x_1, \ldots, x_n)$, $Y^* = (y_1, \ldots, y_m)$ is a mixed Nash equilibrium with component $x_k > 0$. We show that $P = (p_{ij})$, with $p_{ij} = x_i y_j$ is a correlated equilibrium. Since (X^*, Y^*) is a Nash equilibrium we know that

$$v_I = E_1(X^*, Y^*) = E_1(k, Y^*)$$

for every k for which $x_k > 0$. Thus,

$$E_1(k, Y^*) \geq E_1(q, Y^*), \qquad \text{for all } q = 1, 2, \ldots, n,$$

which says

$$\sum_{j=1}^m a_{kj} y_j \geq \sum_{j=1}^m a_{qj} y_j, \qquad q = 1, 2, \ldots, n.$$

Multiplying both sides by $x_k > 0$ gives

$$\sum_{j=1}^m a_{kj} x_k y_j \geq \sum_{j=1}^m a_{qj} x_k y_j, \qquad q = 1, 2, \ldots, n.$$

This is true for any component of X^* which is positive, but it is also true for any component which is zero (because then both sides are zero). Similarly,

$$\sum_{i=1}^n b_{ij} x_i y_j \geq \sum_{i=1}^n b_{ir} x_i y_j, \qquad r = 1, 2, \ldots, m.$$

This says that $P = (x_i, y_j)$ is indeed a correlated equilibrium.

3.43 Consider the following game of chicken:

I/II	Avoid	Straight
Avoid	(4, 4)	(1, 5)
Straight	(5, 1)	(0, 0)

Show that the following are all correlated equilibria of this game:

$$P_1 = \begin{bmatrix} 0 & 1 \\ 0 & 0 \end{bmatrix}, \quad P_2 = \begin{bmatrix} 0 & 0 \\ 1 & 0 \end{bmatrix},$$

$$P_3 = \begin{bmatrix} \frac{1}{4} & \frac{1}{4} \\ \frac{1}{4} & \frac{1}{4} \end{bmatrix}, \quad P_4 = \begin{bmatrix} 0 & \frac{1}{2} \\ \frac{1}{2} & 0 \end{bmatrix}, \quad P_5 = \begin{bmatrix} \frac{1}{3} & \frac{1}{3} \\ \frac{1}{3} & 0 \end{bmatrix}.$$

Show also that P_4 gives a bigger social welfare than P_1, P_2, or P_3.

3.43 Answer: This chicken game has three Nash equilibria. There are two pure Nash equilibria at $(5, 1)$, $(1, 5)$. There is a mixed Nash equilibrium $X^* = Y^* = (\frac{1}{2}, \frac{1}{2})$.

Now P_1 and P_2 correspond to the two pure Nash equilibria. Under P_1 and P_2 the social welfare is 6. Also P_3 corresponds to the mixed Nash equilibrium with social welfare 5. Since $P = X^{*T} Y^* = (x_i y_j)$, if (X^*, Y^*) is a Nash equilibrium, then P_i, $i = 1, 2, 3$, is a correlated equilibrium.

Next consider P_4. We have

$$
\begin{aligned}
{}_1 A \cdot_1 P &= \tfrac{1}{2} \geq_2 A \cdot_1 P = 0, \\
{}_2 A \cdot_2 P &= \tfrac{5}{2} \geq_1 A \cdot_2 P = 2, \\
B_1 \cdot P_1 &= \tfrac{1}{2} \geq B_2 \cdot P_1 = 0, \\
B_2 \cdot P_2 &= \tfrac{5}{2} \geq B_1 \cdot P_2 = 2.
\end{aligned}
$$

Thus, P_4 is a correlated equilibrium. The social welfare under P_4 is 6.

Finally, consider P_5. We have

$$
\begin{aligned}
{}_1 A \cdot_1 P &= \tfrac{5}{3} \geq_2 A \cdot_1 P = \tfrac{5}{3}, \\
{}_2 A \cdot_2 P &= \tfrac{5}{3} \geq_1 A \cdot_2 P = \tfrac{4}{3}, \\
B_1 \cdot P_1 &= \tfrac{5}{3} \geq B_2 \cdot P_1 = \tfrac{5}{3}, \\
B_2 \cdot P_2 &= \tfrac{5}{3} \geq B_1 \cdot P_2 = \tfrac{4}{3}.
\end{aligned}
$$

Thus, P_5 is also a correlated equilibrium. The social welfare under P_5 is $\tfrac{20}{3}$.

P_5 is the correlated equilibrium which maximizes the social welfare. This comes from the solution of

```
Maximize[{Sum[Sum[(A[[i,j]]+B[[i,j]])Q[[i,j]],{j,1,2}],{i,1,2}],
A[[1]].Q[[1]]>=A[[2]].Q[[1]],A[[2]].Q[[2]]>=A[[1]].Q[[2]],
B[[All,1]].Q[[All,1]]>=B[[All,2]].Q[[All,1]],
B[[All,2]].Q[[All,2]]>=B[[All,1]].Q[[All,2]],
q11+q12+q21+q22==1,
q11>=0,q22>=0,q12>=0,q21>=0},
{q11,q12,q21,q22}]
```

3.44 Consider the game in which each player has two strategies Wait and Go. The game matrix is

I/II	Wait	Go
Wait	1, 1	$1 - \varepsilon, 2$
Go	2, $1 - \varepsilon$	0, 0

(a) Find the correlated equilibria corresponding to the Nash equilibria.

3.44.a Answer: The Nash equilibria are $(X_1 = (0, 1), Y_1 = (1, 0))$, $(X_2 = (1, 0), Y_2 = (0, 1))$, and $X_3 = (\tfrac{1-\varepsilon}{2-\varepsilon}, \tfrac{1}{2-\varepsilon}) = Y_3$. The corresponding correlated equilibria are

$$
P_1 = \begin{bmatrix} 0 & 0 \\ 1 & 0 \end{bmatrix}, \quad
P_2 = \begin{bmatrix} 0 & 1 \\ 0 & 0 \end{bmatrix}, \quad
P_3 = \frac{1}{(2-\varepsilon)^2} \begin{bmatrix} (1-\varepsilon)^2 & 1-\varepsilon \\ 1-\varepsilon & 1 \end{bmatrix}.
$$

(b) Find the correlated equilibrium corresponding to maximizing the social welfare for any $0 < \varepsilon < 1$.

3.44.b Answer: The linear program we have to solve is

$$\text{Maximize } 2p_{11} + (p_{12} + p_{21})(3 - \varepsilon)$$

Subject to

$$p_{12}(1 - \varepsilon) \geq p_{11}$$
$$p_{21} \geq (1 - \varepsilon)p_{22}$$
$$p_{21}(1 - \varepsilon) \geq p_{11}$$
$$p_{12} \geq (1 - \varepsilon)p_{22}$$
$$p_{11} + p_{12} + p_{21} + p_{22} = 1$$
$$p_{ij} \geq 0.$$

Considering the objective function, since $2 \leq 3 - \varepsilon$, the maximum is achieved at $p_{11} = 0$, $p_{12} + p_{21} = 1$, if the rest of the constraints are satisfied.

If $p_{11} = 0$, $p_{12} + p_{21} = 1$, then $p_{22} = 0$. The constraints simply reduce to $p_{ij} \geq 0$. Thus, there is a collection of correlated equilibria

$$P = \begin{bmatrix} 0 & p_{12} \\ p_{21} & 0 \end{bmatrix}, \qquad p_{12} + p_{21} = 1,$$

all giving the maximum social welfare $3 - \varepsilon$. In particular, P_1 and P_2 are both correlated equilibria maximizing the social welfare. One red light should do the trick.

3.6 Choosing Among Several Nash Equilibria

Problems

3.45 In the following games solved earlier, determine which, if any, Nash equilibrium is payoff dominant, risk-dominant, and Pareto-optimal.

(a) The Game of Chicken

I/II	Avoid	Straight
Avoid	(1, 1)	(−1, 2)
Straight	(2, −1)	(−3, −3)

(b) Arms Control

	Cooperate	Don't
Cooperate	(3, 3)	(−1, −3)
Don't	(3, −1)	(1, 1)

(c)

$$\begin{bmatrix} (1, 2) & (0, 0) & (2, 0) \\ (0, 0) & (2, 3) & (0, 0) \\ (2, 0) & (0, 0) & (-1, 6) \end{bmatrix}$$

3.45 Answer:

(a) The Nash equilibria are as follows:

$$X_1 = (1, 0), \quad Y_1 = (0, 1), \quad E_\mathrm{I} = -1, E_\mathrm{II} = 2,$$
$$X_2 = (0, 1), \quad Y_2 = (1, 0), \quad E_\mathrm{I} = 2, E_\mathrm{II} = -1,$$
$$X_3 = \left(\tfrac{2}{3}, \tfrac{1}{3}\right), \ Y_3 = \left(\tfrac{2}{3}, \tfrac{1}{3}\right), \ E_\mathrm{I} = \tfrac{1}{3}, E_\mathrm{II} = \tfrac{1}{3}.$$

They are all Pareto-optimal because it is impossible for either player to improve their payoff without simultaneously decreasing the other player's payoff, as you can see from the figure:

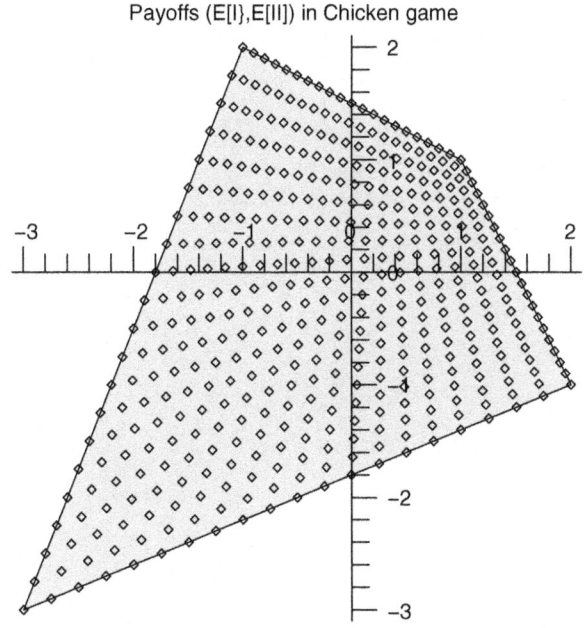

Payoffs (E[I},E[II]) in Chicken game

None of the Nash equilibria are payoff-dominant. The mixed Nash (X_3, Y_3) risk dominates the other two.

(b) The Nash equilibria are as follows:

$$X_1 = \left(\frac{1}{4}, \frac{3}{4}\right), Y_1 = (1, 0), E_1 = 3, E_2 = 0,$$
$$X_2 = (0, 1), Y_2 = (0, 1), E_1 = E_2 = 1,$$
$$X_3 = (1, 0) = Y_3, E_1 = E_2 = 3.$$

(X_3, Y_3) is payoff-dominant and Pareto-optimal.

(c)

$$X_1 = \left(\frac{1}{2}, \frac{1}{3}, \frac{1}{6}\right), Y_1 = \left(\frac{6}{13}, \frac{5}{13}, \frac{2}{13}\right), E_I = \frac{10}{13}, E_{II} = 1.$$

$$X_2 = \left(\frac{3}{4}, 0, \frac{1}{4}\right) = Y_2, E_I = \frac{5}{4}, E_{II} = \frac{3}{2}.$$

$$X_3 = Y_3 = (0, 1, 0), E_I = 2, E_{II} = 3.$$

Clearly X_3, Y_3 is payoff-dominant and Pareto-optimal. Neither (X_1, Y_1) nor (X_2, Y_2) are Pareto-optimal relative to the other Nash equilibria, but they each risk dominate (X_3, Y_3).

CHAPTER FOUR

Games in Extensive Form: Sequential Decision Making

4.1 Introduction to Game Trees—Gambit

Problems

4.1 Convert the following extensive form game to a strategic form game. Be sure to list all of the pure strategies for each player and then use the matrix to find the Nash equilibrium.

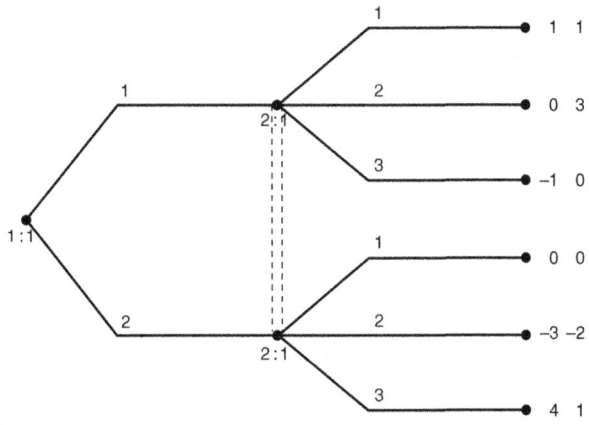

4.1 Answer: Player I has two pure strategies: $1 =$ take action 1 at node $1:1$, and $2 =$ take action 2 at node $1:1$. Player II has three pure strategies: $1 =$ take action 1 at information set $2:1$, $2 =$ take action 2 at $2:1$, and $3 =$ take action 3 at $2:1$.

The game matrix is

I/II	1	2	3
1	1, 1	0, 3	−1, 0
2	0, 0	−3, −2	4, 1

There is a pure Nash at (4, 1).

Solutions Manual to Accompany Game Theory: An Introduction, Second Edition. E.N. Barron.
© 2013 John Wiley & Sons, Inc. Published 2013 by John Wiley & Sons, Inc.

4.2 Given the game matrix draw the tree. Assume the players make their choices simultaneously. Solve the game.

I/II	a	b
A	3, 3	0, 4
B	4, 0	2, −1

4.2 Answer: The solution of the matrix game is the Nash equilibrium at row 2, column 1, giving payoffs $(4, 0)$. The equivalent tree form of the game is

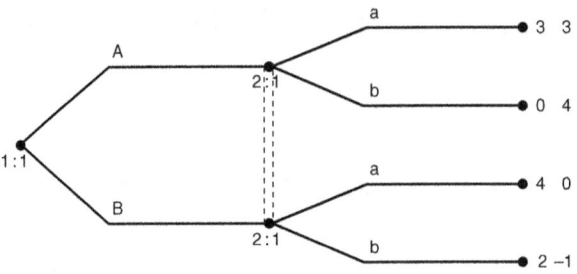

Matrix to tree.

Since the players make their choices simultaneously, the information set $2 : 1$ reflects that player II makes her choice without knowing player I's choice.

4.3 Consider the three player game in which each player can choose Up or Down. In matrix form, the game is

I/II	III plays Up		I/II	III plays Down	
	Up	Down		Up	Down
Up	$(1, 2, −1)$	$(−2, 1, 3)$	Up	$(1, −2, 4)$	$(1, −1, −3)$
Down	$(−1, 3, 0)$	$(2, −1, 4)$	Down	$(−1, 3, −1)$	$(2, 2, 1)$

(a) Draw the extensive form of the game assuming the players choose simultaneously.

4.3.a Answer: The extensive form along with the solution is shown in the figure.

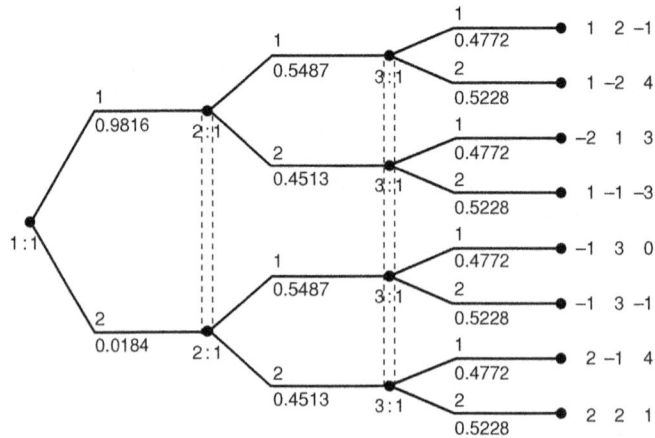

Three player game tree.

The probability a player takes a branch is indicated below the branch.

(b) Use Gambit to solve the game and show that the mixed Nash Equilibrium is
$X^* = (0.98159, 0.01841)$, $Y^* = (0.54869, 0.45131)$, $Z^* = (0.47720, 0.5228)$.

4.4 In Problem 2.45, we considered the game in which Sonny is trying to rescue family members Barzini has captured. With the same description as in that problem, draw the game tree and solve the game.

4.4 Answer: The game tree is in the figure. Note that we have one information set for Barzini because he does not know whether Sonny is headed for Brooklyn or Queens. This is a zero sum game.

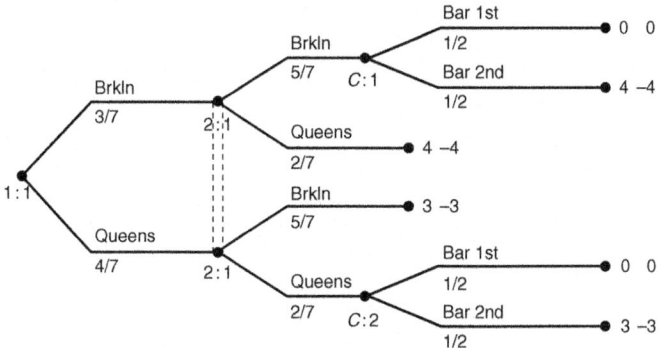

The solution of the game is $X^* = (\frac{3}{7}, \frac{4}{7})$, $Y^* = (\frac{5}{7}, \frac{2}{7})$, and the value is $v = \frac{18}{7}$.

4.5 Aggie and Baggie are each armed with a single coconut cream pie. Since they are the friends of Larry and Curly, naturally, instead of eating the pies they are going to throw them at each other. Aggie goes first. At 20 paces she has the option of hurling the pie at Baggie or passing. Baggie then has the same choices at 20 paces but if Aggie hurls her pie and misses, the game is over because Aggie has no more pies. The game could go into a second round but now at 10 paces, and then a third round at 0 paces. Aggie and Baggie have the same probability of hitting the target at each stage: $\frac{1}{3}$ at 20 paces, $\frac{3}{4}$ at 10 paces, and 1 at 0 paces. If Baggie gets a pie in the face, Aggie scores $+1$, and if Aggie gets a pie in the face Baggie scores $+1$. This is a zero sum game.

(a) Draw the game tree and convert to a matrix form game.

4.5.a Answer: The game tree follows the description of the problem in a straightforward way. It is considerably simplified if you use the fact that if a player fires the pie and misses, all the opponent has to do is wait until zero paces and winning is certain. Here is the tree.

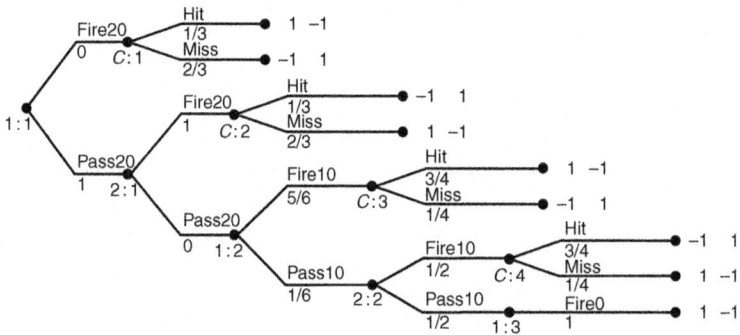

Aggie has three pure strategies:

(a1) Fire at 20 paces;
(a2) Pass at 20 paces; If Baggie passes at 20, Fire at 10 paces;
(a3) Pass at 20 paces; If Baggie passes at 20, Pass at 10 paces; If Baggie passes at 10, Fire at 0 paces.

Baggie has three pure strategies:

(b1) If Aggie passes, Fire at 20 paces;
(b2) If Aggie passes, Pass at 20 paces; If Aggie passes at 10, Fire at 10 paces;
(b3) If Aggie passes, Pass at 20 paces; If Aggie passes at 10, Pass at 10 paces.

The game matrix is then

Aggie/Baggie	(b1)	(b2)	(b3)
(a1)	$-\frac{1}{3}$	$-\frac{1}{3}$	$-\frac{1}{3}$
(a2)	$\frac{1}{3}$	$\frac{1}{2}$	$\frac{1}{2}$
(a3)	$\frac{1}{3}$	$-\frac{1}{2}$	1

To see where the entries come from, consider (a1) versus (b1). Aggie is going to throw her pie and if Baggie doesn't get hit, then Baggie will throw her pie. Aggie's expected payoff will be

$$(+1)\frac{1}{3} + (-1)\frac{2}{3} = -\frac{1}{3}$$

because Aggie hits Baggie with probability $\frac{1}{3}$ and gets $+1$, but if she misses, then she is going to get a pie in the face.

If Aggie plays (a2) and Baggie plays (b3), then Aggie will pass at 20 paces; Baggie will also pass at 20; then Aggie will fire at 10 paces. Aggies' expected payoff is then

$$(+1)\frac{3}{4} + (-1)\frac{1}{4} = \frac{1}{2}.$$

(b) Solve the game.

4.5.b Answer: The solution of the game can be obtained from the matrix using dominance. The saddle point is for Aggie to play (a2) and Baggie plays (b1). In words, Aggie will pass at 20 paces, Baggie will fire at 20 paces, with value $v = \frac{1}{3}$ to Aggie. Since Baggie moves second, her best strategy is to take her best shot as soon as she gets it.

Here is a slightly modified version of this game in which both players can miss their shot with a resulting score of 0. The value of the game is still $\frac{1}{3}$ to Aggie. It is still optimal for Aggie to pass at 20, and for Baggie to take her shot at 20. If Baggie misses, she is certain to get a pie in the face.

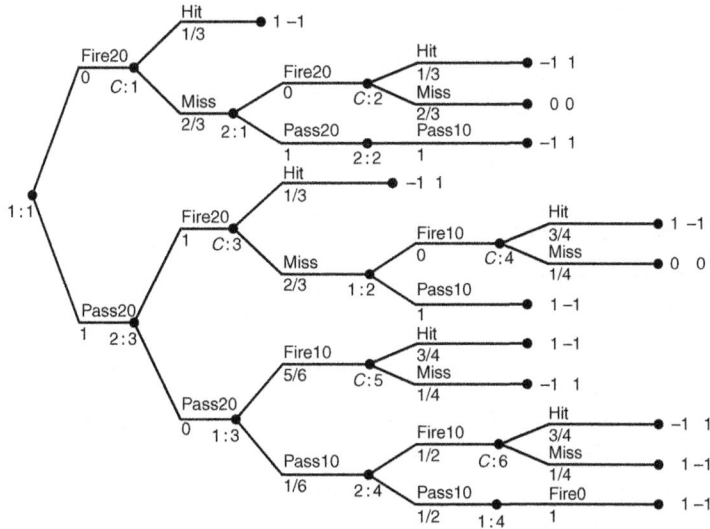

4.6 One possibility not accounted for in a Battle of the Sexes game is the choice of player I to go to a bar instead of the Ballet or Wrestling. Suppose the husband, who wants to go to Wrestling, has a 15% chance of going to a bar instead. His wife, who wants to go to the Ballet, also has a 25% chance of going to the same bar instead. If they both meet at Wrestling, the husband gets 4 and the wife gets 2. If they both meet at the Ballet, the wife gets 4 and the husband gets 1. If they both meet at the bar, the husband gets 4 and the wife also gets 4. If they decide to go to different places they each get 2. Assume the players make their decisions without knowledge of the other's choice. Draw the game and solve using Gambit.

4.6 Answer: This problem further exhibits the use of chance moves in games and modeled in Gambit. We begin by flipping a biased coin to see if the husband will go to the Bar or to Ballet–Wrestling. Then the wife also flips a biased coin to determine if she will go to the Bar. The information set for the wife must account for the fact that she is unaware of the husband's choice. Here is the figure.

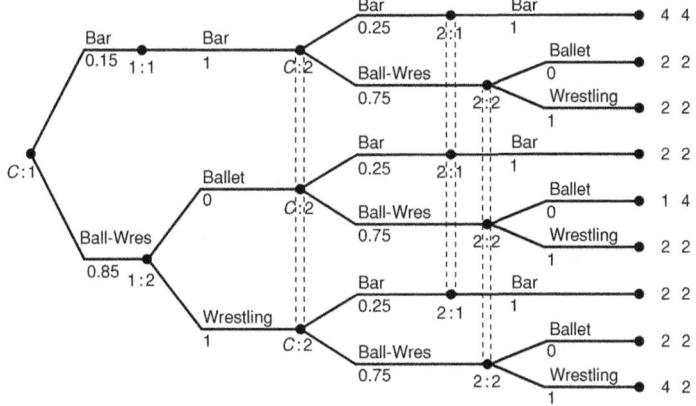

Bar or ballet or wrestling.

The solution of the game is
- If Husband's coin comes up Ballet–Wrestling, the husband should go to wrestling.
- If Wife's coin comes up Ballet–Wrestling, the wife should go to wrestling.

The expected payoff to the husband is $\frac{67}{20}$ and to the wife $\frac{83}{40}$.

4.2 Backward Induction and Subgame Perfect Equilibrium

Problems

4.7 Modify Solomon's decision problem so that each Woman actually loses the value she places on the child if she is not awarded the child. The value to the true mother is assumed to satisfy $C_T > 2C_F$. Assume that Solomon knows both C_T and C_F and so can set $V > 2C_F$. Apply backward induction to find the Nash equilibrium and verify that the true mother always gets the child.

4.7 Answer: Assume first that woman 1 is the true mother. We look at the matrix for this game.

1/2	Agree	Object
Mine	$(C_T, -C_F)$	$(-C_T - W, C_F - V)$
Hers	$(-C_T, C_F)$	$(-C_T, C_F)$

Since $V > 2C_F$, there are two pure Nash equilibria: (1) at $(C_T, -C_F)$ and (2) at $(-C_T, C_F)$. If we consider the tree, we see that woman 2 will always Agree if woman 1 claims Mine. This eliminates the branch Object by woman 2. Next, since $C_T > -_T$, woman 1 will always call Mine, and hence woman 1 will end up with the child. The Nash equilibrium $(C_T, -C_F)$ is the equilibrium that will be played by backward induction. Note that we can't obtain that conclusion just from the matrix.

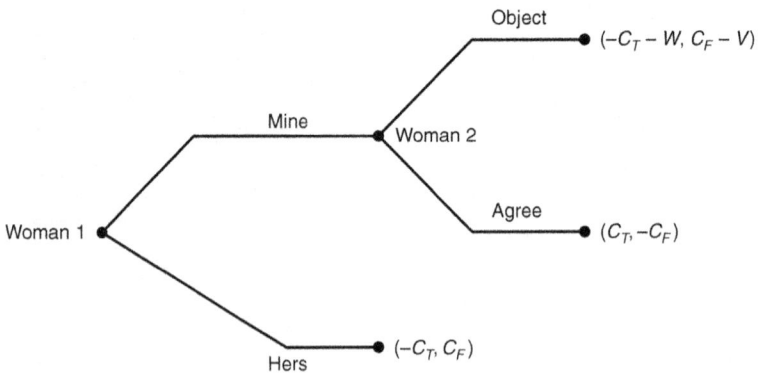

Woman 1 the true mother.

Similarly, assume that woman 2 is the true mother. Then the tree becomes

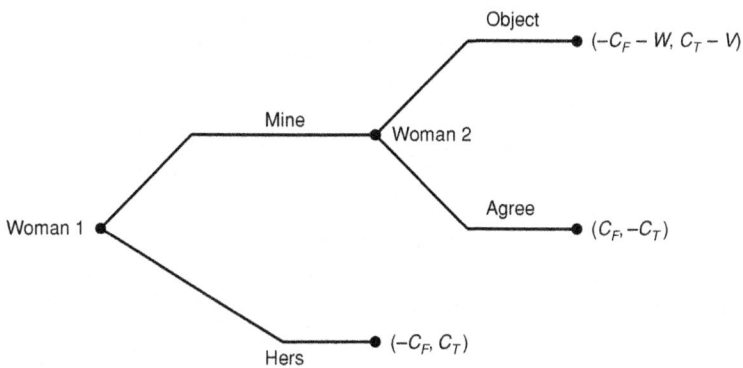

Woman 2 the true mother.

Since $C_T - V > -C_T$, woman 2 always Objects if woman 1 calls Mine. Eliminating the Agree branch for woman 2, we now compare $(-C_F, C_T)$ and $(-C_F - W, C_T - V)$. Since $-C_F > -C_F - W$, we conclude that woman 1 will always call Hers. Backward induction gives us that woman 1 always calls Hers (and if woman 1 calls Mine, then woman 2 Objects). Either way, woman 2, the true mother, ends up with the child.

4.8 Find the Nash equilibrium using backward induction for the tree:

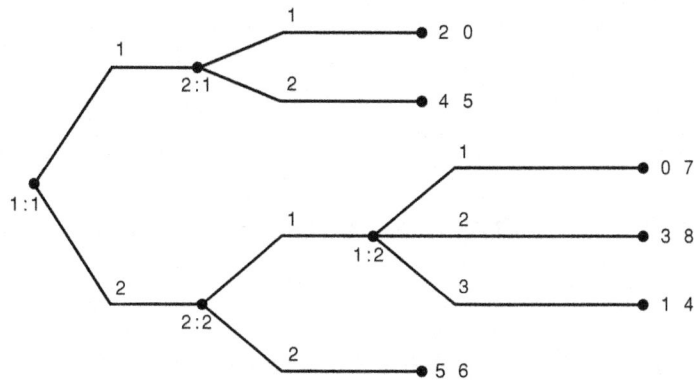

4.8 Answer: The backward Nash is that player 1 plays branch 1 and player 2 (from node 2 : 1) plays branch 2. The payoff is $(4, 5)$.

4.9 Two players ante \$1. Player I is dealt an Ace or a Two with equal probability. Player I sees the card but player II does not. Player I can now decide to Raise, or Fold. If he Folds, then player II wins the pot if the card is a Two, while player I wins the pot if the card is an Ace. If player I Raises, he must add either \$1 or \$2 to the pot. Player II now decides to Call or Fold. If she Calls and the card was a Two, then II wins the pot. If the card was an Ace, I wins the pot. If she Folds, player I takes the pot. Model this as an extensive game and solve it using Gambit.

4.9 Answer: This is a simple model to see if bluffing is ever optimal. Here is the figure.

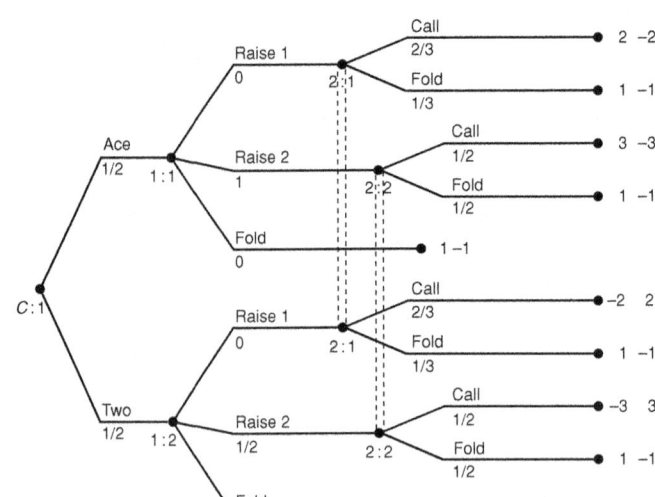

Ace–Two Raise or Fold.

Player I has perfect information but player II only knows if player I has raised $1 or $2, or has folded. This is a zero sum game with value $\frac{1}{2}$ to player I and $-\frac{1}{2}$ to player II. The Nash equilibrium is the following.

If player I gets an Ace, she should always Raise $2, while if player I gets a Two, she should Raise $2 50% of the time and Fold 50% of the time. Player II's Nash equilibrium is that if player I Raises $1, player II should Call with probability $\frac{2}{3}$ and Fold with probability $\frac{1}{3}$; if player I Raises $2, then player II should Call 50% of the time and Fold 50% of the time. A variant equilibrium for player II is that if player I Raises $1, then II should Call 100% of the time.

4.2.2 EXAMPLES OF EXTENSIVE GAMES USING GAMBIT

Problems

4.10 Consider the following game in extensive form depicted in the figure

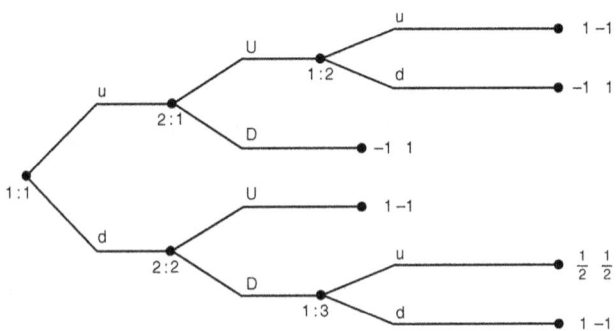

(a) Find the strategic form of the game.

4.10.a Answer: In Gambit form, the strategic form of the game is

I/II	11	12	21	22
11_	1, −1	1, −1	−1, 1	−1, 1
12_	−1, 1	−1, 1	−1, 1	−1, 1
2_1	1, −1	$\frac{1}{2}, \frac{1}{2}$	1, −1	$\frac{1}{2}, \frac{1}{2}$
2_2	1, −1	1, −1	1, −1	1, −1

An underscore character in a strategy means that no decision is necessary at the info set associated with the underscore. The strategies for this game are

Player I

1. 11_- = u at $1:1$, u at $1:2$, no decision at $1:3$
2. 12_- = u at $1:1$, d at $1:2$, no decision at $1:3$
3. 2_1 = d at $1:1$, no decision at $1:2$, u at $1:3$
4. 2_2 = d at $1:1$, no decision at $1:2$, d at $1:3$

Player II

1. 11 = U at $2:1$, U at $2:2$
2. 12 = U at $2:1$, D at $2:2$
3. 21 = D at $2:1$, U at $2:2$
4. 22 = D at $2:1$, D at $2:2$

(b) Find all the Nash equilibria.

4.10.b Answer: There is only one Nash equilibrium $X^* = (0, 0, 0, 1) = Y^*$, (or I plays 2_2, II plays 22) giving payoff 1 to player I and −1 to player II. That is, player I always chooses d at $1:1$, and then d at $1:3$. Player II should always choose D. Note that 212 versus 22 is dud versus DD yields $(1, -1)$ and 222 versus 22 is ddd versus DD and this also yields $(1, -1)$, that is, once player I chooses d at $1:1$, player I knows that info set $1:2$ is never reached, and therefore, no decision for player I at node $1:2$ is necessary.

(c) Find the subgame perfect equilibrium using backward induction.

4.10.c Answer: At the terminal nodes, player I chooses u from $1:2$ and d from $1:3$ resulting in payoff 1 to I and −1 to II. Then II plays D from $2:1$ and anything from $2:2$ because the payoff is the same no matter what II chooses. Finally, from $1:1$ player I chooses d no matter what. The subgame perfect Nash equilibrium is that I plays d from $1:1$ and d from $1:3$. Player II plays D from $2:1$ and either U or D from $2:2$. This includes the Nash equilibrium found earlier, and we conclude $X^* = (0, 0, 0, 1) = Y^*$ is subgame perfect.

4.11 BAT (British American Tobacco) is thinking of entering a market to sell cigarettes currently dominated by PM (Phillip Morris). PM knows about this and can choose to either be passive and give up market share, or be tough and try to prevent BAT from gaining any market share. BAT knows that PM can do this and then must make a decision of whether to fight back, enter the market but act passively, or just stay out entirely. Here is the tree:

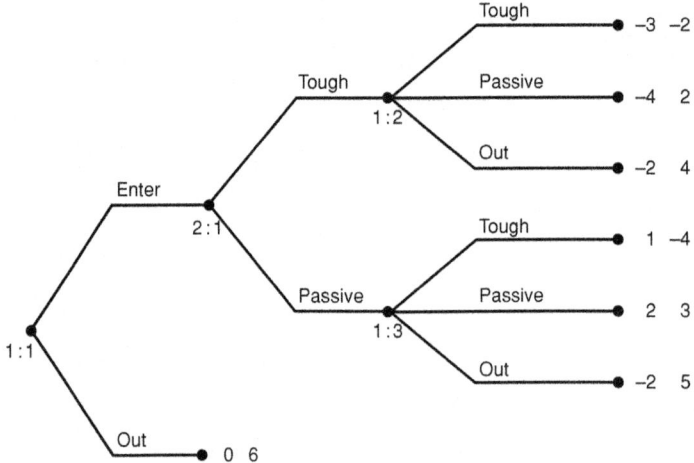

Find the subgame perfect Nash equilibrium.

4.11 Answer: If we use Gambit to solve this game, we get the game matrix

BAT/PM	1	2
111	$-3, -2$	$1, -4$
112	$-3, -2$	$2, 3$
113	$-3, -2$	$-2, 5$
121	$-4, 2$	$1, -4$
122	$-4, 2$	$2, 3$
123	$-4, 2$	$-2, 5$
131	$-2, 4$	$1, -4$
132	$-2, 4$	$2, 3$
133	$-2, 4$	$-2, 5$
2__	$0, 6$	$0, 6$

By dominance, the matrix reduces to the simple game

BAT/PM	1	2
132	$-2, 4$	$2, 3$
2__	$0, 6$	$0, 6$

This game has a Nash equilibrium at $(0, 6)$. We will see it is subgame perfect.

The subgame perfect equilibrium is found by backward induction. At node $1:2$, BAT will play Out with payoff $(-2, 4)$ for each player. At $1:3$, BAT plays Passive with payoffs $(2, 3)$. At $2:1$, PM has the choice of either payoff 4 or payoff 3 and hence plays Tough at $2:1$. Now player BAT at $1:1$ compares $(0, 6)$ with $(-2, 4)$ and chooses Out. The subgame perfect equilibrium is thus BAT plays Out, and PM has no choices to make. The Nash equilibrium at $(0, 6)$ is subgame perfect.

4.12 In a certain card game, player 1 holds two Kings and one Ace. He discards one card, lays the other two face down on the table, and announces either 2 Kings or Ace King.

Ace King is a better hand than 2 Kings. Player 2 must either Fold or Bet on player 1's announcement. The hand is then shown and the payoffs are as follows:

1. If player 1 announces the hand truthfully and player 2 Folds, player 1 wins $1 from player 2.
2. If player 1 lies that the hand is better than it is and player 2 Folds, player 1 wins $2 from player 2.
3. If player 1 lies that the hand is worse than it is and player 2 Folds, player 2 wins $2 from player 1.
4. If player 2 Bets player 1 is lying and if player 1 is actually lying the above payoffs are doubled and reversed. If player 1 is not lying, player 1 wins $2 from player 2.

(a) Use Gambit to draw a game tree.

4.12.a Answer: The tree with the probabilities of playing each branch is the following:

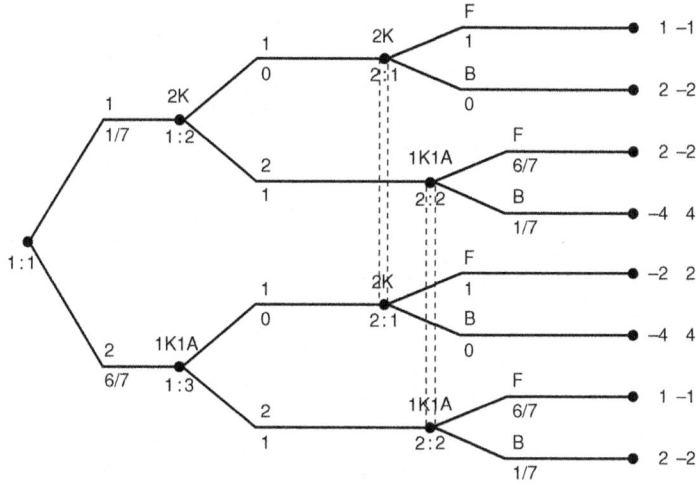

Call 2K's or 1A1K.

(b) Give a complete list of the pure strategies for each player as given by Gambit, and write down the game matrix.

4.12.b Answer: The game matrix is

1/2	11	12	21	22
11_	$1, -1$	$1, -1$	$2, -2$	$2, -2$
12_	$2, -2$	$-4, 4$	$2, -2$	$-4, 4$
2_1	$-2, 2$	$-2, 2$	$4, -4$	$4, -4$
2_2	$1, -1$	$2, -2$	$1, -1$	$2, -2$

The strategies for each player are as follows:

Player 1
1. 11_–Discard Ace, Call 2K;
2. 12_–Discard Ace, Call 1K1A;
3. 2_1–Discard K, Call 2K;
4. 2_2–Discard K, Call 1K1A.

Player 2
1. 11–Fold if player 1 calls 2K, Fold if 1 calls 1K1A;
2. 12–Fold if player 1 calls 2K, Bet if 1 calls 1K1A;
3. 21–Bet if player 1 calls 2K, Fold if 1 calls 1K1A;
4. 22–Bet if player 1 calls 2K, Bet if 1 calls 1K1A.

(c) Find the value of the game and the optimal strategies for each player.

4.12.c Answer: The value of this zero sum game is $v = \frac{8}{7}$. It is hardly a fair game. Player 1 should use 12_ with probability $\frac{1}{7}$ and 2_2 with probability $\frac{6}{7}$. In words, player 1 discards a K $\frac{6}{7}$ of the time and then tells the truth. With probability $\frac{1}{7}$, player 1 should discard the A and then call 1K1A, when he really has 2K. For player 2, if player 1 calls 2K, player 2 should Fold; if player 1 calls 1K1A, player 2 should Fold with probability $\frac{6}{7}$, and Bet with probability $\frac{1}{7}$.

(d) Modify the game so that player 1 first chooses one of the three cards randomly and tosses it. Player 1 is left with two cards—either an A and K, or two K's. Player 1 knows the cards he has but player 2 does not and the game proceeds as before. Solve the game.

4.12.d Answer: The modified game tree is given in the figure.

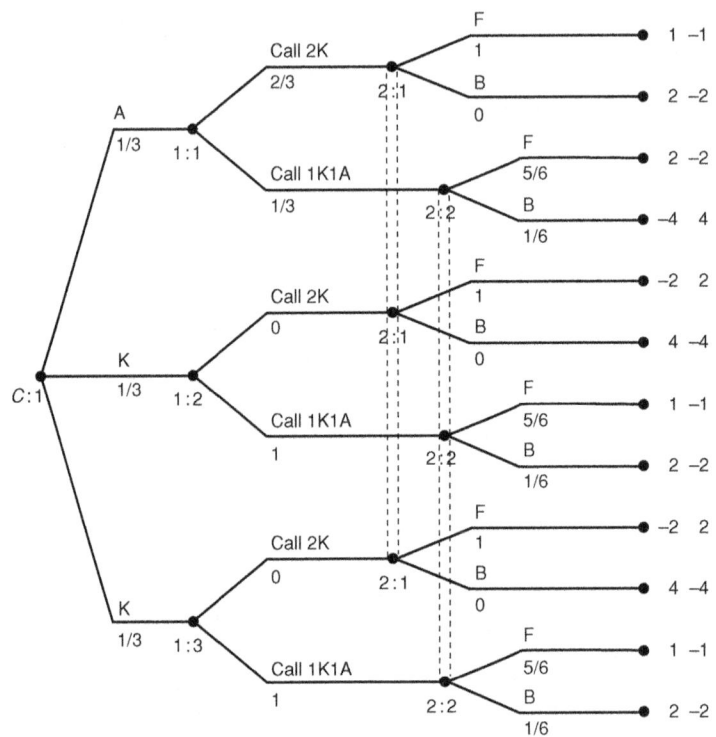

Call 2K's or 1A1K, random draw.

The value of this game is $v = \frac{10}{9} < \frac{8}{7}$. Player 1 should call 2K with probability $\frac{2}{3}$ if the Ace was discarded, and call 1K1A with probability $\frac{1}{3}$. If a K was discarded, player 1 should be truthful and call 1K1A. If player 1 calls 2K, player 2 should always Fold; if player 1 calls 1K1A, player 2 should Fold with probability $\frac{5}{6}$ and Bet with probability $\frac{1}{6}$.

(e) As in the previous part but assume that player 1 does not know the outcome of the random discard. That is, he does not actually know if he has 1A1K or 2K. The payoffs remain as before replacing *lying* with *incorrect* and *truthfully* with *correct*.

4.12.e Answer: This is now a fair game since $v = 0$. Player 1 should always call 1K1A and player 2 should always Bet. Lack of knowledge by player 1 as to which card is discarded makes the game dull but fair. Here is the tree:

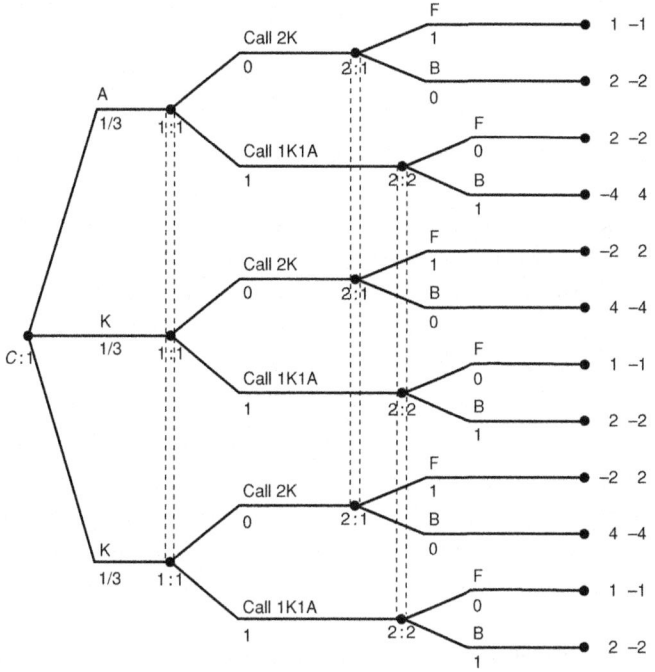

Call 2K's or 1A1K, unknown random draw.

4.13 Solve the Beer or Quiche game assuming Curly has probability $\frac{1}{10}$ of being weak and probability $\frac{9}{10}$ of being strong.

4.13 Answer: The game tree is given in the figure:

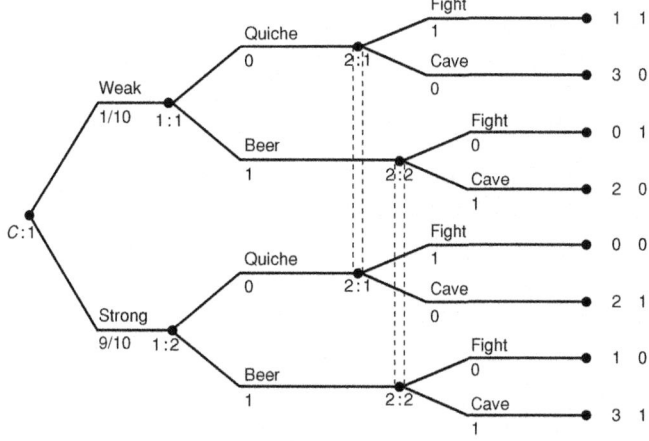

Beer or Quiche Signaling game–weak probability 0.1.

Let's first consider the equivalent strategic form of this game. The game matrix is

Curly/Moe	11	12	21	22
11	0.1, 0.1	0.1, 0.1	2.1, 0.9	2.1, 0.9
12	1, 0.1	2.8, 1	1.2, 0	3, 0.9
21	0, 0.1	0.2, 0	1.8, 1	2, 0.9
22	0.9, 0.1	2.9, 0.9	0.9, 0.1	2.9, 0.9

First note that the first column for player II (Moe) is dominated strictly and hence can be dropped. But that's it for domination.

There are four Nash equilibria for this game. In the first Nash, Curly always chooses to order quiche, while Moe chooses to cave. This is the dull game giving an expected payoff of 2.1 to Curly and 0.9 to Moe. An alternative Nash is Curly always chooses to order a beer and Moe chooses to cave, giving a payoff of 2.9 to Curly and 0.9 to Moe. It seems that Moe is a wimp.

4.14 Curly has two safes, one at home and one at the office. The safe at home is a piece of cake to crack and any thief can get into it. The safe at the office is hard to crack and a thief has only a 15% chance of getting at the gold if it is at the office. Curly has to decide where to place his gold bar (worth 1). On the other hand, the thief has a 50% chance of getting caught if he tries to hit the office and a 20% chance if he hits the house. If he gets the gold he wins 1; if he gets caught he not only doesn't get the gold but he goes to the joint (worth -2). Find the Nash equilibrium and expected payoffs to each player.

4.14 Answer: We will give two solutions, one requires some basic probability and the other doesn't.

Solution 1:

If Curly uses his safe at home and the thief hits the house,

$$E(Curly) = 0 \times 0.2 + (-1) \times 0.8 = -0.8,$$
$$E(Thief) = (-2) \times 0.2 + (1) \times 0.8 = 0.4$$

This assumes the thief can crack the safe at home and he has a 20% chance of getting caught once he's broken into the safe.

If Curly has the gold at home but the thief hits the safe at the office,

$$E(Curly) = 0, \quad E(Thief) = (-2) \times 0.5 + (0 \times 0.15 + 0 \times 0.85) = -1$$

since the thief has a 50% chance of getting caught at the office and a 15% chance of cracking the safe (to find no gold is there).

If Curly has the gold at the office but the thief hits the house,

$$E(Curly) = 0, \quad E(Thief) = (-2) \times 0.2 + 0 \times 0.8 = -0.4.$$

If Curly has the gold at the office and the thief hits the office,

$$E(Curly) = 0 \times 0.5 + 0.5((-1) \times 0.15 + 0 \times 0.85) = -0.075,$$
$$E(Thief) = (-2) \times 0.5 + 0.5((1) \times 0.15 + 0 \times 0.85) = -0.925.$$

With these expected payoffs, we have the following tree:

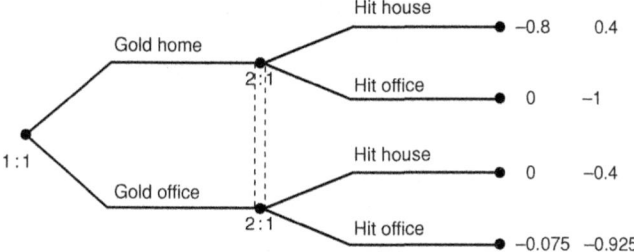

This Nash equilibrium will be subgame perfect. The game matrix is

$$\begin{bmatrix} -\frac{4}{5}, \frac{2}{5} & 0, -1 \\ 0, -\frac{2}{5} & -\frac{3}{40}, -\frac{37}{40} \end{bmatrix}.$$

By dominance, we see that the unique Nash equilibrium is that Curly should always keep the gold at the office and the Thief should always try to hit the House. The payoffs are 0 to Curly and -0.4 to the Thief.

Solution 2:

We let Gambit do the probability work and we simply model the game directly from the description. Here is the tree:

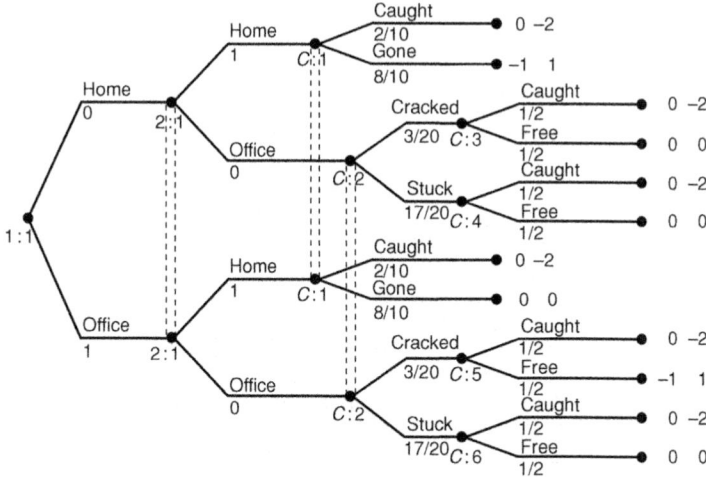

The matrix is exactly as before and the Nash equilibrium is the same.

4.15 Jack and Jill are commodities traders who go to bar after the market closes.[1] They are trying to determine who will buy the next round of drinks. Jack has caught a fly, and he also has a realistic fake fly. Jill, who is always prepared has a fly swatter. They are going to play the following game. Jack places either the real or fake fly on the table

---

[1]Versions of the next three problems appeared on a London School of Economics exam.

covered by his hand. When Jack says "ready," he uncovers the fly (fake or real) and as he does so Jill can either swat the fly or pass. If Jill swats the real fly, Jack buys the drinks; if she swats the fake fly, Jill buys the drinks. If Jill thinks the real fly is the fake and she passes on the real fly, the fly flies away, the game is over, and Jack and Jill split the round of drinks. If Jill passes on the fake fly, then they will give it another go. If in the second round Jill again passes on the fake fly, then Jack buys the round and the game is over. A round of drinks costs $2.

(a) Use Gambit to solve the game.

4.15.a Answer: The game tree is in the figure:

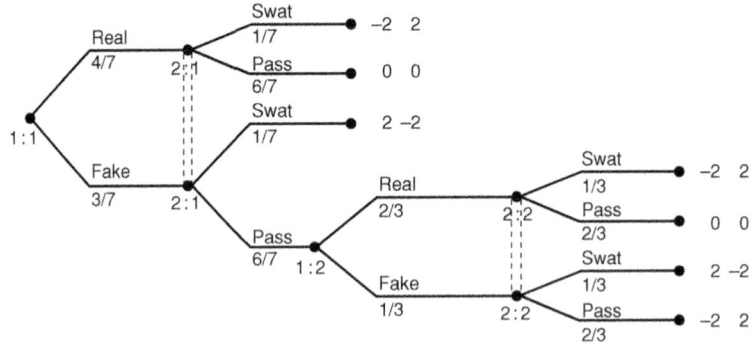

There are three pure strategies for Jack: (1) Real, (2) Fake, then Real, and (3) Fake, then Fake. Jill also has three pure strategies: (1) Swat, (2) Pass, then Swat, and (3) Pass, then Pass.

Jack/Jill	(1)	(2)	(3)
(1)	−2	0	0
(2)	2	−2	0
(3)	2	2	−2

We can use Gambit to solve the problem, but we will use the invertible matrix theorem instead. We get

$$v = \frac{1}{J_3 A^{-1} J_3^T} = -\frac{2}{7}, \quad X^* = v\, J_3 A^{-1} = \left(\frac{4}{7}, \frac{2}{7}, \frac{1}{7}\right),$$
$$Y^* = v \cdot A^{-1} J_3^T = \left(\frac{1}{7}, \frac{2}{7}, \frac{4}{7}\right).$$

(b) What happens if Jill only has a 75% chance of nailing the real fly when she swats at it?

4.15.b Answer: Jack and Jill have the same pure strategies, but the solution of the game is quite different. The matrix is

Jack/Jill	(1)	(2)	(3)
(1)	−1	0	0
(2)	2	−1	0
(3)	2	2	−2

For example, (1) versus (1) means Jack will put down the real fly and Jill will try to swat the fly. Jack's expected payoff is

$$(-2)\frac{3}{4} + (2)\frac{1}{4} = -1.$$

The game tree from the first part is easily modified noting that the probability of hitting or missing the fly only applies to the real fly. Here is the figure with Gambit's solution below the branches.

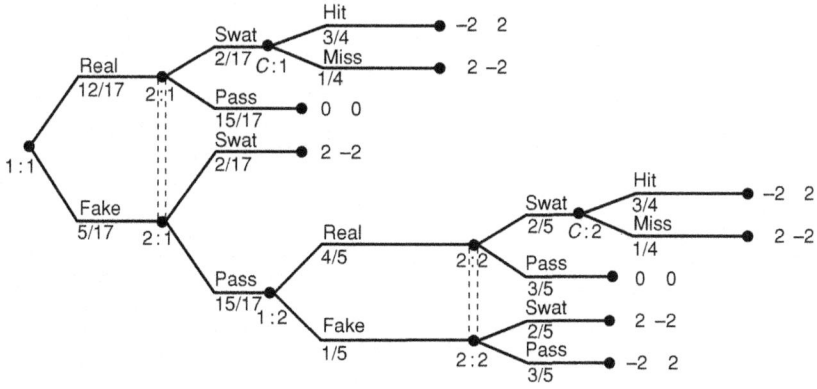

The solution of the game is

$$v = \frac{1}{J_3 A^{-1} J_3^T} = -\frac{2}{17}, \quad X^* = v J_3 A^{-1} = \left(\frac{12}{17}, \frac{4}{17}, \frac{1}{17}\right),$$

$$Y^* = v \cdot A^{-1} J_3^T = \left(\frac{2}{17}, \frac{6}{17}, \frac{9}{17}\right).$$

4.16 Two opposing navys (French and British) are offshore an island while the admirals are deciding whether or not to attack. Each navy is either strong or weak with equal probability. The status of their own navy is known to the admiral but not to the admiral of the opposing navy. A navy captures the island if it either attacks while the opponent does not attack, or if it attacks as strong while the opponent is weak. If the navys attack and they are of equal strength, then neither captures the island.

The island is worth 8 if captured. The cost of fighting is 3 if it is strong and 6 if it is weak. There is no cost of attacking if the opponent does not attack and there is no cost if no attack takes place. What should the admirals do?

4.16 Answer: The information sets must be set up so that each admiral knows the strength of his own navy but not that of his opponent. With that in mind, here is the game tree:

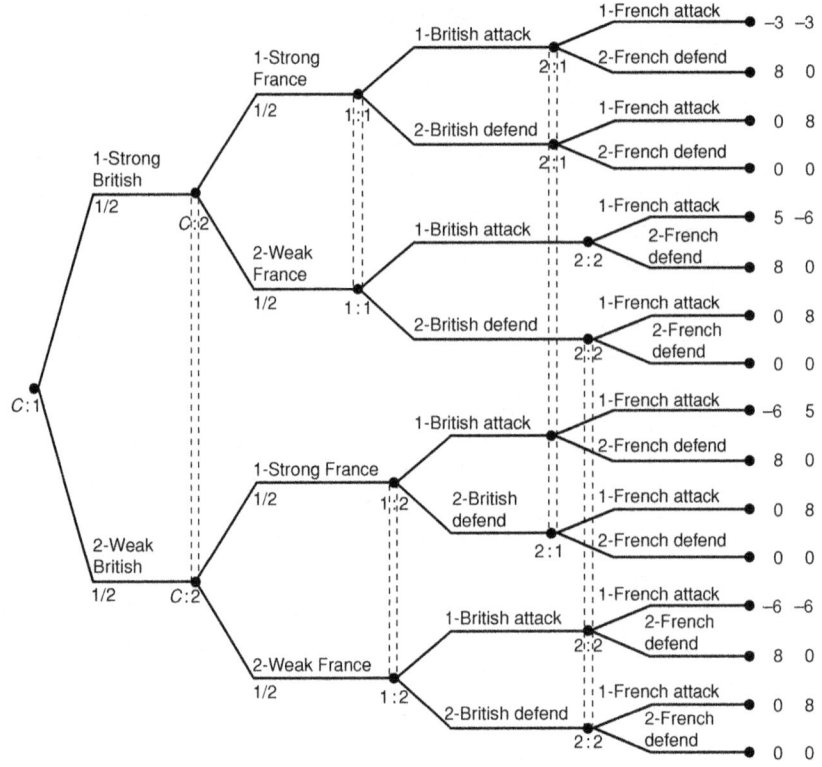

French and British Navys.

The game matrix becomes

Fr/Brit	11	12	21	22
11	$-\frac{5}{2}, -\frac{5}{2}$	$\frac{7}{4}, \frac{1}{2}$	$\frac{15}{4}, -3$	$8, 0$
12	$\frac{1}{2}, \frac{7}{4}$	$\frac{5}{4}, \frac{5}{4}$	$\frac{13}{4}, \frac{1}{2}$	$4, 0$
21	$-3, \frac{15}{4}$	$\frac{1}{2}, \frac{13}{4}$	$\frac{1}{2}, \frac{1}{2}$	$4, 0$
22	$0, 8$	$0, 4$	$0, 4$	$0, 0$

By dominance, we may drop rows 3 and 4 and columns 3 and 4. Then we have a simple 2×2 game to solve.

Fr/Brit	11	12
11	$-\frac{5}{2}, -\frac{5}{2}$	$\frac{7}{4}, \frac{1}{2}$
12	$\frac{1}{2}, \frac{7}{4}$	$\frac{5}{4}, \frac{7}{4}$

We get the following Nash equilibria:
1. $X_1 = Y_1 = (\frac{1}{7}, \frac{6}{7}, 0, 0)$, payoff French $= \frac{8}{7}$, British $= \frac{8}{7}$. When the British and French are strong they should Attack, but if they are Weak they should Attack with probability $\frac{1}{7}$. This is similar to bluffing. The expected payoff to the British and the French is then $\frac{8}{7}$.

2. $X_2 = (1, 0, 0, 0)$, $Y_2 = (0, 1, 0, 0)$, payoff French $= \frac{1}{2}$, British $= \frac{7}{4}$. The British play 11, the French play 12, with probability 1. That is, the British Attack if they are strong or weak, the French Attack if they are strong, but do not attack if they are weak.

3. $X_3 = (0, 1, 0, 0)$, $Y_3 = (1, 0, 0, 0)$, payoff French $= \frac{7}{4}$, British $= \frac{1}{2}$. The French always Attack, while the British Attack only if they are strong.

In Problem 7.3, we will prove that equilibrium 1 is evolutionary stable, which means that this strategy will be played in the long run if there are a series of battles.

4.17 Solve the six-stage centipede game by finding a correlated equilibrium. Since the objective function is the sum of the expected payoffs, this would be the solution assuming that each player cares about the total, not the individual. Use Maple/Mathematica for this solution.

4.17 Answer: Using Maple we have

```
>restart:A:=Matrix([[1,1,1,1],[0,3,3,3],[0,2,5,5],
 [0,2,4,7]]);
>B:=Matrix([[0,0,0,0],[3,2,2,2],[3,5,4,4],[3,5,7,6]]);
>P:=Matrix(4,4,symbol=p);
>Z:=add(add((A[i,j]+B[i,j])*p[i,j],i=1..4),j=1..4):
>with(simplex):
>cnsts:={add(add(p[i,j],i=1..4),j=1..4)=1}:
>for i from 1 to 4 do
    for k from 1 to 4 do
    cnsts:=cnsts union {add(p[i,j]*A[i,j],j=1..4)
>= add(p[i,j]*A[k,j],j=1..4)}:
    end do:
end do:

>for j from 1 to 4 do
    for k from 1 to 4 do
    cnsts:=cnsts union {add(p[i,j]*B[i,j],i=1..4)
>= add(p[i,j]*B[i,k],i=1..4)}:
    end do:
end do:

>maximize(Z,cnsts,NONNEGATIVE);
>with(Optimization):
>LPSolve(Z,cnsts,assume=nonnegative,maximize);
```

The result is $p_{1,1} = 1$ and $p_{i,j} = 0$ otherwise. The correlated equilibrium also gives the action that each player Stops.

4.18 There are three gunfighters A, B, C. Each player has 1 bullet and they will fire in the sequence A then B then C, assuming the gunfighter whose turn comes up is still alive. The game ends after all three players have had a shot. If a player survives, that player gets 2, while if the player is killed, the payoff to that player is -1.

(a) First, assume that all the gunfighters have probability $\frac{1}{2}$ of actually hitting the person they are shooting at. (Hitting the player results in a kill.) Find the extensive game and as many Nash equilibria as you can using Gambit.

4.18.a Answer: The game tree is straightforward because every node has a new information set, that is, it is perfect information.

Here is the tree:

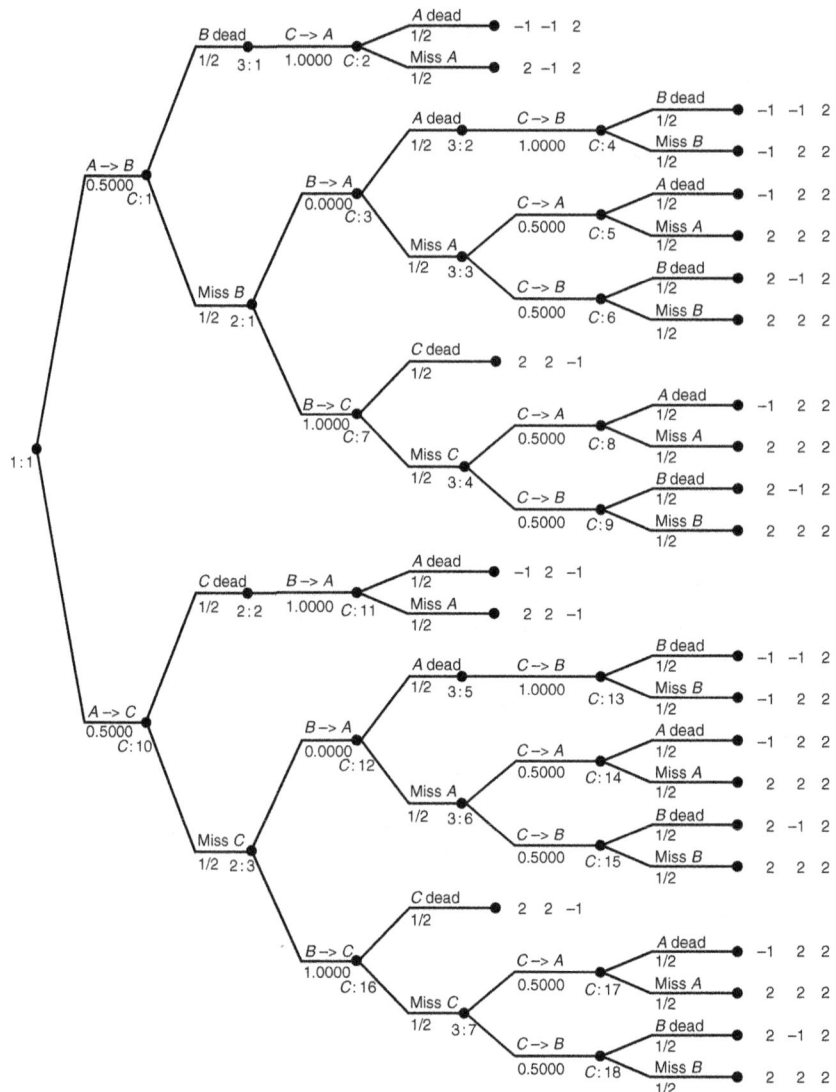

If you look at the strategic game Gambit gives you, it is clearly not feasible to write down all the pure strategies by hand. The Nash equilibrium we get is as follows:

1. A should fire at B or C with probability $\frac{1}{2}$.
2. If A fires at B and B is dead, then C fires at A.
3. If A fires at B and misses, then B should fire at C.
4. If B misses C then C fires at A or B with equal probability.
5. The other case is A fires at C.
6. If A fires at C and kills C, then B fires at A.

7. If A fires at C and misses C, then B should fire at C.

8. If C survives, then C fires at A or B with equal probability.

The expected payoff to A and B is 1.0625; the expected payoff to C is 0.5.

(b) Since the game is perfect information and perfect recall we know there is a subgame perfect equilibrium. Find it.

4.18.b Answer: We use backward induction. We may use Gambit to help in the calculations on this.

At node $3:7$ the payoff to each player is the **node value** given in Gambit as $(1.25, 1.25, 2)$. What this means is that if we arrive at node $3:7$, then the eventual payoff to C is 2, the payoff to A is

$$0.5 \times (0.5(-1) + 0.5(2)) + 0.5 \times (0.5(2) + 0.5(2)) = 1.25$$

and similarly for B. Working backward, we calculate the payoffs when C makes the final move. Then we replace the subtrees in which C has a decision to make, with the payoffs calculated. The result is the tree.

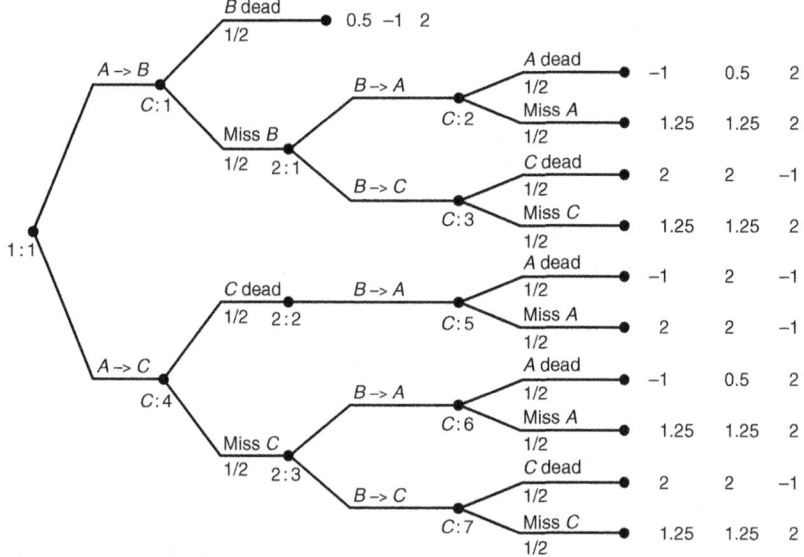

Next, we replace all the subtrees in which B makes the final move. The payoffs are replace with the node values in the preceding tree. The next tree we get is

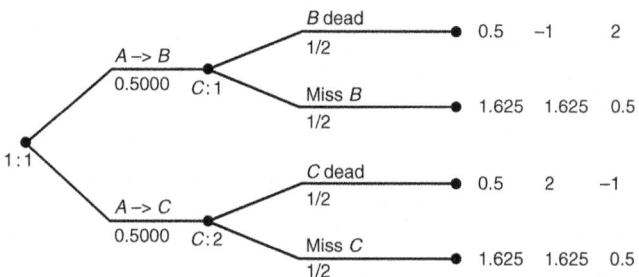

Finally, A makes the last decision. Since the expected payoff to A is the same whether A fires at B or C, A fires at either one with probability $\frac{1}{2}$.

The subgame perfect equilibrium is as follows:

1. A fires at either B or C with probability $\frac{1}{2}$.
2. If A fires at B and B survives, then B should fire at C (since the node value for B is 1.625 at C:3 versus 0.875 at C:2).
3. If A fires at C and C survives, then B should fire at C (node value for B is 1.625 at C:7 versus 0.875 at C:6).
4. If C has survived the first two rounds, then it doesn't matter if C fires at B or A (whoever is still in the game), because the payoff to C will be 2 no matter what.

(c) Now assume that gunfighters have different accuracies. A's accuracy is 40%, B's accuracy is 60% and C's accuracy is 80%. Solve this game.

4.18.c Answer: All we have to do is modify the probabilities at the chance nodes. The Nash equilibrium is as follows:

1. A fires at C.
2. B fires at C.
3. If C survives the first two shots, C fires at A or B with probability $\frac{1}{2}$.

The expected payoffs are 0.992 to A, 1.712 to B, and -0.28 to C. C is the most accurate so eliminating C gives A and B the best chance of survival.

4.19 Moe, Larry, and Curly are in a two round truel. In round 1, Larry fires first, then Moe, then Curly. Each stooge gets one shot in round 1. Each stooge can **choose to fire at one of the other players, or to deliberately miss by firing in the air.** In round 2, all survivors are given a second shot, in the same order of play.

Suppose Larry's accuracy is 30%, Moe's is 80%, and Curly's is 100%. For each player, the preferred outcome is to be the only survivor, next is to be one of two survivors, next is the outcome of no deaths, and the worst outcome is that you get killed. Assume an accurate shot results in death of the one shot at. Who should each player shoot at and find the probabilities of survival under the Nash equilibrium.

4.19 Answer: This problem can be tricky because it quickly becomes unmanageable if you try to solve it by considering all possibilities. The key is to recognize that after the first round, if all three stooges survive it is as though the game starts over. If only two stooges survive, it is like a duel between the survivors with only one shot for each player and using the order of play Larry, then Moe, then Curly (depending on who is alive).

According to the instructions in the problem we take the payoffs for each player to be 2, if they are sole survivor, $\frac{1}{2}$ if there are two survivors, 0 if all three stooges survive, and -1 for the particular stooge who is killed. The payoffs at the end of the first round are -1 to any killed stooge in round 1, and the expected payoffs if there are two survivors, determined by analyzing the two person duels first. This is very similar to using backward induction and subgame perfect equilibria.

Begin by analyzing the two person duels Larry versus Moe, Moe versus Curly, and Curly versus Larry. We assume that round one is over and only the two participants in the duel survive. We need that information in order to determine the payoffs in the two person duels. Thus, if Larry kills Curly, Larry is sole survivor and he gets 2, while Curly gets -1. If Larry misses, he's killed by Curly and Curly is sole survivor.

1. Curly versus Larry:

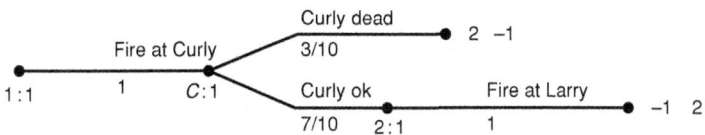

Larry versus Curly.

In this simple game, we easily calculate the expected payoff to Larry is $-\frac{1}{10}$ and to Curly is $\frac{11}{10}$.

2. Moe versus Curly:

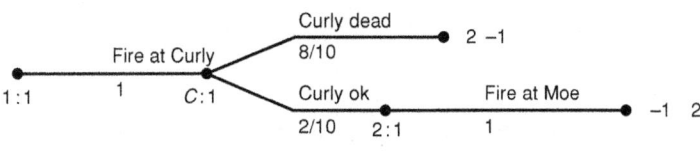

Moe versus Curly.

The expected payoff to Moe is $\frac{7}{5}$, and to Curly is $-\frac{2}{5}$.

3. Larry versus Moe:

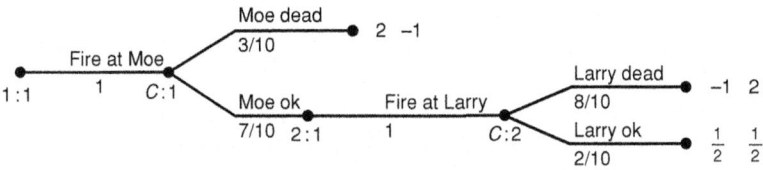

Larry versus Moe.

In this case, if Larry shoots Moe, Larry gets 2 and Moe gets -1, but if Larry misses Moe, then Moe gets his second shot. If Moe misses Larry, then there are two survivors of the truel, and each gets $\frac{1}{2}$. The expected payoff to Moe is $\frac{89}{100}$, and to Curly is $-\frac{11}{100}$.

Next, consider the tree for the first round incorporating the expected payoffs for the second round.

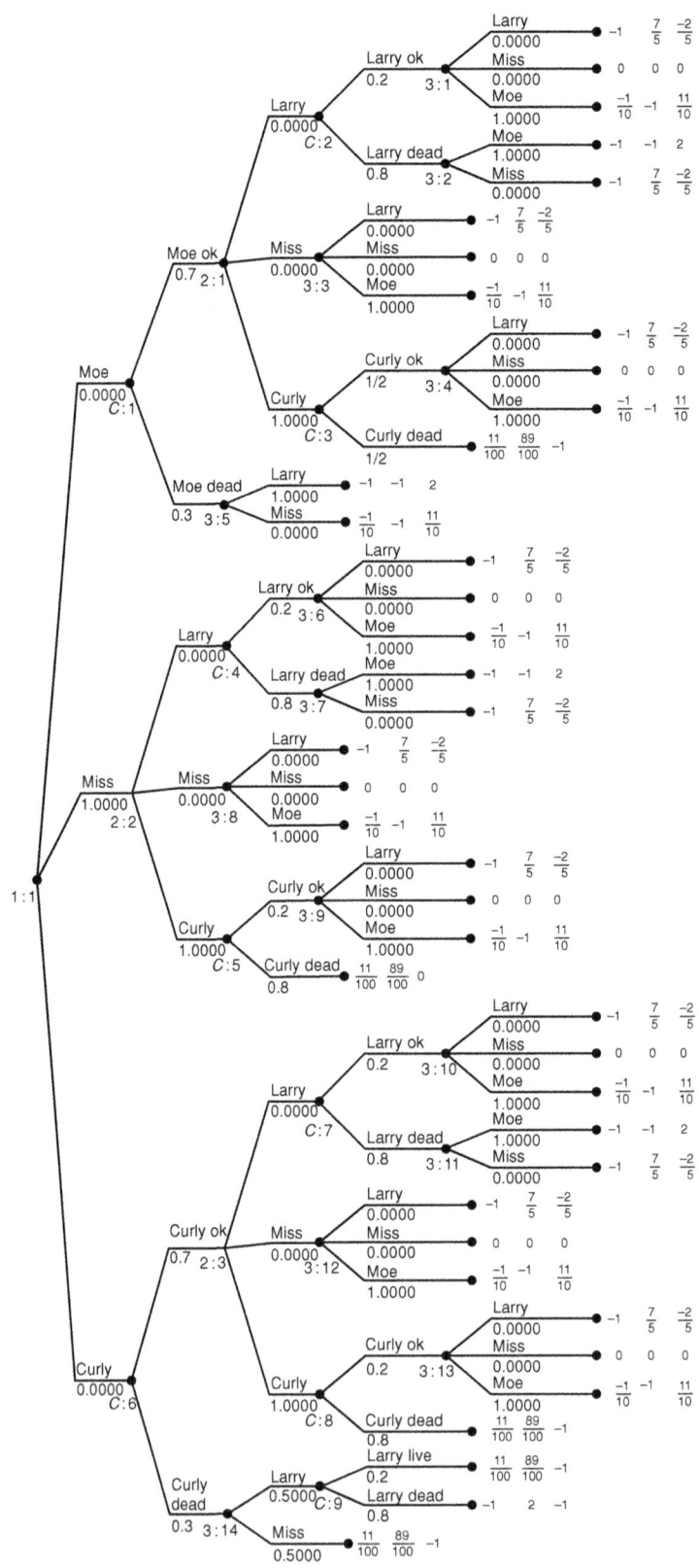

First round tree using second round payoffs.

The expected payoff to Larry is 0.68, Moe is 0.512, and Curly is 0.22. The Nash equilibrium says that Larry should deliberately miss with his first shot. Then Moe should fire at Curly; if Curly lives, then Curly should fire at Moe—and Moe dies. In the second round, Larry has no choice but to fire at Curly, and if Curly lives, then Curly will kill Larry.

Does it make sense that Larry will deliberately miss in the first round? Think about it. If Larry manages to kill Moe in the first round, then surely Curly kills Larry when he gets his turn. If Larry manages to kill Curly, then with 80% probability, Moe kills Larry. His best chance of survival is to give Moe and Curly a chance to shoot at each other. Since Larry should deliberately miss, when Moe gets his shot he should try to kill Curly because when it's Curly's turn, Moe is a dead man.

To find the probability of survival for each player assuming the Nash equilibrium is being played we calculate:

1. For Larry:

$Prob$(Larry Survives)

$= Prob$(Moe kills Curly in round 1 $\cap$ Larry kills Moe in round 2)

$= Prob$(Moe kills Curly in round 1) $\times Prob$(Larry kills Moe in round 2)

$= 0.8 \times 0.3 = 0.24.$

2. For Moe:

$Prob$(Moe Survives)

$= Prob$(Moe kills Curly in round 1 $\cap$ Moe kills Larry in round 2)

$= Prob$(Moe kills Curly in round 1) $\times Prob$(Moe kills Larry in round 2)

$= 0.8 \times 0.8 = 0.64.$

3. For Curly:

$Prob$(Curly Survives)

$= Prob$(Moe misses Curly in round 1 $\cap$ Curly kills Larry in round 2)

$= Prob$(Moe misses Curly in round 1) $\times Prob$(Curly kills Larry in round 2)

$= 0.7 \times 1 = 0.7.$

4.20 Two players will play a Battle of the Sexes game with bimatrix

I/II	A	B
A	3, 1	0, 0
B	0, 0	1, 3

Before the game is played player I has the option of burning \$1 with the result that it would drop I's payoff by 1 in every circumstance.

(a) Draw the extensive form of this game where player II can observe whether or not player I has burned the dollar. Find the strategic form of the game.

4.20.a Answer: The game tree is the following figure:

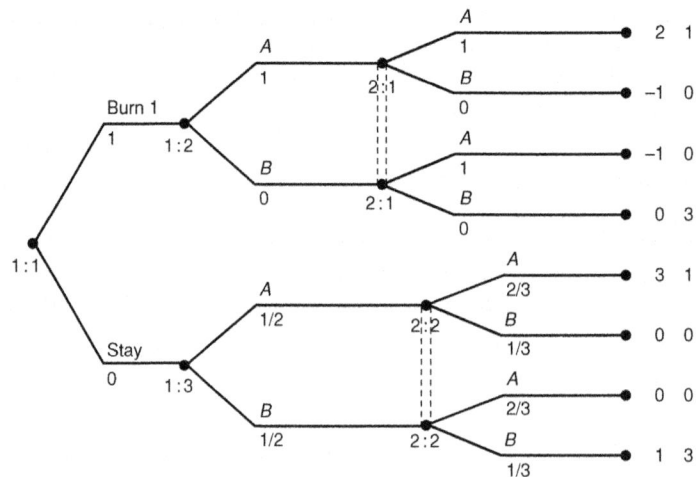

II observes I burn $1.

The game matrix is

I/II	11	12	21	22
11_	2, 1	2, 1	−1, 0	−1, 0
12_	−1, 0	−1, 0	0, 3	0, 3
2_1	3, 1	0, 0	3, 1	0, 0
2_2	0, 0	1, 3	0, 0	1, 3

Strategy 11_ for example, means player I burn's the dollar and then chooses A.

(b) Find all the Nash equilibria of the game and determine which are subgame perfect.

4.20.b Answer: There are four pure Nash equilibria that are easily found from the matrix and they are given in the table

Strategy	Player I	Player II	Payoff to I	Payoff to II
1	2_2	22	1	3
2	2_1	11	3	1
3	2_1	21	3	1
4	11_	12	2	1

There are six more mixed Nash equilibria found by Gambit:

Strategy	Player I	Player II	Payoff to I	Payoff to II
5	$(0, 0, \frac{3}{4}, \frac{1}{4})$	$(0, \frac{7}{12}, \frac{1}{4}, \frac{1}{6})$	$\frac{3}{4}$	$\frac{3}{4}$
6	$(0, 0, \frac{3}{4}, \frac{1}{4})$	$(\frac{1}{4}, \frac{1}{3}, 0, \frac{5}{12})$	$\frac{3}{4}$	$\frac{3}{4}$
7	$(0, 0, \frac{3}{4}, \frac{1}{4})$	$(\frac{1}{4}, 0, 0, \frac{3}{4})$	$\frac{3}{4}$	$\frac{3}{4}$
8	$(0, 0, \frac{3}{4}, \frac{1}{4})$	$(0, 0, \frac{3}{4}, \frac{1}{4})$	$\frac{3}{4}$	$\frac{3}{4}$
9	$(0, 0, 0, 1)$	$(0, \frac{2}{3}, 0, \frac{1}{3})$	1	3
10	$(1, 0, 0, 0)$	$(\frac{2}{3}, \frac{1}{3}, 0, 0)$	2	1

This game has three subgames. To find the subgame perfect Nash, we need to solve the games in which player I burns the dollar, and in which he does not burn the dollar. In this case, the subgame perfect equilibrium is that player I should always choose to not burn the dollar and then the two players play the regular battle of sexes game.

(c) Solve the game when player II cannot observe whether or not player I has burned the dollar.

4.20.c Answer: When player II has no knowledge of player I's choice to burn or not to burn, the game becomes as follows:

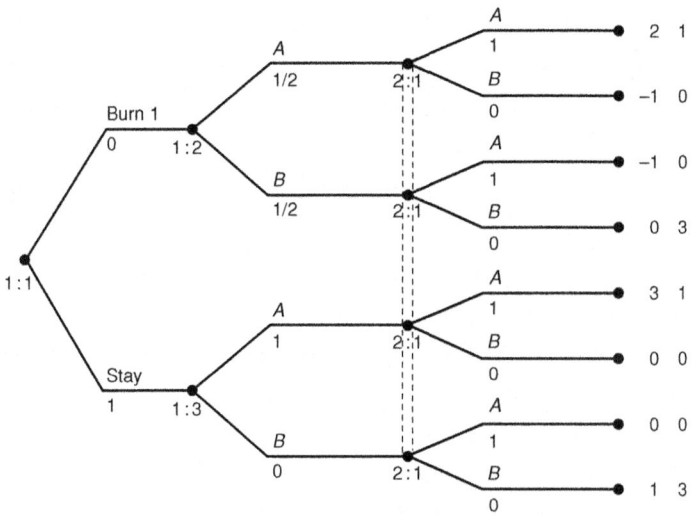

II does not observe I burn $1.

This game has three Nash equilibria all of which are subgame perfect since there are no proper subgames:

1. Player I Stays and then plays B; player II plays B—payoff 1 to I, and 3 to II.
2. Player I Stays and then chooses A with probability $\frac{3}{4}$ and B with probability $\frac{1}{4}$; player II plays A with probability $\frac{3}{4}$ and B with probability $\frac{1}{4}$—payoff $\frac{3}{4}$ to each player.
3. Player I Stays and then plays A; player II plays A—payoff 3 to I, and 1 to II.

4.21 There are two players involved in a two-step centipede game. Each player may be either rational (with probability $\frac{19}{20}$) or an altruist (with probability $\frac{1}{20}$). Neither player knows the other players type.

If both players are rational, each player may either Take the payoff or Pass. Player I moves first and if he Takes, then the payoff is 0.8 to I and 0.2 to II. If I Passes, then II may either Take or Pass. If II Takes, the payoff is 0.4 to I and 1.60 to II. If II Passes the next round begins. Player I may Take or Pass (payoffs 3.20, 0.80 if I Takes) then II may Take or Pass (payoffs 1.60, 6.40 if II Takes and 12.80, 3.20 if II Passes). The game ends.

If player I is rational and player II is altruistic, then player II will Pass at both steps. The payoffs to each player are the same as before if I Takes, or the game is at an end.

The game is symmetric if player II is rational and player I is altruistic.

If both players are altruistic then they each Pass at each stage and the payoffs are 12.80, 3.20 at the end of the second step.

Draw the extensive form of this game and find as many Nash equilibria as you can. Compare the result with the game in which both players are rational.

4.21 Answer: Here's the figure.

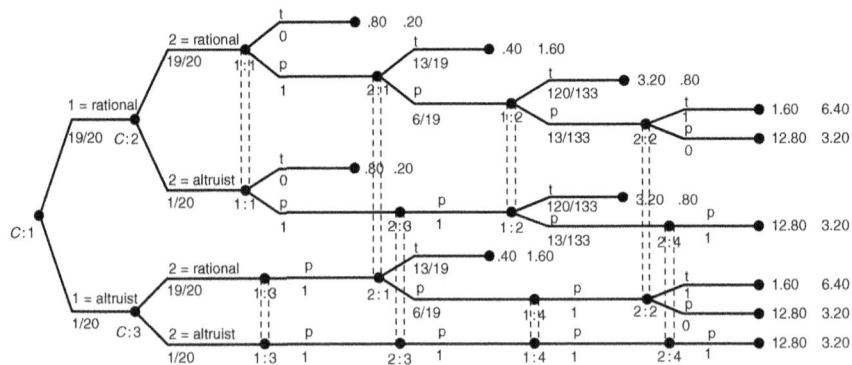

Two-stage centipede with altruism.

The expected payoffs to each player are $\frac{69}{50}$ for I and $\frac{276}{175}$ for II. The Nash equilibrium is the following.

If both players are rational, I should Pass, then II should Pass with probability $\frac{6}{19}$, then I should Pass with probability $\frac{13}{133}$, then player II should Pass with probability 0.

If I is rational and II is an altruist, then I should Pass with probability 1, then II Passes, then I Passes with probability $\frac{13}{133}$, then II Passes.

If I is an altruist and II is rational, then I Passes, II Passes with probability $\frac{6}{19}$, I Passes, and II Passes with probability 0.

If they are both altruists, they each Pass at each step and go directly to the end payoff of 12.80 for I and 3.20 for II.

N-Person Nonzero Sum Games and Games with a Continuum of Strategies

5.1 The Basics

Problems

5.1 This problem makes clear the connection between a pair of numbers (q_1^*, q_2^*) that maximizes both payoff functions and a Nash equilibrium. Suppose that we have a two-person game with payoff functions $u_i(q_1, q_2)$, $i = 1, 2$. Suppose there is a pure strategy pair (q_1^*, q_2^*) that maximizes both u_1 and u_2 as functions of the pair (q_1, q_2).

(a) Verify that (q_1^*, q_2^*) is a Nash equilibrium.

5.1.a Answer: Since (q_1^*, q_2^*) maximizes both u_1 and u_2, we have

$$u_1\left(q_1^*, q_2^*\right) = \max_{(q_1, q_2)} u_1(q_1, q_2) \quad \text{and} \quad u_2\left(q_1^*, q_2^*\right) = \max_{(q_1, q_2)} u_2(q_1, q_2).$$

Thus,

$$u_1\left(q_1^*, q_2^*\right) \geq u_1(q_1, q_2^*) \quad \text{and} \quad u_2\left(q_1^*, q_2^*\right) \geq u_2(q_1^*, q_2),$$

for every $q_1 \neq q_1^*, q_2 \neq q_2^*$. Thus, (q_1^*, q_2^*) automatically satisfies the definition of a Nash equilibrium. A maximum of both payoffs is a much stronger requirement than a Nash point.

(b) Construct an example in which a Nash equilibrium (q_1^*, q_2^*) does *not* maximize both u_1 and u_2 as functions of (q_1, q_2).

5.1.b Answer: Consider $u_1(q_1, q_2) = q_2^2 - q_1^2$ and $u_2(q_1, q_2) = q_1^2 - q_2^2$ and $-1 \leq q_1 \leq 1, -1 \leq q_2 \leq 1$.

Solutions Manual to Accompany Game Theory: An Introduction, Second Edition. E.N. Barron.
© 2013 John Wiley & Sons, Inc. Published 2013 by John Wiley & Sons, Inc.

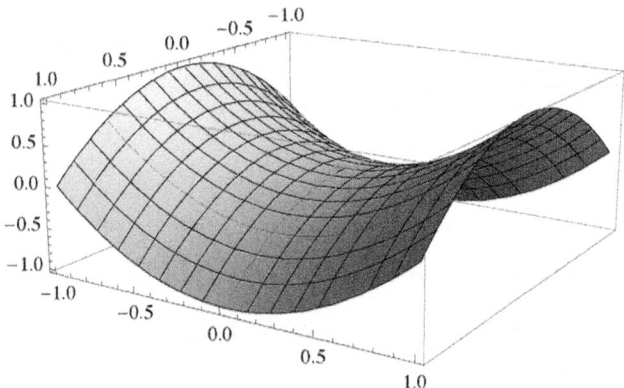

The function $u_1(q_1, q_2) = q_2^2 - q_1^2$.

There is a unique Nash equilibrium at $(q_1^*, q_2^*) = (0, 0)$ since

$$\frac{\partial u_1(q_1, q_2)}{\partial q_1} = -2q_1 = 0 \quad \text{and} \quad \frac{\partial u_2(q_1, q_2)}{\partial q_2} = -2q_2 = 0$$

gives $q_1^* = q_2^* = 0$ and these points provide a maximum of each function with the other variable fixed because the second partial derivatives are $-2 < 0$. On the other hand,

$$u_1\left(q_1^*, q_2^*\right) = 0, \ \max_{(q_1, q_2)} u_1(q_1, q_2) = 1, \quad \text{and} \quad u_2\left(q_1^*, q_2^*\right) = 0, \ \max_{(q_1, q_2)} u_2(q_1, q_2) = 1.$$

Thus, $(q_1^*, q_2^*) = (0, 0)$ maximizes **neither** of the payoff functions. Furthermore, there does not exist a point that maximizes **both** u_1, u_2 at the same point (q_1, q_2).

5.2 Consider the zero sum game with payoff function for player 1 given by $u(x, y) = -2x^2 + y^2 + 3xy - x - 2y, 0 \leq x, y \leq 1$. Show that this function is concave in x and convex in y. Find the saddle point and the value.

5.2 Answer: Take partial derivatives

$$\frac{\partial u}{\partial x} = -4x + 3y - 1 \quad \text{and} \quad \frac{\partial u}{\partial y} = 2y + 3x - 2.$$

Then $\frac{\partial^2 u}{\partial x^2} = -4 < 0$, and $\frac{\partial^2 u}{\partial y^2} = 2 > 0$, which mean u is concave in x and convex in y. Solving the equations where the partial derivatives are zero results in the Nash equilibrium (the same as the saddle point) and value. $x = \frac{4}{17}, y = \frac{11}{17}, u = -\frac{13}{17}$.

5.3 The payoff functions for Curly and Shemp are, respectively,

$$c(x, y) = xy^2 - x^2, \qquad s(x, y) = 8y - xy^2,$$

with $x \in [1, 3], y \in [1, 3]$. Curly chooses x and Shemp chooses y.

(a) Find and graph the rational reaction sets for each player. Recall that the best response for Curly is $BR_c(y) \in \arg\max c(x, y)$ and the rational reaction set for Curly is the graph of $BR_c(y)$. Similarly for Shemp.

5.3.a Answer: By taking derivatives, we get the best response functions

$$BR_c(y) = \begin{cases} 1, & \text{if } 1 \le y \le \sqrt{2}; \\ \dfrac{y^2}{2}, & \text{if } \sqrt{2} \le y \le \sqrt{6}; \\ 3, & \text{if } \sqrt{6} \le y \le 3. \end{cases}$$

$$BR_s(x) = \begin{cases} 3, & \text{if } 1 \le x \le \frac{4}{3}; \\ \dfrac{4}{x}, & \text{if } \frac{4}{3} \le x \le 3. \end{cases}$$

The graphs of the rational reaction sets are as follows:

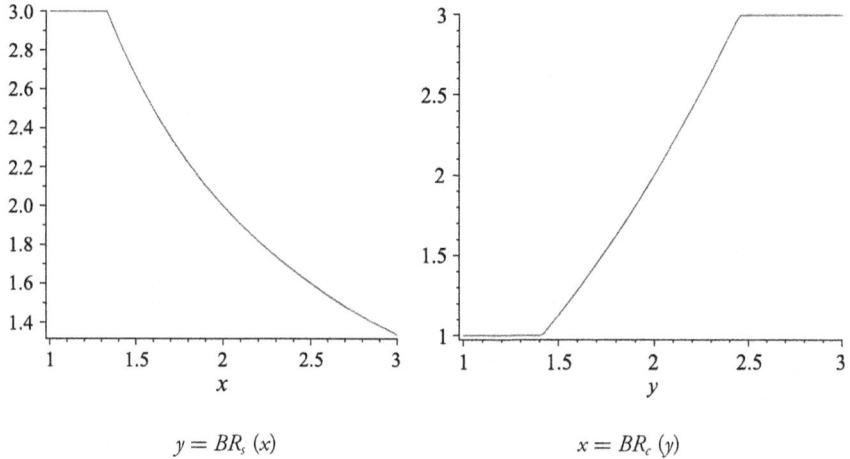

$$y = BR_s(x) \qquad\qquad x = BR_c(y)$$

(b) Find a pure Nash equilibrium.

5.3.b Answer: The rational reaction sets intersect at the unique point $(x^*, y^*) = (2, 2)$, which is the unique pure Nash equilibrium.

5.4 Consider the game in which two players choose nonnegative integers no greater than 1000. Player 1 must choose an even integer, while player 2 must choose an odd integer. When they announce their number, the player who chose the lower number wins the number she announced in dollars. Find the Nash equilibrium.

5.4 Answer: This is similar to the Traveler's Paradox game. The Nash equilibrium is player 1 announces 0 and player 2 announces 1. To see why, suppose player 1 chooses an even number $2n \le 1000$ and player 2 chooses an odd integer $2k + 1 \le 999$. Assume that $n > 0, k > 0$. If $2n > 2k + 1$ then 1 loses and 2 wins $2k + 1$. Player 1 does better by switching to $2k < 2k + 1$. Thus, no strategy in which player 1 picks a bigger integer than player 2 can be part of a Nash equilibrium for player 1. Similarly, no strategy in which player 2 picks a bigger integer than player 1 can be part of a Nash equilibrium for player 2. Thus $n = 0, k = 0$, which means player 1 will call 0 and player 2 will call 1.

5.5 In the tragedy of the commons Example 5.6, we saw that if $N \geq 5$ everyone owning a sheep is a Nash equilibrium. Analyze the case $N \leq 4$.

5.5 Answer: We claim that if $N \leq 4$, then $(0, 0, \ldots, 0)$ is a Nash equilibrium. In fact, $u_1(0, 0, 0, 0) = 0$ but $u_1(1, 0, 0, 0) = 1 - \frac{5}{4} < 0$. Similarly for the remaining cases.

5.6 There are N players each using r_i units of a resource whose total value is $R = \sum_{i=1}^{N} r_i$. The cost to player i for using r_i units of the resource is $f(r_i) + g(R - r_i)$ where we take $f(x) = 2x^2$, $g(x) = x^2$. This says that a player's cost is a function of the amount of the total used by that player and a function of the amount used by the other players. The revenue player i receives from using r_i units is $h(r_i)$, where $h(x) = \sqrt{x}$. Assume that the total resources used by all players is not unlimited so that $0 \leq R \leq R_0 < \infty$.

(a) Find the payoff function for player $i = 1, 2, \ldots, N$.

5.6.a Answer: The payoff function for each player $i = 1, 2, \ldots, N$, is

$$u_i(r_1, \ldots, r_N) = h(r_i) - (f(r_i) + g(R - r_i)) = \sqrt{r_i} - 2r_i^2 - (R - r_i)^2,$$

where $R = r_1 + r_2 + \cdots + r_N$.

(b) Find the Nash equilibrium in general and when $N = 12$.

5.6.b Answer: Taking a partial derivative of u_i with respect to r_i gives

$$\frac{\partial u_i}{\partial r_i} = \frac{1}{2}\frac{1}{\sqrt{r_i}} - 4r_i = 0,$$

which implies $r_i = \frac{1}{4}$. Since $\frac{\partial^2 u_i}{\partial r_i^2} < 0$, we conclude that $(r_1, \ldots, r_N) = (\frac{1}{4}, \ldots, \frac{1}{4})$ is the Nash equilibrium. The total amount of resources used by all the players is then $R = \frac{N}{4}$.

When $N = 12$, the total resources used will be $R = 3$. The payoff to each player when $N = 12$ is

$$u_i\left(\frac{1}{4}, \ldots, \frac{1}{4}\right) = \frac{1}{2} - \frac{1}{8} - \left(\frac{11}{4}\right)^2 = -7.396.$$

(c) Now suppose that the total amount of resources is R and each player will use $\frac{R}{N}$ units of the total. Here, R is unknown and is chosen so that the *social welfare* is maximized, that is,

$$\text{Maximize} \sum_{i=1}^{N} u_i\left(\frac{R}{N}, \ldots, \frac{R}{N}\right),$$

over $R \geq 0$. Practically, this means that the players do not act independently but work together so that the total payoffs to all players is maximized. Find the value of R that provides the maximum, R^s, and find it's value when $N = 12$.

5.6.c Answer: Set

$$F(R) = N\left(h\left(\frac{R}{N}\right) - f\left(\frac{R}{N}\right) - g\left(R - \frac{R}{N}\right)\right) = \sum_{i=1}^{N} u_i\left(\frac{R}{N}, \ldots, \frac{R}{N}\right)$$

We have to find the maximum of F. Take a derivative with respect to R and set to zero:

$$F'(R) = \left(4 - \frac{6}{N} - 2N\right)R + \frac{1}{2\sqrt{\frac{R}{N}}} = 0.$$

After some algebra, we get

$$R^s = \frac{N}{\left(2(4 + 2(N-1))^2\right)^{2/3}}.$$

Since $F''(R) = 4 - \frac{6}{N} - 2N - \frac{\sqrt{N}}{4R^{3/2}} < 0$ we know that R^s provides a maximum.

When $N = 12$ we get $R^s = 0.192547$ and the value of the maximum social welfare is $F(R^s) = 1.14004$.

5.7 Consider the median voter model Example 5.3. Let X denote the preference of a voter and suppose X has density $f(x) = -1.23x^2 + 2x + 0.41, 0 \leq x \leq 1$. Find the Nash equilibrium position a candidate should take in order to maximize their voting percentage.

5.7 Answer: According to the solution of the median voter problem in Example 5.3, we need to find the median of X. To do that we solve for γ in the equation

$$\int_0^\gamma f(x)\, dx = \int_0^\gamma -1.23x^2 + 2x + 0.41\, dx = \frac{1}{2}.$$

This equation becomes $-0.41\gamma^3 + \gamma^2 + 0.41\gamma - 0.5 = 0$ which implies $\gamma = 0.585$. The Nash equilibrium position for each candidate is $(\gamma^*, \gamma^*) = (0.585, 0.585)$. With that position, each candidate will get 50% of the vote.

5.8 In the arbitration game Example 5.4, we have seen that the payoff function is given by

$$u(x, y) = xG\left(\frac{x+y}{2}\right) + y\left(1 - G\left(\frac{x+y}{2}\right)\right),$$

where G is a cumulative distribution function.

(a) Assume that $G' = g$ is a continuous density function with $g(z) > 0$. Show that any critical point of $u(x, y)$ must be a saddle point of u in the sense that

$$\det \begin{bmatrix} u_{xx} & u_{xy} \\ u_{xy} & u_{yy} \end{bmatrix} < 0.$$

5.8.a Answer: We have

$$u_{xx} = g\left(\frac{x+y}{2}\right) + \frac{1}{4}(x-y)g'\left(\frac{x+y}{2}\right),$$

$$u_{yy} = -g\left(\frac{x+y}{2}\right) + \frac{1}{4}(x-y)g'\left(\frac{x+y}{2}\right),$$

$$u_{xy} = \frac{1}{4}(x-y)g'\left(\frac{x+y}{2}\right).$$

Consequently the determinant of the Hessian is

$$u_{xx}u_{yy} - u_{xy}^2 = -g\left(\frac{x+y}{2}\right)^2 < 0.$$

(b) Derive the optimal offers for the exponential distribution with $\lambda > 0$

$$G(z) = 1 - e^{-\lambda z}, z \geq 0 \Rightarrow g(z) = \lambda e^{-\lambda z}, z > 0.$$

Assume that the minimum offer must be at least $a > 0$ and find the smallest possible $a > 0$ so that the offers are positive. In other words, the range of offers is $X + a$ and X has an exponential distribution.

5.8.b Answer: If X has an exponential distribution with parameter λ, the median is $m = \frac{\ln 2}{\lambda}$. This means $X + a$ has median $m + a$. Thus, we solve the linear system

$$y^* - x^* = \frac{1}{g(m)} = \frac{2}{\lambda} \quad \text{and} \quad y^* + x^* = 2m = \frac{2\ln 2}{\lambda} + 2a$$

to get the offers

$$y^* = \frac{1 + \ln 2}{\lambda} + a \quad \text{and} \quad x^* = \frac{\ln 2 - 1}{\lambda} + a.$$

The problem requires that $x^* > 0$ that means that $a \geq \frac{1-\ln 2}{\lambda} > 0$.

5.9 Consider the game in which player I chooses a nonnegative number x, and player II chooses a nonnegative number y (x and y are not necessarily integers). The payoffs are as follows:

$$u_1(x, y) = x(4 + y - x) \text{ for player I}, \quad u_2(x, y) = y(4 + x - y) \text{ for player II}.$$

Determine player I's best response $x(y)$ to a given y, and player II's best response $y(x)$ to a given x. Find a Nash equilibrium, and give the payoffs to the two players.

5.9 Answer: The best responses are easily found by solving $\frac{\partial u_1}{\partial x} = 0$ and $\frac{\partial u_2}{\partial y} = 0$. The result is

$$x(y) = \frac{4 + y}{2}, \qquad y(x) = \frac{4 + x}{2}$$

and solving for x and y results in $x = 4, y = 4$ as the Nash equilibrium. The payoffs are $u_1(4, 4) = u_2(4, 4) = 16$.

5.10 Two players decide on the amount of effort they each will exert on a project. The effort level of each player is $q_i \geq 0, i = 1, 2$. The payoff to each player is

$$u_1(q_1, q_2) = q_1(c + q_2 - q_1) \quad \text{and} \quad u_2(q_1, q_2) = q_2(c + q_1 - q_2),$$

where $c > 0$ is a constant. This choice of payoff models a synergistic effect between the two players.

(a) Find each player's best response function.

5.10.a Answer: Each payoff is a quadratic in the variable the player controls when the opponent's variable is fixed. The maximum will be achieved where the derivative is zero:

$$\frac{\partial u_1}{\partial q_1} = c + q_2 - 2q_1 = 0 \Rightarrow q_1 = \frac{c + q_2}{2} = BR_1(q_2),$$

and

$$\frac{\partial u_2}{\partial q_2} = c + q_1 - 2q_2 = 0 \Rightarrow q_2 = \frac{c + q_1}{2} = BR_2(q_1).$$

(b) Find the Nash equilibrium.

5.10.b Answer: As best responses to each other we must solve $q_1 = BR_1(\frac{c+q_1}{2})$. The result is $q_1^* = q_2^* = c$.

5.11 Suppose two citizens are considering how much to contribute to a public playground, if anything. Suppose their payoff functions are given by

$$u_i(q_1, q_2) = q_1 + q_2 + w_i - q_i + (w_i - q_i)(q_1 + q_2), \quad i = 1, 2,$$

where $w_i =$ wealth of player i. This payoff represents a benefit from the total provided for the playground $q_1 + q_2$ by both citizens, the amount of wealth left over for private benefit $w_i - q_i$, and an interaction term $(w_i - q_i)(q_1 + q_2)$ representing the benefit of private money with the amount donated. Assume $w_1 = w_2 = w$ and $0 \le q_i \le w$. Find a pure Nash equilibrium.

5.11 Answer: Calculate the derivatives and set to zero: $\frac{\partial u_1}{\partial q_1} = w_1 - 2q_1 - q_2 = 0 \Rightarrow$ $q_1(q_2) = \frac{w_1 - q_2}{2}$ is the best response. Similarly, $q_2(q_1) = \frac{w_2 - q_1}{2}$. The Nash equilibrium is then $q_1^* = \frac{2w_1 - w_2}{3}, q_2^* = \frac{2w_2 - w_1}{3}$, assuming that $2w_1 > w_2, 2w_2 > w_1$. If $w_1 = w_2 = w$, then $q_1 = q_2 = \frac{w}{3}$ and each citizen should donate one-third of their wealth.

5.12 Consider the war of attrition game in Example 5.11. Verify that

$$BR_1(t_2) = \begin{cases} 0, & \text{if } t_2 > \frac{v}{c}; \\ (t_2, \infty), & \text{if } t_2 < \frac{v}{c}; \\ \{0\} \cup (t_2, \infty), & \text{if } t_2 = \frac{v}{c}; \end{cases}$$

and

$$BR_2(t_1) = \begin{cases} 0, & \text{if } t_1 > \frac{v}{c}; \\ (t_1, \infty), & \text{if } t_1 < \frac{v}{c}; \\ \{0\} \cup (t_1, \infty), & \text{if } t_1 = \frac{v}{c}. \end{cases}$$

5.12 Answer: The payoffs are

$$u_1(t_1, t_2) = \begin{cases} v - c\, t_2, & \text{if } t_1 > t_2;\ 2 \text{ quits before } 1; \\ -c\, t_1, & \text{if } t_1 \le t_2;\ 1 \text{ quits before } 2; \end{cases}$$

and

$$u_2(t_1, t_2) = \begin{cases} v - c\, t_1, & \text{if } t_2 > t_1;\ 1 \text{ quits before } 2; \\ -c\, t_2, & \text{if } t_2 \leq t_1;\ 2 \text{ quits before } 1. \end{cases}$$

We consider only u_1 since the analysis is similar for u_2. We need to find $t_1^*(t_2)$ which satisfies

$$u_1(t_1^*(t_2), t_2) = \max_{t_1 \geq 0} u_1(t_1, t_2), \quad t_1^*(t_2) = \arg\max_{t_1} u_1(t_1, t_2).$$

Fix $t_2 > 0$ and consider the graph of $u_1(t_1, t_2)$ as t_1 varies:

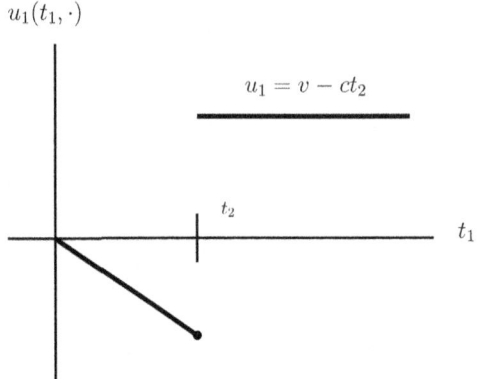

War of attrition–assuming $v - c\, t_2 > 0$.

In the graph, we have assumed first $v - c\, t_2 > 0$ so the horizontal line is above the t_1 axis. If instead we assumed $v - c\, t_2 < 0$ or $= 0$, we would draw a different picture. We look for the maximum of this function.

In the case when $v - c\, t_2 > 0$, we can see from the graph that the largest this function could be is achieved at any $t_1 > t_2$ (not including t_2 because that point is at the bottom of the line through the origin). Thus $t_1^*(t_2) = (t_2, \infty)$. This holds when $v - c\, t_2 > 0 \Leftrightarrow t_2 < \frac{v}{c}$.

In the case when $v - c\, t_2 < 0$, the maximum of the function will be achieved at the unique point $t_1 = 0$ because then the horizontal line will be drawn below the t_1 axis and $(0, 0)$ will be the highest point. Thus $t_1^*(t_2) = \{0\}$. This holds when $v - c\, t_2 < 0 \Leftrightarrow t_2 > \frac{v}{c}$.

In the case when $v - c\, t_2 = 0$, the maximum of the function will be achieved at any $t_1 > t_2$ and at $t_1 = 0$, because then the horizontal line will coincide with the t_1 axis. Thus, $t_1^*(t_2) = \{0\} \cup (t_2, \infty)$. This holds when $v - c\, t_2 = 0 \Leftrightarrow t_2 = \frac{v}{c}$.

(a) Verify that

$$BR_1(t_2) \cap BR_2(t_1) = \left\{ \left(t_1 \geq \frac{v}{c}, t_2 = 0 \right), \left(t_1 = 0, t_2 \geq \frac{v}{c} \right) \right\}.$$

5.12.a Answer: Graph $t_1 = BR_1(t_2)$ and $t_2 = BR_2(t_1)$ on the same graph using $t_1 - t_2$ axes. The intersection of the graphs is the desired set of Nash equilibria.

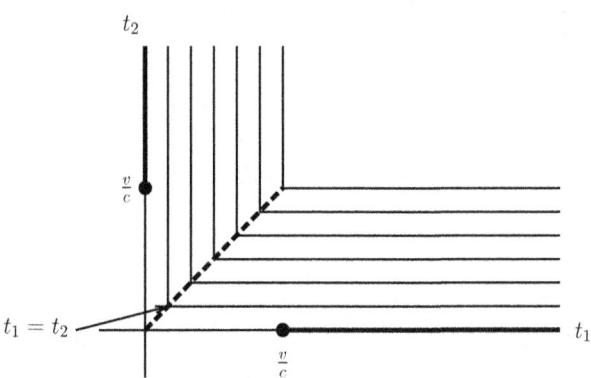

Best response sets.

(b) Verify that $t_1^* = \frac{v}{c}$, $t_2^* = 0$ is a pure Nash equilibrium.

5.12.b Answer: $u_1(t_1^*, t_2^*) = u_1(\frac{v}{c}, 0) = v$ and $u_1(t_1, 0) = v$ if $t_1 > 0$, while $u_1(0, 0) = 0$ and this shows $u_1(t_1^*, t_2^*) \geq u_1(t_1, t_2^*)$ for any $t_1 \geq 0$.

(c) We assumed the cost of continuing the war was $c\, t$. Suppose instead the cost is $c\, t^2$. Find the Nash equilibrium in pure strategies.

5.12.c Answer: Using the same technique as in the previous problem, it is easy to see that

$$BR_1(t_2) \cap BR_2(t_1) = \left\{ \left(t_1 \geq \sqrt{\frac{v}{c}}, t_2 = 0 \right), \left(t_1 = 0, t_2 \geq \sqrt{\frac{v}{c}} \right) \right\}.$$

This means a Nash equilibrium is $t_1^* = 0$, $t_2^* = \sqrt{\frac{v}{c}}$ as well as $t_1^* = \sqrt{\frac{v}{c}}$, $t_2^* = 0$.

(d) Suppose we modify the payoff function to include a cost if country 1 quits after country 2 and vice versa. In addition, we slightly change what happens when they quit at the same time. Take the payoff for country 1 to be

$$u_1(t_1, t_2) = \begin{cases} -c\, t_1, & \text{if } t_1 < t_2; \\ v - c\, t_2 - c\, t_1, & \text{if } t_1 \geq t_2. \end{cases}$$

Find the payoff for country 2 and find the pure Nash equilibrium.

5.12.d Answer: The payoff for country 2 is

$$u_2(t_1, t_2) = \begin{cases} -c\, t_2, & \text{if } t_2 < t_1; \\ v - c\, t_2 - c\, t_1, & \text{if } t_2 \geq t_1. \end{cases}$$

If we graph $t_1 \mapsto u_1(t_1, t_2)$ for fixed $t_2 > 0$, assuming that $v - c t_2 - c t_2 = v - 2c t_2 > 0$, we get the figure

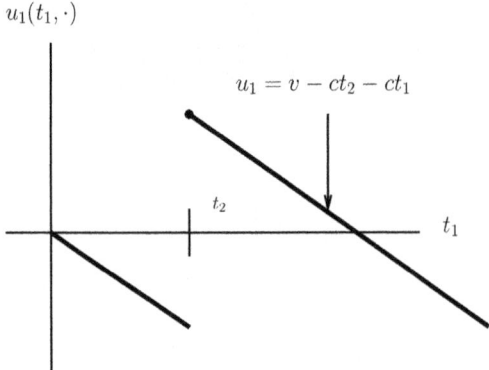

War of attrition–cost to continue.

To calculate the maximum, we see from the graph depending on where t_2 is in relation with $\frac{v}{2c}$, that

$$t_1^*(t_2) = BR_1(t_2) = \begin{cases} \{t_2\}, & \text{if } t_2 \leq \frac{v}{2c}; \\ \{0\}, & \text{if } t_2 > \frac{v}{2c}; \\ \{0\} \cup \{t_2\}, & \text{if } t_2 = \frac{v}{2c}. \end{cases}$$

Similarly,

$$t_2^*(t_1) = BR_2(t_1) = \begin{cases} \{t_1\}, & \text{if } t_1 \leq \frac{v}{2c}; \\ \{0\}, & \text{if } t_1 > \frac{v}{2c}; \\ \{0\} \cup \{t_2\}, & \text{if } t_1 = \frac{v}{2c}. \end{cases}$$

Here is a graph of $BR_2(t_1)$ and $BR_1(t_2)$.

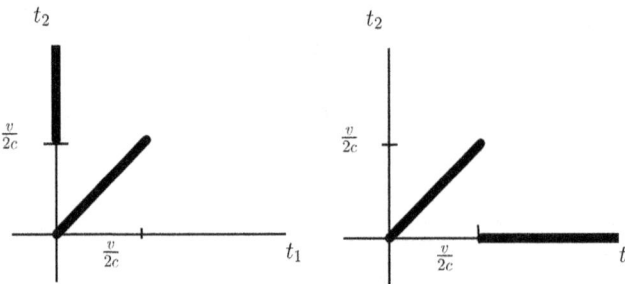

Best response functions.

Overlaying these two (set valued) function on the same set of axes, we see that

$$BR_1(t_2) \cap BR_2(t_1) = \left\{ (x, x) \mid 0 \leq x \leq \frac{v}{2c} \right\}.$$

To verify this we will check that $(x^*, x^*), 0 \leq x^* \leq \frac{v}{2c}$ is a Nash equilibrium directly. We have $u_1(x^*, x^*) = v - 2cx^*$ and

$$u_1(t_1, x^*) = \begin{cases} -ct_1, & \text{if } t_1 < x^*; \\ v - cx^* - ct_1, & \text{if } t_1 > x^*. \end{cases}$$

Clearly, $-ct_1 < u(x^*, x)$ and, if $t_1 > x^*$, then $v - cx^* - ct_1 < v - cx^* - cx^* = v - 2cx^* = u_1(x^*, x^*)$. Similarly, $u_2(x^*, x^*) \geq u_2(x^*, t_2)$, $t_2 \neq x^*$ and we conclude (x^*, x^*) is a Nash equilibrium. Thus, each country should stop at the same time, at any time before $\frac{v}{2c}$.

5.13 This problem considers the War of Attrition but with differing costs of continuing the war and differing values placed upon the resource over which the war is fought. Country 1 places value v_1 on the land, and country 2 values it at v_2. The players choose the time at which to concede the land to the other player, but there is a cost for letting time pass. Suppose that the cost to each country is $c_i, i = 1, 2$ per unit of time. The first player to concede yields the land to the other player at that time. If they concede at the same time, each player gets half the land. Determine the payoffs to each player and determine the pure Nash equilibria.

5.13 Answer: The payoffs are as follows:

$$u_1(t_1, t_2) = \begin{cases} v_1 - c_1 t_2 & \text{if } t_1 > t_2; \\ -c_1 t_1 & \text{if } t_1 < t_2; \\ \frac{v_1}{2} - c_1 t_1 & \text{if } t_1 = t_2; \end{cases} \qquad u_2(t_1, t_2) = \begin{cases} v_2 - c_2 t_1 & \text{if } t_2 > t_1; \\ -c_2 t_2 & \text{if } t_2 < t_1; \\ \frac{v_2}{2} - c_2 t_2 & \text{if } t_1 = t_2. \end{cases}$$

First, calculate $\max_{t_1 \geq 0} u_1(t_1, t_2)$, and $\max_{t_2 \geq 0} u_2(t_1, t_2)$. For example,

$$\max_{t_1 \geq 0} u_1(t_1, t_2) = \begin{cases} v_1 - c_1 t_2, & \text{if } t_2 < \frac{v_1}{c_1}; \\ 0, & \text{if } t_2 > \frac{v_1}{c_1}; \\ 0, & \text{if } t_2 = \frac{v_1}{c_1}. \end{cases}$$

and the maximum is achieved at the set-valued function

$$t_1^*(t_2) = \begin{cases} (t_2, \infty), & \text{if } t_2 < \frac{v_1}{c_1}; \\ 0, & \text{if } t_2 > \frac{v_1}{c_1}; \\ 0 \cup (q_2, \infty), & \text{if } t_2 = \frac{v_1}{c_1}. \end{cases}$$

This is the best response of country 1 to t_2. Next, calculate the best response to t_1 for country 2, $t_2^*(t_1)$:

$$t_2^*(t_1) = \begin{cases} (t_1, \infty), & \text{if } t_1 < \frac{v_2}{c_2}; \\ 0, & \text{if } t_1 > \frac{v_2}{c_2}; \\ 0 \cup (t_1, \infty), & \text{if } t_1 = \frac{v_2}{c_2}. \end{cases}$$

If you now graph these sets on the same set of axes, the Nash equilibria are points of intersection of the sets. The result is

$$(t_1^*, t_2^*) = \begin{cases} \text{either } t_1^* = 0 \text{ and } t_2^* \geq \frac{v_1}{c_1} \\ \text{or } t_2^* = 0 \text{ and } t_1^* \geq \frac{v_2}{c_2}. \end{cases}$$

For example, this says that either (1) country 1 should concede immediately and country 2 should wait until first time $\frac{v_1}{c_1}$, or (2) country 2 should concede immediately and country 1 should wait until first time $\frac{v_2}{c_2}$.

5.14 Consider the general war of attrition as described in this section with given functions α, β. The payoff functions are as follows:

$$u_1(t_1, t_2) = \begin{cases} \beta(t_2), & \text{if } t_1 > t_2; \text{ 2 quits before 1;} \\ \alpha(t_1), & \text{if } t_1 < t_2; \text{ 1 quits before 2;} \\ \frac{\alpha(t_1)+\beta(t_1)}{2}, & \text{if } t_1 = t_2; \end{cases}$$

and

$$u_2(t_1, t_2) = \begin{cases} \beta(t_1), & \text{if } t_2 > t_1; \text{ 1 quits before 2;} \\ \alpha(t_2), & \text{if } t_2 < t_1; \text{ 2 quits before 1;} \\ \frac{\alpha(t_1)+\beta(t_1)}{2}, & \text{if } t_1 = t_2. \end{cases}$$

Verify that if we take α and β to be continuous, decreasing functions with $\alpha(t) < \beta(t)$ and $\alpha(0) > \beta(1)$, then the general war of attrition has two pure Nash equilibria given by $\{(t_1 = \tau^*, t_2 = 0), (t_1 = 0, t_2 = \tau^*)\}$ where $\tau^* > 0$ is the first time where $\beta(\tau^*) = \alpha(0)$.

5.14 The following figure depicts the setup.

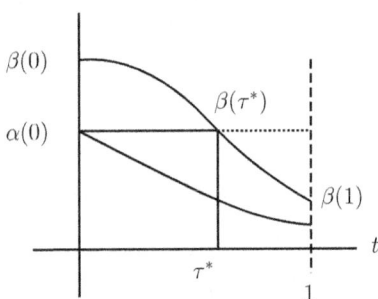

α and β–General game of timing.

We check that $(t_1 = 0, t_2 = \tau^*)$ is a Nash equilibrium. First, $u_1(0, \tau^*) = \alpha(0)$. Now for any $t_1 > 0$ we have

$$u_1(t_1, \tau^*) = \begin{cases} \beta(\tau^*), & \text{if } t_1 > \tau^*; \\ \alpha(t_1), & \text{if } \tau^* > t_1; \\ \frac{\alpha(\tau^*)+\beta(\tau^*)}{2}, & \text{if } t_1 = \tau^*. \end{cases}$$

If $t_1 > \tau^*$, $u_1(0, \tau^*) = \alpha(0) = \beta(\tau^*) = u_1(t_1, \tau^*)$.

If $t_1 < \tau^*$, $u_1(t_1, \tau^*) = \alpha(t_1) < \alpha(0) = u_1(0, \tau^*)$ since α is decreasing implies $\alpha(0) > \alpha(t_1)$ if $t_1 > 0$.

If $t_1 = \tau^*$,

$$u_1(\tau^*, \tau^*) = \frac{\alpha(\tau^*) + \beta(\tau^*)}{2} = \frac{\alpha(\tau^*) + \alpha(0)}{2} < \alpha(0)$$

since $\alpha(\tau^*) < \alpha(0)$. We have in all cases $u_1(t_1, \tau^*) \le \alpha(0) = u_1(0, \tau^*)$.
  Next, $u_2(0, \tau^*) = \beta(0)$.
  If $t_2 > 0$, $u_2(0, \tau^*) = \beta(0) = u_2(0, t_2)$.
  If $t_2 < 0$, there is nothing to check.
  If $t_2 = 0$,

$$u_2(0, 0) = \frac{\alpha(0) + \beta(0)}{2} \le \frac{\beta(0) + \beta(0)}{2} = \beta(0),$$

since $\beta(0) > \alpha(0)$. We have in all cases $u_2(0, \tau^*) = \beta(0) \ge u_2(0, t_2)$. Thus, $(0, \tau^*)$ is a Nash equilibrium.

5.15 This is an exercise on Braess's paradox. Consider the traffic system in which travelers want to go from $A \to C$. The travel time for each car depends on the total traffic.

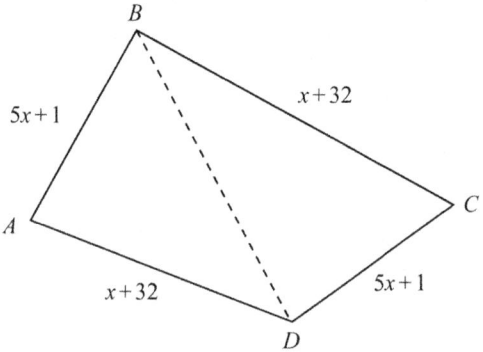

Braess' paradox.

(a) Suppose the total number of cars is 6 and each car can decide to take $A \to B \to C$ or $A \to D \to C$. Show that 3 cars taking each path is a Nash equilibrium.

5.15.a Answer: If three players take $A \to B \to C$, their travel time is 51. No player has an incentive to switch. If four players take this path, their travel time is 57 while the two players on $A \to D \to C$ travel only 45, clearly giving an incentive for a player on $A \to B \to C$ to switch.

(b) Suppose a new road $B \to D$ is built which has travel time zero no matter how many cars are on it. Cars can now take the paths $A \to B \to D \to C$ and $A \to D \to B \to C$ in addition to the paths from the first part. Find the Nash equilibrium and show that the travel time is now worse for all the players.

5.15.b Answer: If three cars take the path $A \to B \to D \to C$, their travel time is 32, while if the other three cars stick with $A \to D \to C$ or $A \to B \to C$, the travel time for them is still 51, so they have an incentive to switch. If they all switch to $A \to B \to D \to C$, their travel time is 62. If 5 cars take $A \to B \to D \to C$, their travel time is 52, while the one car that takes $A \to D \to C$ or $A \to B \to C$ has travel time $33 + 31 = 64$. Note that no player would want to take $A \to D \to B \to C$ because the travel time for even one car is 66. Continuing this way, we see that the only path that gives no player an incentive to switch is $A \to B \to D \to C$. That is the Nash equilibrium and it results in a travel time which is greater than before the new road was added.

5.16 We have a game with $N \geq 2$ players. Each player i must simultaneously decide whether to join the team or not. If player i joins then $x_i = 1$ and otherwise $x_i = 0$. Let $T = \sum_{i=1}^{N} x_i$ denote the size of the team. If player i doesn't join the team, then player i receives a payoff of zero. If player i does join the team then i pays a cost of c. If all N players join the team, so that $T = N$, then each player enjoys a benefit of v. Hence, player i's payoff is

$$u_i(x_1, \ldots, x_N) = \begin{cases} v - c, & \text{if } T = \sum_{i=1}^{N} x_i = N; \\ -x_i c, & \text{if } T < N. \end{cases}$$

Suppose that $v > c > 0$.

(a) Find all of the pure-strategy Nash equilibria.

5.16.a Answer: There are two pure Nash equilibria: (1) Every player joins the team, $x_i = 1, i = 1, 2, \ldots, N$; (2) No player joins the team, $x_i = 0, i = 1, 2, \ldots, N$. To see why,

$$u_i(1, 1, \ldots, 1) = v - c > 0, \text{ while } u_i(1, 1, \ldots, 1, 0, 1, \ldots, 1) = 0$$

if player i chooses to not join. Thus (1) is a Nash equilibrium. Similarly, if no one joins the team $x_i = 0$ for each $i = 1, 2, \ldots, N$, and

$$u_i(0, 0, \ldots, 0) = 0, \text{ while } u_i(0, 0, \ldots, 0, 1, 0, \ldots, 0) = -c < 0,$$

if player i chooses to join but the other players stick with not joining. Thus (2) is also a Nash equilibrium.

(b) Take $N = 2$. Find a mixed Nash equilibrium.

5.16.b Answer: The matrix for this game is The matrix for this game is

I/II	Don't	Join
Don't	0, 0	0, $-c$
Join	$-c, 0$	$v - c, v - c$

Clearly there are two pure Nash equilibria at payoffs $0, 0$ and $v - c, v - c$. By equality of payoffs, if $Y^* = (y, 1 - y)$, we get $0 = -cy + (v - c)(1 - y)$ which results in $y = \frac{v-c}{v}$. By symmetry $X^* = Y^* = (\frac{v-c}{v}, \frac{c}{v})$. The expected payoffs for this Nash equilibrium is $E(X^*, Y^*) = 0$ for each player.

5.17 Two companies are at war over a market with a total value of $V > 0$. Company 1 allocates an effort $x > 0$, and company 2 allocates an effort $y > 0$ to obtain all or a portion of V. The portion of V won by company 1 if they allocate effort x is $(\frac{x}{x+y})V$ at cost $C_1 x$, where $C_1 > 0$ is a constant. Similarly, the portion of V won by company

2 if they allocate effort y is $(\frac{y}{x+y})V$ at cost $C_2 y$, where $C_2 > 0$ is a constant. The total reward to each company is then

$$u_1(x, y) = V\frac{x}{x+y} - C_1 x \quad \text{and} \quad u_2(x, y) = V\frac{y}{x+y} - C_2 y, \quad x > 0, y > 0.$$

Show that these payoff functions are concave in the variable they control and then find the Nash equilibrium using calculus.

5.17 Answer: We have the derivatives

$$\frac{\partial u_1}{\partial x} = -C_1 - V\frac{x}{(x+y)^2} + V\frac{1}{x+y}, \quad \frac{\partial u_2}{\partial y} = -C_2 - V\frac{y}{(x+y)^2} + V\frac{1}{x+y},$$

and

$$\frac{\partial^2 u_1}{\partial x^2} = -2V\frac{y}{(x+y)^3} < 0, \quad \frac{\partial^2 u_2}{\partial y^2} = -2V\frac{x}{(x+y)^3} < 0,$$

and hence concavity in the appropriate variable for each payoff function. Solving the first derivatives set to zero gives

$$x^* = \frac{C_2 V}{(C_1 + C_2)^2}, \quad y^* = \frac{C_1 V}{(C_1 + C_2)^2}.$$

Then

$$u_1(x^*, y^*) = \frac{C_2^2 V}{(C_1 + C_2)^2} \quad \text{and} \quad u_2(x^*, y^*) = \frac{C_1^2 V}{(C_1 + C_2)^2}.$$

5.18 Suppose that a firm's output depends on the number of laborers they have hired. Let p = number of workers, w = worker compensation (per unit time) and assume

$$u_f(p, w) = f(p) - p\,w, \quad f(p) = \begin{cases} p(1000 - p), & \text{if } p \le 500, \\ 25{,}000, & \text{if } p > 500, \end{cases}$$

is the payoff to the firm if they hire p workers and pay them w. Assume that the union payoff is $u_u(p, w) = p\,w$.

Find the best response for the firm to a wage demand assuming $w_m \le w \le W$. Then find w to maximize $u_u(p(w), w)$. This scheme says that the firm will hire the number of workers that maximizes its payoff for a given wage. Then the union, knowing that the firm will use its best response, will choose a wage demand that maximizes its payoff assuming the firm uses $p(w)$.

5.18 Answer: If $p \le 500$ we have $u_f(p, w) = p(1000 - p - w)$ and the maximum is achieved at $p = \frac{1000-w}{2}$, which requires $1000 \ge w$. If $p > 500$, then $u_f(p, w) = 25000 - pw$, which has a maximum achieved at $p = 0$. Thus,

$$p = BR_f(w) = \begin{cases} \frac{1000-w}{2}, & \text{if } w_m \le w \le 1000; \\ 0, & \text{otherwise.} \end{cases}$$

For the union, $u_u(p, w) = pw$, and w will be chosen to maximize

$$u_u(p(w), w) = u_u(BR_f(w), w) = \begin{cases} \frac{1000-w}{2}w, & \text{if } w_m \leq w \leq 1000; \\ 0w = 0, & \text{if } w > 1000. \end{cases}$$

The maximum of u_u is $u_u(p(w^*), w^*) = 125,000$ achieved at $w^* = 500$. This w^* maximizes $\frac{1}{2}(1000 - w)w$. Then the best response to this wage demand by the firm is

$$p(w^*) = \frac{1000 - w^*}{2} = 250.$$

Thus, a wage demand of $w^* = 500$ will be accepted with the firm employing $p^*(w^*) = 250$ workers and a resulting payoff to the firm of $u_f(250, 500) = 62,500$. The union payoff will be $u_u(250, 500) = 125,000$

5.19 Suppose that $N > 2$ players choose an integer in $\{1, 2, \ldots, 100\}$. The payoff to each player is 1 if that player has chosen an integer which is closest to $\frac{2}{3}$ of the average of the numbers chosen by all the players. If two or more players choose the closest integer, then they equally split the 1. The payoff of other players is 0. Show that the Nash equilibrium is for each player to choose the number 1.

5.19 Answer: The payoff function can be written as

$$u_i(x_1, \ldots, x_{100}) = \begin{cases} 1, & \text{if } x_i \text{ closest to } \frac{2}{3}\overline{x}; \\ \frac{1}{2}, & \text{if } x_i = x_j, i \neq j, \ x_i \text{ closest to } \frac{2}{3}\overline{x}; \\ \frac{1}{3}, & \text{if } x_i = x_j = x_k, i \neq j \neq k, \ x_i \text{ closest to } \frac{2}{3}\overline{x}; \\ \vdots, & \\ \frac{1}{100}, & \text{if } x_1 = x_2 = \cdots = x_{100}. \end{cases}$$

We need to show that

$$u_i(1, 1, 1, \cdots, 1) \geq u_i(x_i, 1_{-i}), \quad x_i = 2, \ldots, 100.$$

Recall that 1_{-i} denotes that all the players except player i are playing 1. Note that since $\overline{x} = 1, u_i(1, 1, \ldots, 1) = \frac{1}{100}$ and $|1 - \frac{2}{3}\overline{x}| = \frac{1}{3}$. What is a better x_i for player i to switch to? In order to be better it must satisfy

$$\left| x_i - \frac{2}{3}\overline{x} \right| = \left| x_i - \frac{2}{3}\frac{99 + x_i}{100} \right| < \frac{1}{3}.$$

Note that $x_i - \frac{2}{3}\frac{1}{100}x_i - \frac{2}{3}\frac{99}{100} = x_i\frac{298}{300} - \frac{198}{300} > 0$. Hence, we must have

$$0 < x_i\frac{298}{300} - \frac{198}{300} < \frac{1}{3} \Rightarrow x_i\frac{298}{300} < \frac{298}{300} \Rightarrow x_i < 1.$$

That's impossible. Hence, $(1, 1, \cdots, 1)$ is a Nash equilibrium.

5.20 Two countries share a long border. The pollution emitted by one country affects the other. If country $i = 1, 2$ pollutes a total of Q_i tons per year, they will emit p_i tons per year into the atmosphere, and clean up $Q_i - p_i$ tons per year (before it is emitted) at cost c dollars per ton. Health costs attributed to the atmospheric pollution are proportional to the square of the total atmospheric pollution in each country. The total atmospheric pollution in country 1 is $p_1 + kp_2$, and in country 2 it is $p_2 + kp_1$. The constant $0 < k < 1$ represents the fraction of the neighboring country's pollution entering the atmosphere. Assume the proportionality constants and cost of cleanup is the same for both countries.

(a) Find a function giving the total cost of polluting for each country as a function of p_1, p_2.

5.20.a Answer: We take the payoff functions for each country as

$$u_1(p_1, p_2) = a(p_1 + kp_2)^2 + c(Q_1 - p_1),$$
$$u_2(p_1, p_2) = a(p_2 + kp_1)^2 + c(Q_2 - p_2).$$

(b) Find the Nash equilibrium level of pollution emitted for each country. Note here that we want to minimize the payoff functions in this problem, not maximize.

5.20.b Answer: If we take partial derivatives we get

$$\frac{\partial u_1}{\partial p_1} = -c + 2a(p_1 + kp_2) = 0,$$

$$\frac{\partial u_2}{\partial p_2} = -c + 2a(p_2 + kp_1) = 0.$$

Solving we see that

$$p_1^* = \frac{c}{2a(1 + k)}, \qquad p_2^* = \frac{c}{2a(1 + k)}$$

and each point does provide a minimum payoff because the second partials are positive. Therefore, (p_1^*, p_2^*) is a Nash equilibrium. Interestingly, this solution does not depend on Q_1, Q_2.

(c) Find the optimal levels of pollution and payoffs if $c = 1000, a = 10, k = \frac{1}{2}, Q_1 = Q_2 = 300$.

5.20.c Answer: Using the Nash equilibrium formulas, we get

$$p_1^* = p_2^* = \frac{c}{2a(1 + k)} = \frac{1000}{2 \cdot 10(1 + \frac{1}{2})} = \frac{100}{3}.$$

Then, each country emits $\frac{100}{3}$ tons and cleans up $Q_i - p_i^* = \frac{800}{3}$ tons.

$$u_1\left(\frac{100}{3}, \frac{100}{3}\right) = u_1\left(\frac{100}{3}, \frac{100}{3}\right) = \frac{875,000}{3}.$$

5.21 Corn is a food product in high demand but also enjoys a government price subsidy. Assume that the demand for corn (in bushels) is given by $D(p) = 150,000(15 - p)^+$, where p is the price per bushel. The government program guarantees that $p \geq 2$. Suppose that there are three corn producers who have reaped 1 million bushels each. They each have the choice of how much to send to market and how much to use for feed (at no profit). Find the Nash equilibrium. What happens if one farmer sends the entire crop to market?

5.21 Answer: The price subsidy kicks in if $p = 15 - \frac{q_1+q_2+q_3}{150,000} \leq 2$, which implies that the total quantity shipped, $Q = q_1 + q_2 + q_3 \leq 1,950,000$. The price per bushel is given by

$$p(q_1, q_2, q_3) = \begin{cases} 15 - \frac{q_1+q_2+q_3}{150,000}, & \text{if } Q < 1,950,000; \\ 2, & \text{if } Q \geq 1,950,000. \end{cases}$$

Each farmer has the payoff function

$$u_i(q_1, q_2, q_3) = p(q_1, q_2, q_3)q_i, \quad i = 1, 2, 3.$$

Assume that $Q < 1,950,000$. Take a partial derivative and set to zero to get $q_i = 562,500$ bushels each. Then $Q = 1,687,500 < 1,950,000$. So an interior pure Nash equilibrium consists of each farmer sending 562,500 bushels each to market and using 437,500 bushels for feed. The price per bushel will be $p = 15 - \frac{3 \times 562,500}{150,000} = 3.75$, which is greater than the government-guaranteed price.

If each producer ships 562,500 bushels, the profit for each producer is

$$u_i(562,500, 562,500, 562,500) = 3.75 \times 562,500 = 2,109,375.$$

Now suppose producer 1 ships his entire crop of $q_1 = 10^6$ bushels. Assuming the other producers are not aware that producer 1 is shipping 10^6 bushels, if the other producers still ship 562,500 bushels the total shipped will be $2,125,000 > 1,950,000$ and the price per bushel would be subsidized at 2. The revenue to each producer is then

$$u_1(10^6, 562,500, 562,500) = 2,000,000,$$
$$u_2(10^6, 562,500, 562,500) = 1,125,000,$$
$$u_3(10^6, 562,500, 562,500) = 1,125,000.$$

This means all the producers make less money but 2 and 3 make significantly less.

It's very unlikely that 2 and 3 would be happy with this solution. We need to find 2 and 3's best responses to 1's shipment of 10^6. Assume first that $10^6 + q_2 + q_3 \leq 1,950,000$. We calculate

$$\max_{q_2} u_2(10^6, q_2, q_3) = \max_{q_2} q_2 \left(15 - \frac{1}{150,000}(10^6 + q_2 + q_3) \right)$$

The problem for q_3 is similar. Taking a derivative, setting to zero, and solving for q_2, q_3 results in $q_2 = q_3 = (1.25 \times 10^6)/3 = 416,667$. That is the best response of producers 2 and 3 to producer 1 shipping his entire crop assuming $Q \leq 1,950,000$.

The price per bushel at total output 1, 833, 334 bushels will be \$2.78 and the profit for each producer will be $u_1 = 2,780,000$, $u_2 = u_3 = 1,158,334$.

Now assume that $Q \geq 1,950,000$. In this case $p = 2$ and the profit for each producer is $u_1 = 2,000,000$, $u_2 = 2 \times q_2$, $u_3 = 2 \times q_3$. The maximum for producers 2 and 3 is achieved with $q_2 = q_3 = 10^6$ and all 3 producers ship their entire crop to market. This is clearly the best response of producers 2 and 3 to $q_1 = 10^6$.

5.22 This is known as the **division of land game**. Suppose that there is a parcel of land as in the following figure:

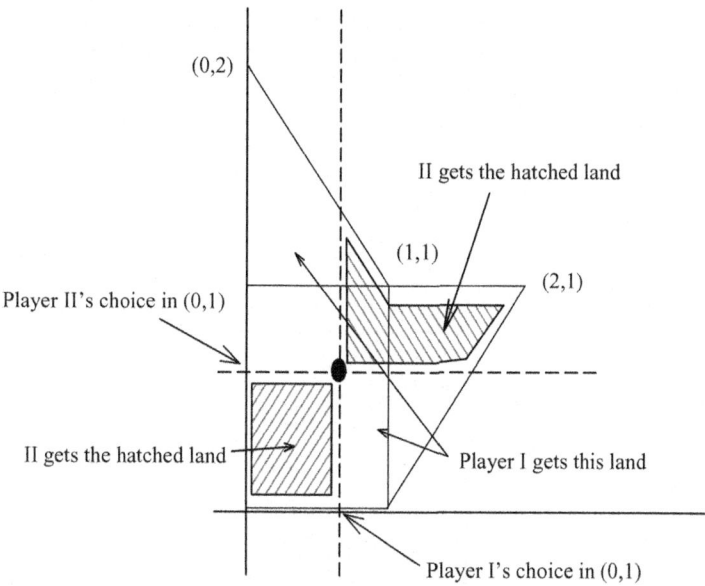

Division of land.

Player I chooses a vertical line between $(0, 1)$ on the x-axis and player II chooses a horizontal line between $(0, 1)$ on the y-axis. Player I gets the land below II's choice and right of I's choice as well as the land above II's choice and left of I's line. Player II gets the rest of the land. Both players want to choose their line so as to maximize the amount of land they get.

(a) Formulate this as a game with continuous strategies and solve it.

5.22.a Answer: This is actually a constant sum game since the portions of land that each player gets must add to the total, which is 2. The formula for the area of a trapezoid is $\frac{1}{2}(b_1 + b_2)h$, where b_1, b_2 are the two parallel sides and h is the height. Using this formula, we can calculate the area for player 1 assuming she chooses $0 \leq x \leq 1$ and player 2 chooses $0 \leq y \leq 1$. The area for player I is

$$A(x, y) = x - 2xy + y + x - \frac{x^2}{2} + \frac{y^2}{2},$$

and the area for player II is

$$B(x, y) = 2xy + 2 - 2x - y + \frac{x^2}{2} - \frac{y^2}{2}.$$

The Nash equilibrium is found by taking partial derivatives of A (with respect to x) and B (with respect to y), setting to zero, and solving. The result is $x^* = \frac{4}{5}, y^* = \frac{3}{5}$. The point $(x^* = \frac{4}{5}, y^* = \frac{3}{5})$ is indeed a Nash equilibrium because $A_{xx}(x, y) = -1 < 0$ and $B_{yy}(x, y) = -1 < 0$ so that the functions are concave down in the variable they control. Finally, the areas each player gets are $A(x^*, y^*) = \frac{11}{10}, B(x^*, y^*) = \frac{9}{10}$. Surprisingly, they are not equal areas.

(b) Now consider a seemingly easier parcel of land to analyze. The parcel of land is the triangle with vertices at $(0, 0)$, $(0, 1)$, $(1, 1)$. Given a choice of pure strategy $0 \le x_0 \le 1, 0 \le y_0 \le 1, y_0 \le x_0$, player I gets the land in the triangle with $x \le x_0, y \le y_0$ as well as the land $x \ge x_0, y \ge y_0$ in the triangle. Player II gets the remainder of the land. Find a Nash equilibrium.

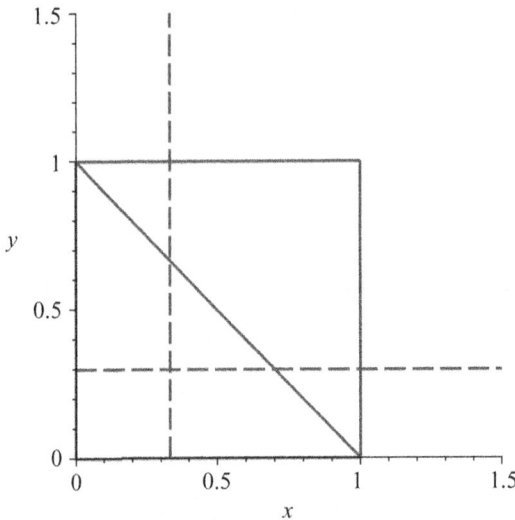

5.22.b Answer:
We have for $x + y \le 1, 0 \le x \le 1, 0 \le y \le 1$,

$$u_1(x, y) = \frac{1}{2}xy + \frac{1}{2}(1 - x - y)^2, \quad u_2(x, y) = \frac{1}{2} - u_1(x, y).$$

Taking derivatives and setting to zero gives $x = y = \frac{1}{3}$. Then $u_1\left(\frac{1}{3}, \frac{1}{3}\right) = \frac{1}{6}, u_2\left(\frac{1}{3}, \frac{1}{3}\right) = \frac{1}{3}$. However, the point found by taking derivatives is NOT the Nash equilibrium. In fact, take second derivatives and you find $\frac{\partial^2 u}{\partial x^2} = 1 > 0$, so u is not concave in x but rather convex. This is the wrong direction. Furthermore, $u\left(\frac{1}{3}, \frac{1}{3}\right) = 0.1667, u\left(\frac{1}{2}, \frac{1}{2}\right) = 0.25$, and $u\left(\frac{1}{2}, \frac{1}{3}\right) = 0.1805$. This means it is not true that $u\left(\frac{1}{3}, \frac{1}{3}\right) > u\left(\frac{1}{2}, \frac{1}{3}\right)$, and we have exhibited explicitly, that $\left(\frac{1}{3}, \frac{1}{3}\right)$ is not a Nash equilibrium.

It seems reasonable to guess and from geometry it looks obvious that the Nash Equilibrium should be $x = \frac{1}{2}, y = \frac{1}{2}$. To prove it we have to show $u\left(\frac{1}{2}, \frac{1}{2}\right) \geq u\left(x, \frac{1}{2}\right)$ for all $0 \leq x < \frac{1}{2}$. Consider

$$u_1\left(x, \frac{1}{2}\right) = \frac{1}{8} + \left(\frac{1}{2}\right)x^2 \leq u_1\left(\frac{1}{2}, \frac{1}{2}\right) = \frac{1}{4},$$

since, for any $0 \leq x \leq \frac{1}{2}$ the max of $u_1\left(x, \frac{1}{2}\right)$ occurs at $x = \frac{1}{2}$. Similarly, $u_2\left(\frac{1}{2}, \frac{1}{2}\right) \geq u_2\left(\frac{1}{2}, y\right)$ for all $0 \leq y \leq \frac{1}{2}$.

We should be a little concerned about the fact that the x's and y's are not chosen independently as is the usual case in Nash equilibria. The problem here is that we require that $x + y \leq 1$ which means the order in which the players choose a point in $[0, 1]$ makes a difference. To avoid this problem we change the problem slightly by defining the payoff

$$u(x, y) = \begin{cases} xy + \frac{1}{2}(x + y - 1)^2, & \text{if } x + y < 1; \\ xy, & \text{if } x + y = 1; \\ (1 - x)(1 - y) + \frac{1}{2}(x + y - 1)^2, & \text{if } x + y > 1; \end{cases}$$

with $x \in [0, 1]$, $y \in [0, 1]$. This payoff gives the same area as before if $x + y \leq 1$, and if $x + y > 1$, it gives the area to the right of x and above y as well as the area of the triangle formed left of x, below y but above $x + y = 1$. In other words, the same area as before for the figures symmetric about the line $x + y = 1$. Here is a plot of $u(x, y)$.

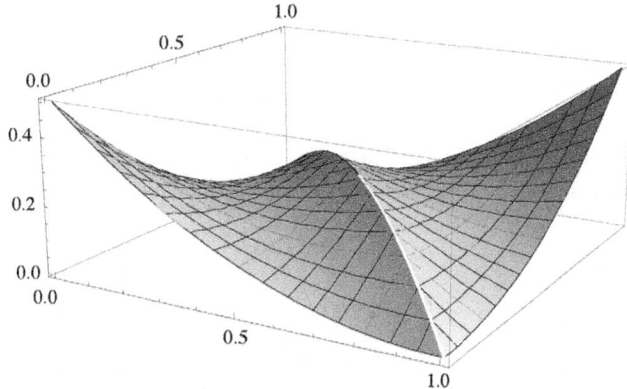

You can see the function is not differentiable along $x + y = 1$. The Nash equilibrium cannot be found by taking derivatives and setting to zero. The Nash equilibrium is at $(\frac{1}{2}, \frac{1}{2})$ with equal areas of $\frac{1}{4}$.

5.23 A famous economics problem is called the Samaritan's Dilemma. In one version of this problem, a citizen works in period 1 at a job giving a current annual net income of I. Out of her income she saves x for her retirement and earns $r\%$ interest per period. When she retires, considered to occur at period 2, she will receive an annual amount y from the government. The payoff to the citizen is

$$u_C(x, y) \equiv u_1(I - x) + u_2(x(1 + r) + y)\delta,$$

where u_1 is the first-period utility function, and is increasing but concave down, and u_2 is the second-period utility function, that is also increasing and concave down. These utility functions model increasing utility but at a diminishing rate as time progresses. The constant $\delta > 0$, called a discount rate, finds the present value of dollars that are not delivered until the second period. The government has a payoff function:

$$u_G(x, y) = u(T - y) + \alpha u_C(x, y) = u(T - y) + \alpha(u_1(I - x) + u_2(x(1 + r) + y)\delta),$$

where $u(t)$ is the government's utility function for income level t, such as tax receipts, and $\alpha > 0$ represents the factor of benefits received by the government for a happy citizen, called an **altruism factor**.

(a) Assume that the utility functions u_1, u_2, u are all logarithmic functions of their variables. Find a Nash equilibrium.

5.23.a Answer: If we take the first derivatives, we get

$$\frac{\partial u_C(x, y)}{\partial x} = -\frac{1}{I - x} + \frac{\delta(1 + r)}{y + (1 + r)x}, \quad \frac{\partial u_C(x, y)}{\partial y} = -\frac{1}{T - y} + \frac{\alpha\delta}{y + (1 + r)x}.$$

The second partials give us

$$\frac{\partial^2 u_C(x, y)}{\partial x^2} = -\frac{1}{(I - x)^2} - \frac{\delta(1 + r)^2}{(y + (1 + r)x)^2},$$

$$\frac{\partial^2 u_C(x, y)}{\partial y^2} = -\frac{1}{(T - y)^2} - \frac{\alpha\delta}{(y + (1 + r)x)^2}$$

both of which are < 0. That means that we can find the Nash equilibrium where the first partials are zero. Solving, we get the general solution

$$x^* = \frac{(1 + \alpha\delta)I(1 + r) - \alpha T}{(1 + \alpha + \alpha\delta)(1 + r)},$$

$$y^* = \frac{-I(1 + r) + \alpha(1 + \delta)T}{1 + \alpha + \alpha\delta},$$

and that is the Nash equilibrium as long as $0 \le x^* < I, 0 \le y^* < T$.

(b) Given $\alpha = 0.25$, $I = 22{,}000$, $T = 100{,}000$, $r = 0.15$, and $\delta = 0.03$, find the optimal amount for the citizen to save and the optimal amount for the government to transfer to the citizen in period 2.

5.23.b Answer: For the data given in the problem $x^* = 338.66$, $y^* = 357.85$.

5.24 Suppose we have a zero sum game and mixed densities X_0 for player I and Y_0 for player II.

(a) Suppose there are constants c and b so that $u(X_0, y) = c$ for all pure strategies y for player II and $u(x, Y_0) = b$, for all pure strategies x for player II. Show that (X_0, Y_0) is a saddle point in mixed strategies for the game and we must have $b = c$.

5.24.a Answer: Let Y be any mixed density for player II. Then

$$u(X_0, Y) = \int_{Q_2} u(X_0, y) Y(y) \, dy = c$$

since $\int_{Q_2} Y(y) \, dy = 1$. Similarly, $u(X, Y_0) = b$ for any density X for player I. This immediately implies $b = c$ since $u(X_0, Y_0) = b = c$.

(b) Suppose there is a constant c so that $u(X_0, y) \geq c$, for all pure x, and $u(x, Y_0) \leq c$, for all pure y. Show that (X_0, Y_0) is a saddle point in mixed strategies.

5.24.b Answer: We have $u(X_0, Y_0) = \int_{Q_2} u(X_0, y) Y_0(y) \, dy \geq c$. Similarly, $u(X_0, Y_0) = \int_{Q_1} u(x, Y_0) X_0(x) \, dx \leq c$. Thus, $u(X_0, Y_0) = c$. Further, if Y is any mixed density,

$$u(X_0, Y) = \int_{Q_2} u(X_0, y) Y(y) \, dy \geq c \int_{Q_2} Y(y) \, dy = c = u(X_0, Y_0).$$

Similarly, $u(X_0, Y_0) \geq u(X, Y_0)$ for any X.

5.25 Two investors choose investment levels from the unit interval. The investor with the highest level of investment wins the game, which has payoff 1 but costs the level of investment. If the same level of investment is chosen, they split the market. Take the investment levels to be $x \in [0, 1]$ for player I and $y \in [0, 1]$ for player II. Investor I's payoff function is

$$u_1(x, y) = \begin{cases} 1 - x, & \text{if } x > y; \\ \frac{1}{2} - x, & \text{if } x = y; \\ -x, & \text{if } x < y. \end{cases}$$

(a) What is player II's payoff function?

5.25.a Answer:

$$u_2(x, y) = \begin{cases} -y, & \text{if } x > y; \\ \frac{1}{2} - y, & \text{if } x = y; \\ 1 - y, & \text{if } x < y. \end{cases}$$

(b) Show that there is no pure Nash equilibrium in this game.

5.25.b Answer: If $y < 1$, then investor I would like to choose an $x \in (y, 1)$ (in particular we want to choose $x = y$ but $y \notin (y, 1)$) and therefore no best response exists. If $x < 1$, then investor II would like to choose a $y \in (x, 1)$ and II has no best response either. The last case is $x = y = 1$. In this case, both investors have best responses $x = 0$ or $y = 0$.

(c) Now use mixed strategies. Let $f(x)$ be the density for player I and $g(y)$ the density for player II. The equality of payoffs theorem would say $v_I = E_I(x, Y) = E_I(x', Y)$ for any $x, x' \in [0, 1]$ with $f(x), f(x') > 0$. This says

$$E_I(x, Y) = \int_0^1 u_1(x, y) g(y) \, dy$$

is a constant, v_I, independent of x (assuming $f(x) > 0$). Take the derivative of both sides with respect to x and solve for the density $g(x)$ for player II. Do a similar procedure to find the density for player I. Now find v_I and v_{II}.

5.24.c Answer: First observe that $u_1(x, y) = u_2(y, x)$ which means that if X is a Nash equilibrium for player I, then it is also a Nash equilibrium for player II. From the definition of $v_I, v_I = \int_0^x (1 - x) g(y) \, dy + \int_x^1 (-x) g(y) \, dy$ so that taking derivatives,

$$0 = -\int_0^x g(y) \, dy + (1 - x) g(x) - \int_x^1 g(y) \, dy + x g(x) = g(x) - \int_0^1 g(y) \, dy.$$

Now since $\int_0^1 g(y) \, dy = 1$, we conclude that $g(x) = 1, 0 \leq x \leq 1$. This is the density for a uniformly distributed random variable on $[0, 1]$. Thus the Nash equilibrium is that each investor chooses an investment level at random.

Finally, we will see that $v_I = v_{II} = 0$. But that is immediate from

$$v_I = \int_0^x (1 - x) \, dy + \int_x^1 (-x) \, dy = 0.$$

5.2 Economics Applications of Nash Equilibria

Problems

5.26 In the Cournot model, we assumed there were two firms in competition. Instead suppose that there is only one firm, a monopolist, producing at quantity q. Formulate a model to determine the optimal production quantity. Compare the optimal production quantity for the monopolist with the quantities derived in the Cournot duopoly assuming $c_1 = c_2 = c$. Does competition produce lower prices?

5.26 Answer: If there is only one firm, the payoff function is given by $u(q) = q P(q) - c q, P(q) = (\Gamma - q)^+$. Take a derivative and set to zero to see that the maximum of u is achieved at $q_m^* = \frac{1}{2}(\Gamma - c)$.

Comparing q_m^* with the two firm optimal production quantity $q^* = \frac{1}{3}(2\Gamma - c_1 - c_2)$, taking $c_1 = c_2 = c$, we will have $q_m^* = \frac{\Gamma - c}{2} < q^* = \frac{2}{3}(\Gamma - c)$. Less will be produced if there is no competition and since the price function is a decreasing function of quantity, this means the price with only one firm for a gadget will be higher.

5.27 The Cournot model assumed the firms acted independently. Suppose instead the two firms collude to set production quantities.

(a) Find the optimal production quantity for each firm assuming the unit costs are the same for both firms, $c_1 = c_2 = c$.

5.27.a Answer: If they have the same unit costs, the two firms together should act as a monopolist. This means the optimal production quantities for each firm should be

$$q_1^* = q_2^* = \frac{q_m^*}{2} = \frac{1}{4}(\Gamma - c).$$

The payoffs to each firm is then $u_i(q_1^*, q_2^*) = \frac{1}{8}(\Gamma - c)^2$, which is greater than the profit under competition. Furthermore, the cartel price will be the same as the monopoly price since the total quantity produced is the same.

(b) Show that the optimal production quantities in the cartel solution are unstable in the sense that they are not best responses to each other. Assume $c_1 = c_2 = c$.

5.27.b Answer: The best response of firm 1 to the quantity q_2^* is given by

$$q_1(q_2^*) = \frac{1}{2}(\Gamma - c - q_2^*) = \frac{3}{8}(\Gamma - c) > \frac{1}{4}(\Gamma - c) = q_2^*.$$

Thus firm 1 should produce *more* than firm 2 as a best response. Similarly, firm 2 has an incentive to produce more than firm 1. The cartel collapses.

5.28 Two firms produce identical widgets. The price function is $P(q) = (150 - q)^+$ and the cost to produce q widgets is $C(q) = 120q - \frac{2}{3}q^2$ for each firm.

(a) What are the profit functions for each firm?

5.28.a Answer: $u_i(q_1, q_2) = q_i(150 - q_1 - q_2) - 120q_i + \frac{2}{3}q_i^2$.

(b) Find the Nash equilibrium quantities of production.

5.28.b Answer: Take derivatives, set to zero to see that $q_1 = q_2 = 18$.

(c) Find the price of a widget at the Nash equilibrium level of production as well as the profits for each firm.

5.28.c Answer: $P(18 + 18) = 114$. $u_1(18, 18) = u_2(18, 18) = 108$.

(d) If firm 1 assumes firm 2 will use it's best response function, what is the best production level of firm 1 to maximize it's profits assuming firm 1 will then publicly announce the level. In other words, they will play sequentially, not simultaneously.

5.28.d Answer: The best response functions are $q_1(q_2) = \frac{3}{2}(30 - q_2)$, $q_2(q_1) = \frac{3}{2}(30 - q_1)$. This requires $0 \le q_1 \le 30, 0 \le q_2 \le 30$. Set

$$f(q_1) = u_1(q_1, q_2(q_1)) = q_1\left(\frac{7}{6}q_1 - 15\right).$$

This is a parabola that opens up. **The derivative gives the minimum, not the maximum.** The maximum occurs at the right endpoint, which is where $q_2(q_1) = 0$ or $0 \le q_1 \le 30$ implies that $q_1 = 30$ and then $q_2(30) = 0$. Thus, $u_1(30, 0) = 600$.

5.29 Suppose instead of two firms in the Cournot model with payoff functions 5.1 that there are N firms. Formulate this model and find the optimal quantities each of the N firms should produce. Instead of a duopoly, this is an oligopoly. What happens when the firms all have the same costs and $N \to \infty$?

5.29 Answer: The profit function for firm i is $u_i(q_1, \ldots, q_i, \ldots, q_N) = q_i((\Gamma - \sum_{j=1}^{N} q_j)^+ - c_i)$. If we take derivatives and set to zero, we get

$$\frac{\partial u_i}{\partial q_i} = \Gamma - \sum_{j=1}^{N} q_j - c_i - q_i = 0, \ i = 1, 2, \ldots, N.$$

Since $\frac{\partial^2 u_i}{\partial q_i^2} = -2 < 0$, any critical point will provide a maximum. Set $\alpha_i = \Gamma - c_i$. We have to solve the system of equations $q_i + \sum_{j=1}^{N} q_j = \alpha_i$. In matrix form, we get the system

$$A\vec{q} = \vec{\alpha}, \quad A = \begin{bmatrix} 2 & 1 & 1 & \cdots & 1 \\ 1 & 2 & 1 & \cdots & 1 \\ \vdots & \vdots & \vdots & \vdots & \vdots \\ 1 & 1 & 1 & \cdots & 2 \end{bmatrix}, \quad \vec{q} = \begin{bmatrix} q_1 \\ q_2 \\ \vdots \\ q_N \end{bmatrix}, \quad \vec{\alpha} = \begin{bmatrix} \alpha_1 \\ \alpha_2 \\ \vdots \\ \alpha_N \end{bmatrix},$$

It is easy to check that

$$A^{-1} = \frac{1}{N+1} \begin{bmatrix} N & -1 & -1 & \cdots & -1 \\ -1 & N & -1 & \cdots & -1 \\ \vdots & \vdots & \vdots & \vdots & \vdots \\ -1 & -1 & -1 & \cdots & N \end{bmatrix}$$

and then

$$\vec{q} = A^{-1}\vec{\alpha}.$$

After a little algebra, we see that the optimal quantity that each firm should produce is

$$q_i = \frac{1}{N+1} \left(\Gamma - Nc_i + \sum_{j=1, j \neq i}^{N} c_j \right).$$

If $c_i = c$, $i = 1, 2, \ldots, N$, then $q_i = \frac{\Gamma - c}{N+1} \to 0$, $N \to \infty$. The profits then also approach zero.

5.30 Two firms produce identical products. The market price for total production quantity q is $P(q) = 100 - 2\sqrt{q}$. Firm 1's production cost is $C_1(q_1) = q_1 + 10$, and firm 2's production cost is $C_2(q_2) = 2q_2 + 5$. Find the profit functions and the Nash equilibrium quantities of production and profits.

5.30 Answer: With

$$u_1(q_1, q_2) = q_1(100 - 2\sqrt{(q_1 + q_2)}) - q_1 - 10, \quad \text{and}$$
$$u_2(q_1, q_2) = q_2(100 - 2\sqrt{(q_1 + q_2)}) - 2q_2 - 5,$$

we get the equations

$$99\sqrt{q_1 + q_2} - 2(q_1 + q_2) - q_1 = 0,$$
$$98\sqrt{q_1 + q_2} - 2(q_1 + q_2) - q_2 = 0.$$

Adding and setting $b = q_1 + q_2$, we have $197\sqrt{b} = 5b$, or $197 = 5\sqrt{b}$. The solution is $b = q_1 + q_2 = 1552.36$. Plugging into the equations, we have $99(39.41) - 2(1552.36) = q_1$ and $98(39.41) - 2(1552.35) = q_2$. The conclusion is $q_1^* = 795.88$, $q_2^* = 756.48$, and the resulting profits are $u_1^* = 16066.77$ and $u_2^* = 14519.42$.

5.31 Consider the Cournot duopoly model with a somewhat more realistic price function given by

$$P(q) = \begin{cases} \frac{1}{4}q^2 - 5q + 26, & \text{if } 0 \le q \le 10; \\ 1, & \text{if } q > 10. \end{cases}$$

Take the cost of producing a gadget at $c = 1$ for both firms.

(a) We may restrict productions for both firms q_1, q_2 to $[0, 10]$. Why?

5.31.a Answer: No firm would produce more than 10 because the cost of producing and selling one gadget would essentially be the same as the return on that gadget.

(b) Find the optimal production quantity and the profit at that quantity if there is only one firm making the gadget.

5.31.b Answer: The profit function for the monopolist is

$$u(q) = q\, P(q) - q = \frac{1}{4}q^3 - 5q^2 + 25q.$$

Taking a derivative and setting to zero gives

$$u'(q) = \frac{3}{4}q^2 - 10q + 25 = 0 \Rightarrow q = 10 \text{ or } q = \frac{10}{3}$$

and it is easy to check that $q = \frac{10}{3}$ provides the maximum. Also, $u\left(\frac{10}{3}\right) = \frac{1000}{27}$.

(c) Now suppose there are two firms with the same price function. Find the Nash equilibrium quantities of production and the profits for each firm.

5.31.c Answer: Now we have

$$u_i(q_1, q_2) = q_i\, P(q_1 + q_2) - q_i, \quad i = 1, 2.$$

We can take advantage of the fact that there is symmetry between the two firms since they have the same price function and costs. That means, for firm 1,

$$u_1(q_1, q_2) = u_1(q_1) = q_1 P(2q_1) - q_1.$$

The maximum is achieved at $q_1 = \frac{5}{3}$. Thus, the Nash equilibrium is $q_1 = \frac{5}{3}, q_2 = \frac{5}{3}$ and $u_1 = u_2 = \frac{500}{27}$.

5.32 Compare profits for firm 1 in the model with uncertain costs and the standard Cournot model. Assume $\Gamma = 15$, $c_1 = 4$, $c^+ = 5$, $c^- = 1$, and $p = 0.5$.

5.32 Answer: Consider first the problem with uncertain costs. The optimal production quantities are as follows:

$$q_1^* = \frac{1}{3}[\Gamma - 2c_1 + pc^+ + (1-p)c^-],$$

$$q_2^{+*} = \frac{1}{3}[\Gamma + c_1] - \frac{1}{6}[(1-p)c^- + pc^+] - \frac{1}{2}c^+,$$

$$q_2^{-*} = \frac{1}{3}[\Gamma - 2c^- + c_1] + \frac{1}{6}p(c^- - c^+).$$

For the data of the problem, we get

$$q_1^* = \frac{10}{3}, \quad q_2^{+*} = \frac{10}{3}, \quad q_2^{-*} = \frac{16}{3}.$$

The expected profit for firm 1 is

$$u_1\left(q_1^*, q_2^*\right) = q_1^*(\Gamma - (q_1^* + q_2^{+*}) - c_1)p + q_1(\Gamma - (q_1^* + q_2^{-*}) - c_1)(1-p)$$
$$= 11.11,$$

where Q_2^* denotes the random variable giving the production quantities for firm 2. If, we assume that the costs for firm 2 is $c_2 = 5$, firm 1's profit will be $q_1^*(\Gamma - (q_1^* + q_2^{+*}) - c_1) = 14.44$, while if firm 2's costs are $c_2 = 1$, firm 1's profit would be $q_1(\Gamma - (q_1^* + q_2^{-*}) - c_1) = 7.78$.

5.33 Suppose that we consider the Cournot model with uncertain costs but with three possible costs for firm 2, $Prob(C_2 = c^i) = p_i$, $i = 1, 2, 3$, where $p_i \geq 0$, $p_1 + p_2 + p_3 = 1$.

(a) Solve for the optimal production quantities.

5.33.a Answer: We have to solve the system

$$q_1 = \frac{1}{2}[(\Gamma - q_2^1 - c_1)p_1 + (\Gamma - q_2^2 - c_1)p_2 + (\Gamma - q_2^3 - c_1)p_3],$$

$$q_2^i = \frac{1}{2}[\Gamma - q_1 - c^i], i = 1, 2, 3,$$

Observe that q_1 is the expected value of the best response production quantities when the costs for firm 2 are c^i, $i = 1, 2, 3$ with probability p_i, $i = 1, 2, 3$. The system of best response equations has solution

$$q_1 = \frac{1}{3}\left[\Gamma - 2c_1 + p_1(c^1 - c^3) + p_2(c^2 - c^3)\right],$$

$$q_2^1 = \frac{1}{3}\left[\Gamma + c_1 - \frac{c^3}{2} + \frac{1}{2}p_1(c^3 - c^1) + \frac{p_2}{2}(c^3 - c^2)\right] - \frac{c^1}{2},$$

$$q_2^2 = \frac{1}{3}\left[\Gamma + c_1 - \frac{c^3}{2} + \frac{1}{2}p_1(c^3 - c^1) + \frac{p_2}{2}(c^3 - c^2)\right] - \frac{c^2}{2},$$

$$q_2^3 = \frac{1}{3}\left[\Gamma + c_1 + \frac{1}{2}p_1(c^3 - c^1) + \frac{p_2}{2}(c^3 - c^2)\right] - \frac{2c^3}{3}.$$

(b) Find the explicit optimal production quantities when $p_1 = \frac{1}{2}$, $p_2 = \frac{1}{8}$, $p_3 = \frac{3}{8}$, $\Gamma = 100$, and $c_1 = 2$, $c^1 = 1$, $c^2 = 2$, $c^3 = 5$.

5.33.b Answer: The optimal production quantities with the information given are $q_1 = \frac{749}{24}$ for firm 1, and $q_2^1 = \frac{529}{16}$, $q_2^2 = \frac{521}{16}$, and $q_2^3 = \frac{497}{16}$.

5.34 Suppose that two firms have constant unit costs $c_1 = 2$, $c_2 = 1$, and $\Gamma = 19$ in the Stackelberg model.

(a) How much should firm 2 produce as a function of q_1?

5.34.a Answer: We have

$$q_2(q_1) = \frac{\Gamma - q_1 - c_2}{2} = \frac{18 - q_1}{2}.$$

(b) How much should firm 1 produce?

5.34.b Answer: Firm 1's optimal production quantity is

$$q_1^* = \frac{\Gamma - 2c_1 + c_2}{2} = \frac{19 - 4 + 1}{2} = 8.$$

(c) How much, then, should firm 2 produce?

5.34.c Answer: Firm 2's production quantity should be

$$q_2^* = \frac{\Gamma + 2c_1 - 3c_2}{4} = \frac{19 + 4 - 3}{4} = 5.$$

5.35 Set up and solve a Stackelberg model given three firms with constant unit costs c_1, c_2, c_3 and firm 1 announcing production quantity q_1.

5.35 Answer: First we have to find the best response functions for firm 2 and firm 3. We have from the profit functions and setting derivatives to zero,

$$q_2(q_1) = \frac{\Gamma - q_1 - q_3 - c_2}{2} \quad \text{and} \quad q_3(q_1) = \frac{\Gamma - q_1 - q_2 - c_3}{2}.$$

Solving these two equation results in

$$q_2(q_1) = \frac{\Gamma - q_1 - 2c_2 + c_3}{3}, \qquad q_3(q_1) = \frac{\Gamma - q_1 - 2c_3 + c_2}{3}.$$

Next, firm 1 will choose q_1 to maximize

$$u_1(q_1, q_2(q_1), q_3(q_1)) = q_1(\Gamma - q_1 - q_2(q_1) - q_3(q_1) - c_1).$$

Taking a derivative and setting to zero, we get $q_1 = \frac{\Gamma + c_2 + c_3 - 3c_1}{2}$. Finally, plugging this q_1 into the best response functions q_2, q_3, we get the optimal Stackelberg production quantities

$$q_1 = \frac{\Gamma + c_2 + c_3 - 3c_1}{2},$$

$$q_2 = \frac{\Gamma - 5c_2 + c_3}{6} + \frac{c_1}{2},$$

$$q_3 = \frac{\Gamma + c_2 - 5c_3}{6} + \frac{c_1}{2}.$$

The profits for each firm are as follows:

$$u_1(q_1, q_2, q_3) = \frac{(\Gamma + c_2 + c_3 - 3c_1)^2}{12},$$

$$u_2(q_1, q_2, q_3) = \frac{(\Gamma - 5c_2 + c_3 + 3c_1)^2}{36},$$

$$u_2(q_1, q_2, q_3) = \frac{(\Gamma + c_2 - 5c_3 + 3c_1)^2}{36}.$$

5.36 Gadget prices range in the interval $(0, 100]$. Assume the demand function is simply $D(p) = 100 - p$. If two firms make the gadget, firm 1's profit function is given by

$$u_1(p_1, p_2) = \begin{cases} p_1(100 - p_1), & p_1 < p_2; \\ \frac{p_1(100 - p_1)}{2}, & p_1 = p_2; \\ 0, & p_1 > p_2. \end{cases}$$

(a) What is firm 2's profit function?

5.36.a Answer: Firm 2's profit function is

$$u_2(p_1, p_2) = \begin{cases} p_2(100 - p_2), & p_2 < p_1; \\ \frac{p_2(100 - p_2)}{2}, & p_1 = p_2; \\ 0, & p_2 > p_1. \end{cases}$$

(b) Suppose firm 2 picks a price $p_2 > 50$. What is the best response of firm 1?

5.36.b Answer: If $p_2 > 50$, the best response of firm 1 is $p_1 = 50$ because that value maximizes u_1. If $p_2 \leq 50$, firm 1 has no best response since the best response would be at $p_1 < p_2 - \varepsilon$ but not at $p_1 = p_2$. The same conclusions hold for firm 2.

(c) Suppose firm 2 picks a price $p_2 \leq 50$. Is there a best response of firm 1?

5.36.c Answer: If $p_2 \leq 50$, firm 1 has no best response. Symmetric with firm 1.

(d) Consider the game played in rounds. In round, one eliminate all strategies that are *not* best responses for both firms in the two cases considered in earlier parts of this problem. In round two, now find a better response to what's left from round one. If you keep going what is the conclusion?

5.36.d Answer:
In the second round, $p_1 = 49$ is a better response to firm 2's choice of $p_2 = 50$ in the first round. Similarly for firm 2. This means we may eliminate $(50, 50)$ as a possible Nash equilibrium. Continuing in rounds, we see by elimination of strategies which are the only possible Nash's, that there are no Nash equilibria for this game. Payoffs are not continuous so there is no guarantee that a Nash equilibrium will exist.

5.37 In the Bertrand model of this section, show that if $c_1 = c_2 = c$, then $(p_1^*, p_2^*) = (c, c)$ is a Nash equilibrium but each firm makes zero profit.

5.37 Answer: If $c_1 = c_2 = c$, we have the profit function for firm 2:

$$u_2(p_1, p_2) = \begin{cases} p_2(\Gamma - p_2) - c(\Gamma - p_2), & \text{if } p_2 < p_1; \\ \frac{(p-c)(\Gamma-p)}{2}, & \text{if } p_1 = p_2 = p \geq c; \\ 0, & \text{if } p_2 > p_1. \end{cases}$$

We will show that $u_2(c, c) \geq u_2(c, p_2)$, for any $p_2 \neq c$. Now, $u_2(c, c) = 0$ and

$$u_2(c, p_2) = \begin{cases} (p_2 - c)(\Gamma - p_2) < 0, & \text{if } p_2 < c; \\ \frac{(p-c)(\Gamma-p)}{2} = 0, & \text{if } p_1 = p_2 = p \geq c; \\ 0, & \text{if } p_2 > c. \end{cases}$$

In every case, $u_2(c, c) = 0 \geq u_2(c, p_2)$. Similarly, $u_1(c, c) \geq u_1(p_1, c)$ for every $p_1 \neq c$. This says (c, c) is a Nash equilibrium and both firms make zero profit.

5.38 Determine the entry deterrence level of production for the model in this section for firm 1 given $\Gamma = 100$, $a = 2$, $b = 10$.

5.38 Answer: We have $q_1^0 = \Gamma - 2\sqrt{b} - a = 91.675$ as the entry deterrence level of production to make $u_2 = 0$. The price of a gadget at this level is $2\sqrt{b} + a = 8.324$, and then firm 1's profit at this price is $u_1(q_1^0) = 2\sqrt{b}(\Gamma - a) - 5b = 569.81$

(a) How much profit is lost by setting the price to deter a competitor?

5.38.a Answer: The lost profit is the difference between the monopolist profit and the entry deterrence profit:

$$q_1 = \frac{\Gamma - a}{2} = 49 \Rightarrow u_1(49) = \left(\frac{\Gamma - a}{2}\right)^2 - b = 2391$$

and so the lost profit is $2391 - 569.81 = 1821.19$.

5.39 We could make one more adjustment in the Bertrand model of this section and see what effect it has on the model. What if we put a limit on the total quantity that a firm can produce? This limits the supply and possibly will put a floor on prices. Let $K \geq \frac{\Gamma}{2}$ denote the maximum quantity of gadgets that each firm can produce and recall that $D(p) = \Gamma - p$ is the quantity of gadgets demanded at price p. Find the profit functions for each firm.

5.39 Answer: The profit functions for each firm are as follows:

$$u_1(p_1, p_2) = \begin{cases} (p_1 - c_1) \min\{(\Gamma - p_1), K\}, & \text{if } p_1 < p_2; \\ \frac{(p - c_1)(\Gamma - p)}{2}, & \text{if } p_1 = p_2 = p \geq c_1; \\ 0, & \text{if } p_1 > p_2, p_2 \geq \Gamma - K. \\ (p_1 - c_1)(\Gamma - p_1 - K), & \text{if } p_1 > p_2, p_2 < \Gamma - K. \end{cases}$$

and

$$u_2(p_1, p_2) = \begin{cases} (p_2 - c_2) \min\{(\Gamma - p_2), K\}, & \text{if } p_2 < p_1; \\ \frac{(p - c_2)(\Gamma - p)}{2}, & \text{if } p_1 = p_2 = p \geq c_2; \\ 0, & \text{if } p_2 > p_1, p_1 \geq \Gamma - K. \\ (p_2 - c_2)(\Gamma - p_2 - K), & \text{if } p_2 > p_1, p_2 < \Gamma - K. \end{cases}$$

To explain the profit function for firm 1, if $p_1 < p_2$, firm 1 will be able to sell the quantity of gadgets $q = \min\{\Gamma - p_1, K\}$, the smaller of the demand quantity at price p_1 and the capacity of production for firm 1.

If $p_1 = p_2$, so that both firms are charging the same price, they split the market and since $K \geq \frac{\Gamma}{2}$ there is enough capacity to fill the demand.

If $p_1 > p_2$ in the standard Bertrand model, firm 1 loses all the business since they charge a higher price for gadgets. In the limited capacity model, firm 1 loses all the business only if, in addition, $p_2 \geq \Gamma - K$; that is, if $K \geq \Gamma - p_2$, which is the amount that firm 2 will be able to sell at price p_2, and this quantity is less than the production capacity.

Finally, if $p_1 > p_2$, and $K < \Gamma - p_2$, firm 1 will be able to sell the amount of gadgets that exceed the capacity of firm 2. That is, if $K < \Gamma - p_2$, then the quantity demanded from firm 2 at price p_2 is greater than the production capacity of firm 2, so the residual amount of gadgets can be sold to consumers by firm 1 at price p_1. But note that in this case, the number of gadgets that are demanded at price p_1 is $\Gamma - p_1$, so firm 1 can sell at most $\Gamma - p_1 - K < \Gamma - p_2 - K$.

What about Nash equilibria? Even in the case $c_1 = c_2 = 0$, there is no pure Nash equilibrium even at $p_1^* = p_2^* = 0$. This is known as the Edgeworth paradox.

5.40 Suppose that the demand functions in the Bertrand model are given by

$$q_1 = D_1(p_1, p_2) = (\Gamma - p_1 + bp_2)^+ \quad \text{and} \quad q_2 = D_2(p_1, p_2) = (\Gamma - p_2 + bp_1)^+,$$

where $1 \geq b > 0$. This says that the quantity of gadgets sold by a firm will increase if the price set by the opposing firm is too high. Assume that both firms have a cost of production $c \leq \min\{p_1, p_2\}$. Then, since profit is revenue minus costs the profit functions will be given by

$$u_i(p_1, p_2) = D_i(p_1, p_2)(p_i - c), \quad i = 1, 2.$$

(a) Using calculus, show that there is a unique Nash equilibrium at

$$p_1^* = p_2^* = \frac{\Gamma + c}{2 - b}.$$

5.40.a Answer: Firm i wants to solve

$$\max_{0 \le p_i}[\Gamma - p_i + bp_j][p_i - c].$$

Taking derivatives and solving gives the best response functions $p_i^*(p_j) = \frac{1}{2}(\Gamma + bp_j + c)$, $i = 1, 2$. Since, for example,

$$\frac{\partial u_1}{\partial p_1} = \Gamma - 2p_1 + bp_2 + c \Rightarrow \frac{\partial^2 u_1}{\partial p_1^2} = -2 < 0,$$

we know we have a maximum. Solving the best responses gives the unique Nash equilibrium

$$p_1^* = p_2^* = \frac{\Gamma + c}{2 - b}.$$

(b) Find the profit functions at equilibrium.

5.40.b Answer: We have

$$u_1(p_1^*, p_2^*) = u_2(p_1^*, p_2^*) = \left(\frac{\Gamma + c(b - 1)}{2 - b}\right)^2.$$

Both firms make the same profit at equilibrium.

(c) Suppose the firms have different costs and sensitivities so that

$$q_1 = D_1(p_1, p_2) = (\Gamma - p_1 + b_1 p_2)^+$$

and

$$q_2 = D_2(p_1, p_2) = (\Gamma - p_2 + b_2 p_1)^+,$$

and

$$u_i(p_1, p_2) = D_i(p_1, p_2)(p_i - c_i), \quad i = 1, 2.$$

Find a Nash equilibrium and the profits at equilibrium.

5.40.c Answer: By using the same procedure as before we get the equilibrium prices

$$p_1^* = \frac{2(\Gamma + c_1) + b_1(\Gamma + c_2)}{4 - b_1 b_2}, \qquad p_2^* = \frac{(\Gamma + c_1)b_2 + 2(\Gamma + c_2)}{b_1 b_2 - 4}$$

and equilibrium profits

$$u_1(p_1^*, p_2^*) = \left(\frac{(\Gamma + c_1 b_2 + c_2)b_1 + 2\Gamma - 2c_1}{b_1 b_2 - 4}\right)^2,$$

and

$$u_2(p_1^*, p_2^*) = \left(\frac{(\Gamma + c_2 b_1 + c_1)b_2 + 2\Gamma - 2c_2}{b_1 b_2 - 4}\right)^2.$$

(d) Find the equilibrium prices, production quantities, and profits if $\Gamma = 100$, $c_1 = 5$, $c_2 = 1$, $b_1 = \frac{1}{2}$, $b_2 = \frac{3}{4}$.

5.40.d Answer: Substituting these values into the solution from the previous part, we have

$$p_1^* = 71.86, \quad p_2^* = 77.45, \quad u_1 = 4470.54, \quad \text{and} \quad u_2 = 5844.34.$$

For the production quantities, we have

$$q_1^* = \Gamma - p_1^* + b_1 p_2^* = 66.86 \quad \text{and} \quad q_2^* = \Gamma - p_2 + b_2 p_1^* = 76.45.$$

5.41 Suppose that firm 1 announces a price in the Bertrand model with demand functions

$$q_1 = D_1(p_1, p_2) = (\Gamma - p_1 + bp_2)^+ \quad \text{and} \quad q_2 = D_2(p_1, p_2) = (\Gamma - p_2 + bp_1)^+,$$

where $1 \geq b > 0$. Assume the firms have unit costs of production c_1, c_2.

Construct the Stackelberg model; firm 2 should find the best response $p_2 = p_2(p_1)$ to maximize $u_2(p_1, p_2)$ and then firm 1 should choose p_1 to maximize $u_1(p_1, p_2(p_1))$. Find the equilibrium prices and profits. What is the result when $\Gamma = 100, c_1 = 5, c_2 = 1, b = 1/2$?

5.41 Answer: Start with profit functions

$$u_1(p_1, p_2) = (\Gamma - p_1 + bp_2)(p_1 - c_1) \quad \text{and} \quad u_2(p_1, p_2) = (\Gamma - p_2 + bp_1)(p_2 - c_2).$$

Solve $\frac{\partial u_2}{\partial p_2} = 0$ to get the best response function

$$p_2(p_1) = \frac{1}{2}(c_2 + \Gamma + bp_1).$$

Next solve $\frac{\partial u_1(p_1, p_2(p_1))}{\partial p_1} = 0$ to get

$$p_1^* = \frac{2c_1 + 2\Gamma - c_1 b^2 + bc_2 + b\Gamma}{4 - 2b^2},$$

and then the optimal price for firm 2 is

$$p_2^* = p_2(p_1^*) = \frac{-4c - 2 + c_2 b^2 - 4\Gamma + \Gamma b^2 - 2bc_1 + c_1 b^3 - 2b\Gamma}{4b^2 - 8}.$$

The profit functions become

$$u_1(p_1^*, p_2^*) = \frac{1}{8} \frac{(-2c_1 + b^2 c_1 + 2\Gamma + bc_2 + b\Gamma)^2}{2 - b^2}$$

and

$$u_2(p_1^*, p_2^*) = \frac{1}{16} \frac{\left(-4\Gamma + \Gamma b^2 + 4c_2 - 3c_2 b^2 - 2bc_1 + c_1 b^3 - 2b\Gamma\right)^2}{\left(-2 + b^2\right)^2}.$$

Note that with the Stackelberg formulation and the formulation in the preceding problem the Nash equilibrium has positive prices and profits.

The Maple commands to do these calculations are as follows:

```
> u1:=(p1,p2)->(G-p1+b*p2)*(p1-c1);
> u2:=(p1,p2)->(G-p2+b*p1)*(p2-c2);
> eq1:=diff(u2(p1,p2),p2)=0;
> solve(eq1,p2);
> assign(p2,%);
> f:=p1->u1(p1,p2);
> eq2:=diff(f(p1),p1)=0;
> solve(eq2,p1);
> simplify(%);
> assign(p1,%);
> p2;simplify(%);
> factor(%);
> assign(a2,%);
> u1(p1,a2);
> simplify(%);
> u2(p1,a2);
> simplify(%);
```

For the data of the problem we have

$$p_1^* = 74.07, \quad p_2^* = 69.02, \quad u_1 = 4174.50, \quad u_2 = 4626.43.$$

5.3 Duels

Problems

5.42 Determine the optimal time to fire for each player in the noisy duel with accuracy functions $p_I(x) = \sin(\frac{\pi}{2}x)$ and $p_{II}(x) = x^2$, $0 \le x \le 1$.

5.42 Answer: We have to find $x \in [0, 1]$ that satisfies $p_1(x) + p_2(x) = 1$. The Mathematica commands to do this are as follows:

```
p1[x_] = Sin[Pi/2 x]
p2[x_] = x^2
Solve[p1[x] + p2[x] == 1, 0 <= x <= 1, x]
```

The result is $x = 0.52$.

5.4 Auctions

5.4.1 COMPLETE INFORMATION

Problems

5.43 Verify that in the first-price auction $(b_1, \ldots, b_N) = (v_1, \ldots, v_N)$ is a Nash equilibrium assuming $v_1 = v_2$.

5.43 Answer: The payoff functions are as follows:

$$u_i(b_1, \ldots, b_N) = \begin{cases} 0, & \text{if } b_i < M, \text{ she is not a high bidder;} \\ v_i - b_i, & \text{if } b_i = M, \text{ she is the sole high bidder;} \\ \frac{v_i - b_i}{k}, & \text{if } i \in \{k\}, \text{ she is one of } k \text{ high bidders;} \end{cases}$$

and recall that $v_1 \geq v_2 \cdots \geq v_N$. We have to show that $u_i(v_1, \ldots, v_N)$ gives a larger payoff to player i if player i makes any other bid $b_i \neq v_i$. We assume that $v_1 = v_2$ so the two highest valuations are the same. Now for any player i if she bids less than $v_1 = v_2$ she does not win the object and her payoff is zero. If she bids $b_i > v_1$ she wins the object with payoff $v_i - b_i < v_1 - b_i < 0$. If she bids $b_i = v_1$, her payoff is $\frac{v_i - b_i}{3} = \frac{v_i - v_1}{3} < 0$. In all cases, she is worse off if she deviates from the bid $b_i = v_i$ as long as the other players stick with their valuation bids.

5.44 In a second-price sealed-bid auction with complete information the winner is the high bidder but she pays, not the price she bid, but the second highest bid. If there are ties, then the winner is drawn at random from among the high bidders and she pays the highest bid. Formulate the payoff functions and show that the following rules are optimal:

    (a) Each player bids $b_i \leq v_i$.

    (b) If $v_1 > v_2$, then player 1 wins by bidding any amount $v_2 < b_1 < v_1$.

    (c) If $v_1 = v_2 = \cdots v_k$, then $(v_1, v_2, \ldots, v_N)$ is a Nash equilibrium.

5.44 Answer: The payoff functions are as follows:

$$u_i(b_1, \ldots, b_N) =$$
$$\begin{cases} 0, & \text{if } b_i < \max_{j \neq i} b_j, \text{ she is not a high bidder;} \\ v_i - \max_{j \neq i} b_j, & \text{if } b_i > \max_{j \neq i} b_j, \text{ she is the sole high bidder;} \\ \dfrac{v_i - \max_{j \neq i} b_j}{k}, & \text{if } b_i = \max_{j \neq i} b_j \text{ and } i \in \{k\}, \text{ she is one of } k \text{ high bidders.} \end{cases}$$

The only difference between this and the first-price auction is if she wins the auction she pays the second highest price given by $\max_{j \neq i} b_j$.

    (a) It isn't hard to check that if $b_i > v_i$ then in all cases she can do no worse and sometimes do better by bidding v_i. For instance, if $\max_{j \neq i} b_j < v_i$ then player i wins the auction and she pays $\max_{j \neq i} b_j$ with payoff $v_i - \max_{j \neq i} b_j$. However, she gets the exact same payoff if she bids $b_i = v_i$. All other cases are similar.

    (b) We claim that $(b_1, \ldots, b_N) = (v_1, \ldots, v_N)$ is a Nash equilibrium. If everyone uses that strategy, then $u_1(v_1, \ldots, v_N) = v_1 - v_2$ and player 1 wins the auction, paying v_2 for the object. The other players get zero, If player 1 uses any other bid she cannot do better. For player i with $v_i < v_1$, her payoff will be zero (since she doesn't win the auction with such a bid), and in order get a bigger payoff she must bid $b_i > v_1$. But then her payoff is $v_i - v_1 < 0$. Thus, no player can do better by switching to a bid other than her valuation.

5.45 A homeowner is selling her house by auction. Two bidders place the same value on the house at \$100,000, while the next bidder values the house at \$80,000. Should the homeowner use a first-price or second-price auction to sell the house, or does it matter? What if the highest valuation is \$100,000, the next is \$95,000 and the rest are no more than \$90,000?

5.45 Answer: In the first case, it will not matter if she uses a first or second-price auction. Either way she will sell it for $100,000. In the second case, the winning bid for player 1 is between $95,000 < b_1 < 100,000$ whether it is a first or second price auction. However, in the second price auction, the house will sell for $95,000. Thus, a first-price auction is better for the seller.

5.46 Find the optimal reserve price to set in an auction assuming that the density of the value random variable V is $f(p) = 6p(1 - p), 0 \le p \le 1$.

5.46 Answer: The cumulative distribution function is $F(p) = -2p^3 + 3p^2, 0 < p < 1$. The interior solution of $1 - F(p) - pf(p) = 0$ is $p^* = 0.422$, so the reserve price should be set at 42.2% of the normalized range of prices. Note that even though the density is symmetric around $p = \frac{1}{2}$, the optimal reserve price is not 0.5. The Maple commands to solve are as follows:

```
> restart: f:=x->6*x*(1-x);
> F:=x->int(f(y),y=0..x);
> fsolve(1-F(x)-x*f(x)=0,x);
```

5.4.2 INCOMPLETE INFORMATION

Problems

5.47 In an **all-pay auction**, all the bidders must actually pay their bids to the seller, but only the high bidder gets the object up for sale. This type of auction is also called a **charity auction**. By following the procedure for a Dutch auction, show that the equilibrium bidding function for all players is $\beta(v) = ((N - 1)/N)v^N$, assuming that bidders' valuations are uniformly distributed on the normalized interval $[0, 1]$. Find the expected total amount collected by the seller.

5.47 Answer: The expected payoff of a bidder with valuation v who makes a bid of b is given by

$$u(b) = vProb(b \text{ is high bid}) - b = vF(\beta^{-1}(b))^{N-1} - b = v\beta^{-1}(b)^{N-1} - b.$$

Differentiate, set to zero, and solve to get $\beta(v) = \frac{(N-1)}{N})v^N$.

Since all bidders will actually pay their own bids and each bid is $\beta(v) = \frac{N-1}{N})v^N$, the expected payment from each bidder is

$$E[\beta(V)] = \frac{N - 1}{N} \int_0^1 v^N dv = \frac{N - 1}{N(N + 1)}.$$

Since there are N bidders, the total expected payment to the seller will be $\frac{N-1}{N+1}$.

CHAPTER SIX

Cooperative Games

6.1 Coalitions and Characteristic Functions

Problems

6.1 A stag-hunt game has characteristic function $v(S) = \alpha |S|$, $S \subset N$, $v(N) = 1$, where $\frac{1}{n} > \alpha > 0$.

(a) Find the normalized stag-hunt characteristic function.

6.1.a Answer: The normalized function is $v'(S) = \frac{v(S) - \sum_{i \in S} v(i)}{v(N) - \sum_{i \in N} v(i)} = 0$, $S \subsetneq N$, and $v'(N) = 1$.

(b) Find $C(0)$ using the normalization.

6.1.b Answer: The core using v' is

$$ C(0) = \left\{ \vec{x}' \mid x_i' \geq 0, \sum_{i=1}^{n} x_i' = 1 \right\}. $$

The unnormalized allocations satisfy

$$ x_i = x_i' \left(v(N) - \sum_{i=1}^{n} v(i) \right) + v(i) = x_i' \left(1 - \sum_{i=1}^{n} \alpha \right) + \alpha = x_i'(1 - n\alpha) + \alpha. $$

Since $1 - n\alpha > 0$, $x_i - \alpha \geq 0$ and $\sum_{i=1}^{n} x_i = (1 - n\alpha) + n\alpha = 1$. In terms of the unnormalized allocations, $C(0) = \{\vec{x} \mid x_i \geq \alpha, \sum_{i=1}^{n} x_i = 1\}$.

6.2 A customer wants to buy a bolt and a nut for the bolt. There are three players but player 1 owns the bolt and players 2 and 3 each own a nut. A bolt together with a nut is worth 5 but is worthless otherwise. Also, a nut without a bolt is worthless. Define a characteristic function for this game and verify that it is superadditive.

Solutions Manual to Accompany Game Theory: An Introduction, Second Edition. E.N. Barron.
© 2013 John Wiley & Sons, Inc. Published 2013 by John Wiley & Sons, Inc.

6.2 Answer: We could define a characteristic function for this game as

$$v(123) = 5, \ v(12) = v(13) = 5, \ v(1) = v(2) = v(3) = 0, \ \text{and} \ v(\emptyset) = 0.$$

$v(1) = 0$ because a bolt without a nut is worthless to player 1.

The characteristic function we have defined is superadditive because for any two disjoint coalitions we have $v(S \cup T) \geq v(S) + v(T)$. For instance with $S = \{12\}$, $T = \{3\}$ we have $v(123) = 5 \geq v(12) + v(3)$.

6.3 A river has n pollution-producing factories dumping water into the river. Assume that the factory does not have to pay for the water it uses but it may need to expend money to clean the water before it is suitable for use. Assume the cost of a factory to clean polluted water before it can be used is proportional to the number of polluting factories. Let $c =$ cost per factory. Assume also that a factory may choose to clean the water it dumps into the river at a cost of b per factory. We assume the inequalities $0 < c < b < nc$.

If a coalition S forms, all of its members could agree to pollute with a payoff of $-|S|(nc)$. The other possibility is all of its members could agree to clean the water and the factories not in the coalition pollute the river, which results in a total payoff to coalition S of $-|S|(n - |S|)c - |S|b$. Hence, the characteristic function is

$$v(S) = \begin{cases} \max\{-|S|(nc), -|S|(n - |S|)c - |S|b\}, & \text{if } S \subset N; \\ \max\{-n^2c, -nb\}, & \text{if } S = N. \end{cases}$$

Show that $\vec{x} = (-b, -b, \ldots, -b) \in C(0)$, which means $C(0) \neq \emptyset$. The allocation in which every factory cleans the water before it is dumped in the river is in the core.

6.3 Answer: Let $\vec{x} = (-b, \ldots, -b)$. Calculate for $S \subsetneq N$,

$$e(S, \vec{x}) = v(S) - \vec{x}(S) = \max\{|S|(-nc), |S|(-(n - |S|)c) - |S|b\} + b|S|$$
$$= \max\{|S|(b - nc), |S|(-(n - |S|)c)\}.$$

Since $b < nc$ and $n > |S|$ both terms in the max are negative so $e(S, \vec{x}) \leq 0$, $S \subsetneq N$. If $S = N$,

$$e(N, \vec{x}) = \max\{-n^2c, -nb\} + nb = \max\{n(b - nc), 0\} = 0$$

since $b < nc$. By definition we conclude $\vec{x} \in C(0)$.

6.4 A small research drug company, labeled 1, has developed a drug. It does not have the resources to get FDA (Food and Drug Administration) approval or to market the drug, so it considers selling the rights to the drug to a big drug company. Drug companies 2 and 3 are interested in buying the rights but only if both companies are involved in order to spread the risks. Suppose that the research drug company wants $1 billion, but will take $100 million if only one of the two big drug companies are involved. The profit to a participating drug company 2 or 3 is $5 billion, which they split. Here is a possible characteristic function with units in billions:

$$v(1) = v(2) = v(3) = 0, v(12) = 0.1, v(13) = 0.1, v(23) = 0, v(123) = 5,$$

because any coalition that doesn't include player 1 will be worth nothing.

(a) Find the normalized characteristic function and find the core using the normalized characteristic function.

6.4.a Answer: The normalized characteristic function is simply $v'(S) = \frac{v(S)}{v(N)}$ since $v(i) = 0, i = 1, 2, 3$:

$$v'(i) = 0, v'(12) = 0.02, v'(23) = 0, v'(13) = 0.02, v'(123) = 1.$$

Now we find the core

$$C(0) = \{(x_1, x_2, x_3) \mid x_i \geq 0, x_1 + x_2 \geq 0.02, x_1 + x_3 \geq 0.02, x_2 + x_3 \\ \leq 0, x_1 + x_2 + x_3 = 1\}.$$

We do not use primes to denote the normalized allocations. It is easy to see that the core does not contain a single allocation. For example, $(0.4, 0.4, 0.2)$ and $(0.3, 0.4, 0.3)$ are both in the core.

(b) The least core using normalized allocation is $C(-\frac{1}{3}) = \{(\frac{1}{3}, \frac{1}{3}, \frac{1}{3})\}$. Find the least core in unnormalized allocations.

6.4.b Answer: There is only one normalized allocation in $C(-\frac{1}{3})$ given by $\vec{n} = (\frac{1}{3}, \frac{1}{3}, \frac{1}{3})$. The unnormalized allocation is then

$$x_i = n_i(v(N) - \sum_{j=1}^{3} v(j)) + v(i) = \frac{1}{3}(5 - 0) + 0 = \frac{5}{3}, i = 1, 2, 3.$$

The corresponding first ε that makes the unnormalized core nonempty is $\varepsilon = -\frac{5}{3}$. Thus, the unnormalized core is $C(-\frac{5}{3}) = \{(\frac{5}{3}, \frac{5}{3}, \frac{5}{3})\}$.

6.5 Look back at Example 6.5. Find the normalized characteristic function and the normalized element in the least core.

6.5 Answer: The characteristic function is

$$v(1) = 1, v(2) = 2, v(3) = 3,$$
$$v(12) = 4, v(13) = 5, v(23) = 6, v(123) = 8.$$

The normalized characteristic function is

$$v'(1) = 0, v'(2) = 0, v'(3) = 0,$$
$$v'(12) = \frac{1}{2}, v'(13) = \frac{1}{2}, v'(23) = \frac{1}{2}, v'(123) = 1.$$

The normalized least core is $C(-\frac{1}{6}) = \{(\frac{1}{3}, \frac{1}{3}, \frac{1}{3})\}$. The unnormalized least core is $C(-\frac{1}{3}) = \{(\frac{5}{3}, \frac{8}{3}, \frac{11}{3})\}$.

The normalized least core comes from the inequalities

$$\frac{1}{2} - \varepsilon \leq x_1 + x_2, \; x_1 + x_2 \leq 1 + 2\varepsilon \Rightarrow -\frac{1}{2} \leq 3\varepsilon \Rightarrow -\frac{1}{6} \leq \varepsilon,$$

which decides $\varepsilon = -\frac{1}{6}$ is the first ε. Then $x_1 + x_2 = \frac{2}{3} \Rightarrow x_1 = \frac{1}{3}$. Then $x_1 \leq \frac{1}{2} - \frac{1}{6} = \frac{1}{3}$, $x_2 \leq \frac{1}{2} - \frac{1}{6}$ implies $0 = x_1 - \frac{1}{3} + x_2 - \frac{1}{3}$ and hence $x_1 = x_2 = \frac{1}{3}$.

6.6 Consider the bimatrix game with

$$A = \begin{bmatrix} 4 & 1 \\ 0 & 2 \end{bmatrix} \quad \text{and} \quad B = \begin{bmatrix} 2 & 1 \\ 0 & 4 \end{bmatrix}.$$

(a) Find the characteristic function of this game.

6.6.a Answer: $v(1) = value(A) = \frac{8}{5}$, $v(2) = value(B^T) = \frac{8}{5}$, $v(12) = 6$, $v(\emptyset) = 0$.

(b) Find the core of the game $C(0)$.

6.6.b Answer: The core is

$$C(0) = \left\{ (x_1, x_2) \mid \frac{8}{5} - x_1 \leq 0, \frac{8}{5} - x_2 \leq 0, x_1 + x_2 = 6 \right\}.$$

Eliminating $x_1 = 6 - x_2$, we may simplify to get $C(0) = \{ (6 - x_2, x_2) \mid \frac{8}{5} \leq x_2 \leq \frac{22}{5} \}$. This set does not contain only one allocation.

(c) Find the least core.

6.6.c Answer: The least core is

$$C(\varepsilon) = \left\{ (x_1, x_2) \mid \frac{8}{5} \leq x_1 + \varepsilon, \frac{8}{5} \leq x_2 + \varepsilon, x_1 + x_2 = 6 \right\}.$$

Since

$$\frac{8}{5} \leq x_2 + \varepsilon = 6 - x_1 + \varepsilon \Rightarrow x_1 \leq \frac{22}{5} + \varepsilon$$

combined with $\frac{8}{5} - \varepsilon \leq x_1$ implies $\varepsilon \geq -\frac{7}{5} = \varepsilon^1$ is the first ε that makes $C(\varepsilon) \neq \emptyset$. If $\varepsilon = -\frac{7}{5}$ this implies $x_1 = \frac{22}{5} - \frac{7}{5} = 3$ and then $x_2 = 6 - x_1 = 3$.

The least core is then $C\left(-\frac{7}{5}\right) = \{(3, 3)\}$.

6.7 Given the characteristic function $v(i) = 0$, $i = 1, 2, 3, 4$ and

$$v(12) = 4, v(13) = 4, v(14) = 3, v(23) = 6, v(24) = 2, v(34) = 2,$$
$$v(123) = 10, v(124) = 7, v(134) = 7, v(234) = 8, v(1234) = 13,$$

find the normalized characteristic function. Given the fair allocation

$$\vec{x} = \left(\frac{1}{4}, \frac{33}{104}, \frac{33}{104}, \frac{3}{26} \right)$$

for the normalized game, find the unnormalized allocation.

6.7 Answer: In normalized form since $v(i) = 0$, simply divide each allocation value by $v(1234) = 13 : \vec{x}$ unnormalized $= \left(\frac{13}{4}, \frac{33}{8}, \frac{33}{8}, \frac{3}{2} \right)$.

6.8 Odd Man Out is a three-player coin toss game in which each player chooses H or T. If all three make the same choice, the house pays each player 1; otherwise the odd man out pays the other players 1. Consider this as a three-player nonzero sum game.

(a) Find the characteristic function for the game.

6.8.a Answer: If player 3 plays H the game matrix (with player 1 as the row player and 2 as the column player) is

1/2	H	T
H	$(1,1,1)$	$(1,-2,1)$
T	$(-2,1,1)$	$(1,1,-2)$

If player 3 plays T the game matrix (with player 1 as the row player and 2 as the column player) is

1/2	H	T
H	$(1,1,-2)$	$(-2,1,1)$
T	$(1,-2,1)$	$(1,1,1)$

We now have to solve all the zero sum games 1 versus 23, 2 versus 13, 3 versus 12, and 12 versus 3, 13 versus 2, 23 versus 1. The game matrices in two cases are,

$$1 \text{ versus } 23 = \begin{bmatrix} 1 & 1 & 1 & -2 \\ -2 & 1 & 1 & 1 \end{bmatrix} \quad 12 \text{ versus } 3 = \begin{bmatrix} 2 & 2 \\ -1 & -1 \\ -1 & -1 \\ 2 & 2 \end{bmatrix}$$

The columns in the first game are HH, HT, TH, TT as are the rows in the second game. All the one player coalitions have the same game to solve as do the two player coalitions. The largest payoff sum is 3. We get the characteristic function for the cooperative game is

$$v(1) = v(2) = v(3) = -\frac{1}{2}, \quad v(12) = v(13) = v(23) = 2, \quad v(123) = 3.$$

(b) Find the core.

6.8.b Answer: The core is

$$C(0) = \left\{ -\frac{1}{2} \le x_i, 2 \le x_1 + x_2, 2 \le x_1 + x_3, 2 \le x_2 + x_3, x_1 + x_2 + x_3 = 3 \right\}.$$

It is straightforward to check that this reduces to $C(0) = \{(1,1,1)\}$. In fact, eliminating x_3 gives

$$x_2 \le 1, x_1 \le 1, 2 \le x_1 + x_2 \le \Rightarrow x_1 + x_2 = 2 \Rightarrow x_3 = 1.$$

By symmetry, or similar calculations, $x_1 = x_2 = 1$.

6.9 Find the characteristic function for the following three-player game. Each player has two strategies, A, B. If player 1 plays A the matrix is

$$\begin{bmatrix} (1, 2, 1) & (3, 0, 1) \\ (-1, 6, -3) & (3, 2, 1) \end{bmatrix},$$

while if player 1 plays B the matrix is

$$\begin{bmatrix} (-1, 2, 4) & (1, 0, 3) \\ (7, 5, 4) & (3, 2, 1) \end{bmatrix}.$$

In each matrix, player 2 is the row player and player 3 is the column player. Next, find the normalized characteristic function.

6.9 Answer: $v(\emptyset) = 0$, $v(1) = \frac{3}{5}$, $v(2) = 2$, $v(3) = 1$, $v(12) = 5$, $v(13) = 4$, $v(23) = 3$, $v(123) = 16$.

6.10 Derive the least core for the game with

$$v(123) = 1 = v(12) = v(13) = v(23) \quad \text{and} \quad v(1) = v(2) = v(3) = 0.$$

6.10 Answer: $\varepsilon^1 = \frac{1}{3}$ and $C\left(\frac{1}{3}\right) = \left\{\left(\frac{1}{3}, \frac{1}{3}, \frac{1}{3}\right)\right\}$.

6.11 Given the characteristic function

$$v(1) = 1, v(2) = \frac{1}{4}, v(3) = -1, v(12) = 3, v(13) = -1, v(23) = 1, v(123) = 4,$$

find the least core without normalizing.

6.11 Answer: Writing out the inequalities for $C(\varepsilon)$, we have the inequalities

$$e(23, \vec{x}) - x_2 - x_3 \leq \varepsilon, \ e(12, \vec{x}) - x_1 - x_2 \leq \varepsilon$$

and $x_1 + x_2 + x_3 = 4$. These imply that $3 - \varepsilon \leq x_1 + x_2 \leq 5 + \varepsilon$ and then $-2 \leq 2\varepsilon$. Consequently, we derive that $\varepsilon^1 = -1$ and the least core is $C(-1) = \{\vec{x} = (2, 2, 0)\}$.

6.12 Larger amounts of money invested in things like CDs (certificates of deposit) get a better rate of return. Suppose the rates of return depend on the amount invested as follows:

Invested amount	Rate of return
0–1,000,000	4%
1,000,000–3,000,000	5%
>3,000,000	5.5%

Three companies are going to form an investment partnership to pool their money. Suppose Company 1 will invest $1,800,000, company 2, $900,000, and company 3, $300,000. How should the net amount earned on the total investment be split among the three companies? Define an appropriate characteristic function. Find the core.

6.12 Answer: The characteristic function can be taken to be the amount earned on the investment:

$$v(1) = 90,000, \quad v(2) = 36,000, \quad v(3) = 12,000$$
$$v(12) = 135,000, \quad v(13) = 105,000 \quad v(23) = 60,000$$
$$v(123) = 150,000$$

The core is

$$
\begin{aligned}
C(0) &= \{(x_1, x_2, x_3) \mid 90 \le x_1, 36 \le x_2, 12 \le x_3, \\
&\quad 135 \le x_1 + x_2, 60 \le x_2 + x_3, 105 \le x_1 + x_3, \\
&\quad x_1 + x_2 + x_3 = 150,000\} \\
&= \{(x_1, x_2, 150 - x_1 - x_2) \mid 90 \le x_1, 36 \le x_2, x_1 + x_2 \le 138, \\
&\quad 60 \le 150 - x_1, 105 \le 150 - x_2\} \\
&= \{(x_1 = 90, x_2 = 45, x_3 = 15)\}.
\end{aligned}
$$

This is the fair allocation because the core contains only one element.

6.13 A **constant sum game** is one in which $v(S) + v(N - S) = v(N)$ for all coalitions $S \subset N$. Show that any *essential* constant sum game must have empty core $C(0) = \emptyset$.

6.13 Answer: Suppose $\vec{x} \in C(0)$ so that $e(S, \vec{x}) \le 0, \forall S \subsetneq N$. Take the single player coalition $S = \{i\}$ so $v(i) + v(N - i) = v(N)$. Since the game is essential, $v(N) > \sum_{i=1}^{n} v(i)$. Since $\vec{x}$ is in the core, we have

$$
v(N) > \sum_{i=1}^{n} v(i) = \sum_{i=1}^{n} v(N) - v(N - i) = nv(N) - \sum_{i=1}^{n} v(N - i),
$$
$$\Rightarrow$$
$$
v(N)(n - 1) < \sum_{i=1}^{n} v(N - i) \le \sum_{i=1}^{n} \sum_{j \ne i} x_j = \sum_{i=1}^{n} v(N) - x_i
$$
$$
= nv(N) - \sum_{i=1}^{n} x_i = (n - 1)v(N).
$$

That's a contradiction and hence $C(0) = \emptyset$.

6.14 In this problem, you will see why inessential games are of no interest. Show that an **inessential** game has one and only one imputation and is given by

$$\vec{x} = (x_1, \ldots, x_n) = (v(1), v(2), \ldots, v(n));$$

that is, each player is allocated exactly the benefit of the one-player coalition.

6.14 Answer: Since the game is inessential, $v(N) = \sum_{i=1}^{n} v(i)$. It is obvious that $\vec{x} = (v(1), \ldots, v(n)) \in C(0)$ since we have seen that an inessential game must be additive. That would tell us that $v(S) = \sum_{i \in S} v(i)$.

If there is another $\vec{y} \in C(0)$, $\vec{y} \neq \vec{x}$, there must be one component $y_i < v(i)$ or $y_i > v(i)$. Since $\vec{y} \in C(0)$, the first possibility cannot hold and so $y_i > v(i)$. This is true at any j component of $\vec{y}$ not equal to $v(j)$. Let

$$S = \{i \in N \mid y_i > v(i)\}$$

and we know $S \neq \emptyset$.

But then, $\sum_{i \in S} y_i > \sum_{i \in S} v(i) = v(S)$, which contradicts the fact that $\vec{y} \in C(0)$.

6.15 A player i is a **dummy** if $v(S) = v(S \cup i)$, for every $S \subset N$ with $i \notin S$. It looks like a dummy contributes nothing. Show that if i is a dummy and $v(i) = 0$, then for any $\vec{x} \in C(0)$, it must be true that $x_i = 0$.

6.15 Answer: Suppose $i = 1$. Then

$$x_1 + \sum_{j \neq 1} x_j = v(N) = v(N - 1) \le \sum_{j \neq 1} x_j,$$

and so $x_1 \le 0$. But since $-x_1 = v(1) - x_1 \le 0$, we have $x_1 = 0$.

6.16 Show that a vector $\vec{x} = (x_1, x_2, \ldots, x_n)$ is an imputation if and only if there are nonnegative constants $a_i \ge 0$, $i = 1, 2, \ldots, n$, such that $\sum_{i=1}^{n} a_i = v(N) - \sum_{i=1}^{n} v(i)$, and $x_i = v(i) + a_i$ for each $i = 1, 2, \ldots, n$.

6.16 Answer: In order for $\vec{x} = (x_1, x_2, \ldots, x_n)$ to be an imputation we must have (1) $x_i \ge v(i)$, $i = 1, 2, \ldots, n$, and (2) $\sum_{i=1}^{n} x_i = v(N)$.

1. If $\vec{x}$ is an imputation, set $a_i = x_i - v(i) \ge 0$, $i = 1, 2, \ldots, n$. Then $x_i = a_i + v(i)$ and $\sum_{i=1}^{n} a_i = \sum_{i=1}^{n} x_i - \sum_{i=1}^{n} v(i) = v(N) - \sum_{i=1}^{n} v(i)$.

2. If $a_i \ge 0$, $i = 1, 2, \ldots, n$, such that $\sum_{i=1}^{n} a_i = v(N) - \sum_{i=1}^{n} v(i)$, and $x_i = v(i) + a_i$ for each $i = 1, 2, \ldots, n$, then $x_i \ge v(i)$ and

$$\sum_{i=1}^{n} x_i = \sum_{i=1}^{n} v(i) - \sum_{i=1}^{n} a_i = v(N).$$

Hence, $\vec{x} = (x_1, \ldots, x_n)$ is an imputation.

6.17 Let $\delta_i = v(N) - v(N - i)$. Show that $C(0) = \emptyset$ if $\sum_{i=1}^{n} \delta_i < v(N)$.

6.17 Answer: Let $\vec{x} \in C(0)$. Since $v(N - 1) \le x_2 + \cdots + x_n = v(N) - x_1$, we have $x_1 \le v(N) - v(N - 1)$. In general, $x_i \le v(N) - v(N - i)$, $1 \le i \le n$. Now add these up to get $v(N) = \sum_i x_i \le \sum_i \delta_i < v(N)$, which says $C(0) = \emptyset$.

6.18 Verify the statement: $C(0) \neq \emptyset$ if and only if the linear program

$$\text{Minimize } z = x_1 + \cdots + x_n$$

$$\text{subject to } v(S) \le \sum_{i \in S} x_i \text{ for every } S \subsetneq N$$

has a finite minimum, say z^*, and $z^* \le v(N)$.

6.18 Answer: 1. Suppose the linear program

$$\text{Minimize } z = x_1 + \cdots + x_n$$
$$\text{subject to } v(S) \leq \sum_{i \in S} x_i \text{ for every } S \subsetneq N$$

has a finite minimum, say z^*, and $z^* \leq v(N)$. Let $\vec{x}$ be the solution of this linear program. That means it automatically satisfies $e(S, \vec{x}) \leq$ for every $S \subsetneq N$, and $\sum_{i=1}^n x_i = z^* \leq v(N)$. If $\sum_{i=1}^n x_i = v(N)$, then $\vec{x}$ is an imputation and hence $\vec{x} \in C(0)$. If $\sum_{i=1}^n x_i = z^* < v(N)$ then define

$$y_i = x_i + \frac{v(N) - \sum_{j=1}^n x_j}{n},$$

and $\vec{y} = (y_1, \ldots, y_n)$. It is immediate that $y_i \geq x_i \geq v(i)$ and $\sum_{i=1}^n y_i = v(N)$, which means $\vec{y}$ is an imputation. Furthermore, for any $S \subsetneq N$,

$$v(S) \leq \sum_{i \in S} x_i \leq \sum_{i \in S} y_i \Rightarrow e(S, \vec{y}) \leq 0.$$

Putting this all together, we conclude that $\vec{y} \in C(0) \neq \emptyset$.

2. Suppose $\vec{y} \in C(0) \neq \emptyset$. Then $\vec{y}$ satisfies all the constraints of the linear program. Therefore, the linear program has a finite minimum z^* and $z^* \leq \sum_{i=1}^n y_i = v(N)$.

6.19 Show that in any $n \geq 2$ player cooperative game, each player belongs to exactly 2^{n-1} coalitions.

6.19 Answer: If we remove, say player 1 from N, then $N - 1$ has 2^{n-1} coalitions (including the empty coalition), none of which contain player 1. But N has 2^n coalitions and so the number of coalitions which do not contain player 1 is $2^n - 2^{n-1} = 2^{n-1}$ and consequently there are 2^{n-1} coalitions which do contain player 1.

6.1.1 MORE ON THE CORE AND LEAST CORE

Problems

6.20 In a three-player game, each player has two strategies A, B. If player 1 plays A, the matrix is

$$\begin{bmatrix} (1, 2, 1) & (3, 0, 1) \\ (-1, 6, -3) & (3, 2, 1) \end{bmatrix},$$

while if player 1 plays B, the matrix is

$$\begin{bmatrix} (-1, 2, 4) & (1, 0, 3) \\ (7, 5, 4) & (3, 2, 1) \end{bmatrix}.$$

In each matrix, player 2 is the row player and player 3 is the column player. The characteristic function is $v(\emptyset) = 0$, $v(1) = \frac{3}{5}$, $v(2) = 2$, $v(3) = 1$, $v(12) = 5$, $v(13) = 4$, $v(23) = 3$, $v(123) = 16$. Verify that and then find the core and the least core.

6.20 Answer: The core is

$$C(0) = \left\{ (x_1, x_2, 16 - x_1 - x_2) : \frac{3}{5} \le x_1 \le 13, 2 \le x_2 \le 12, 5 \le x_1 + x_2 \le 15 \right\}.$$

The least core: $\varepsilon^1 = -\frac{62}{15}$, $C(\varepsilon^1) = \left\{ \left(\frac{71}{15}, \frac{92}{15}, \frac{77}{15} \right) \right\}$.

6.21 In the sink problem, we took the characteristic function $v(i) = 0$, $v(12) = 105$, $v(13) = 120$, $v(23) = 135$, and $v(123) = 150$, which models the fact that the players get the number of sinks each coalition can load and anyone not in the coalition would get nothing. The truck capacity and least core allocation in this case was

Player	Truck capacity	$C(20)$ Allocation
Aggie	45	35
Maggie	60	50
Baggie	75	65
Total	180	150

(a) Suppose that each coalition now gets the sinks left over after anyone not in the coalition gets their full truck capacity met first. Curly will help load any one player coalition. What is the characteristic function?

6.21.a Answer: The characteristic function for this game is

$$v(1) = 15, v(2) = 30, v(3) = 45, v(12) = 75, v(13) = 90, v(23) = 105, v(123) = 150.$$

For example, for $S = 1$, players 2 and 3 get their capacity first which is 135; the 15 left over go to player 1.

(b) Show that the core of this game is not empty and find the least core.

6.21.b Answer: If we find the least core and show that $\varepsilon^1 < 0$ then it must be true that $C(\varepsilon^1) \subset C(0) \ne \emptyset$. The least core inequalities we must solve are as follows:

$$15 - x_1 \le \varepsilon, 30 - x_2 \le \varepsilon, 45 - x_3 \le \varepsilon,$$
$$75 - \varepsilon \le x_1 + x_2, 90 - \varepsilon \le x_1 + x_3 = 150 - x_2, 105 - \varepsilon \le x_2 + x_3 = 150 - x_1,$$
$$x_1 + x_2 + x_3 = 150.$$

Hence, $x_2 \le 60 + \varepsilon$, $x_1 \le 45 + \varepsilon \Rightarrow 75 - \varepsilon \le x_1 + x_2 \le 105 + 2\varepsilon$ that gives $\varepsilon^1 = -10$. Then $x_1 + x_2 = 85$, $x_3 = 65$, and since $x_1 \le 35$, $x_2 \le 50$, it must be true that $x_1 = 35$, $x_2 = 50$. The least core is $C(-10) = \{(35, 50, 65)\}$ the exact same allocation as before but with a nonempty core. Every player is ok with this allocation.

6.22 A **weighted majority game** has a characteristic function of the form

$$v(S) = \begin{cases} 1, & \text{if } \sum_{i \in S} w_i > q; \\ 0, & \text{otherwise.} \end{cases}$$

Here, $w_i \geq 0$ are called weights and $q > 0$ is called a quota. Take $q = \frac{1}{2} \sum_{i \in N} w_i$. Suppose that there is one large group with two-fifths of the votes and two equal-sized groups with three-tenths of the vote each.

(a) Find the characteristic function.

6.22.a Answer: We consider the groups as individual players. The large group is 1, and the two equal-sized groups are 2 and 3. Next, the quota is $q = \frac{1}{2} \left(\frac{2}{5} + \frac{3}{10} + \frac{3}{10} \right) = \frac{1}{2}$. The characteristic function is $v(i) = 0$, $v(12) = v(13) = v(23) = 1$, $v(123) = 1$. Any coalition of any two groups wins. Single groups lose.

(b) Find the core and the least core.

6.22.b Answer: The core is empty since $\delta_i = v(N) - v(N - i) = 0$, $\sum_{i=1}^{3} \delta_i = 0 < v(123) = 1$ and Problem (6.17) tells us $C(0) = \emptyset$. This can also be seen directly since any two $x_i's$ must have sum at least 1, but all 3 must add up to 1. That's impossible.

The least core is a straightforward computation dealing with the inequalities. The result is $C(\frac{1}{3}) = \{(\frac{1}{3}, \frac{1}{3}, \frac{1}{3})\}$.

6.23 A classic game is the **garbage game**. Suppose that there are n property owners, each with one bag of garbage that needs to be dumped on somebody's property (one of the n). If n bags of garbage are dumped on a coalition S of property owners, the coalition receives a reward of $-n$. The characteristic function is taken to be the best that the members of a coalition S can do, which is to dump all their garbage on the property of the owners not in S.

(a) Explain why the characteristic function should be $v(S) = -(n - |S|)$, $v(N) = -n$, $v(\emptyset) = 0$, where $|S|$ is the number of members in S.

6.23.a Answer: The payoff for player i is the number of bags of garbage dumped in his yard times -1. The payoff for a coalition is the number of bags of garbage dumped on all the yards of members of S times -1. If there are n players, there are $n - |S|$ bags of garbage that will be dumped on the yards of the coalition S.

(b) Show that the core of the game is empty if $n > 2$.

6.23.b Answer: Suppose $n > 2$ and $C(0) \neq \emptyset$. Since $v(S) = |S| - n \leq \sum_{j \in S} x_j$, we have for any $k = 1, 2, \ldots, n, -1 = v(N - k) \leq \sum_{j \in N - k} x_j$. This says

$$x_2 + x_3 + \cdots + x_n \geq -1,$$
$$x_1 + x_3 + \cdots + x_n \geq -1,$$
$$x_1 + x_2 + \cdots + x_n \geq -1,$$
$$\vdots$$
$$x_1 + x_2 + \cdots + x_{n-1} \geq -1.$$

Add these up and use the fact that $\sum_{i=1}^{n} x_i = v(N) = -n$ to get the requirement

$$(n - 1)(x_1 + x_2 + \cdots + x_n) = -n(n - 1) \geq -n \Rightarrow n \leq 2.$$

With this contradiction we see that $n > 2 \Rightarrow C(0) = \emptyset$.

Another way to see this is to use a result from an earlier Problem 6.17 that gives us a criterion to use to determine when the core is empty. The criterion says

$$\sum_{i=1}^{n} \delta_i = \sum_{i=1}^{n} (v(N) - v(N - i)) < v(N) \Rightarrow C(0) = \emptyset.$$

In the garbage game, we have

$$\sum_{i=1}^{n} [-n - (-(n - |N - i|))] = \sum_{i=1}^{n} [-n + n - (n - 1)]$$
$$= -n(n - 1) < v(N) = -n$$

implies that $n - 1 > 1$ or $n > 2$. We conclude that when $n > 2$, from Problem 6.17, the core of the garbage game is empty.

(c) Recall that an imputation $\vec{y}$ dominates an imputation $\vec{x}$ through the coalition S if $e(S, \vec{y}) \geq 0$ and $y_i > x_i$ for each member $i \in S$. Take $n = 4$ in the garbage game. Find a coalition S so that $\vec{y} = (-1.5, -0.5, -1, -1)$ dominates $\vec{x} = (-2, -1, -1, 0)$.

6.23.c Answer: A coalition that works is $S = \{12\}$.

6.24 In the pollution game we have the characteristic function

$$v(S) = \begin{cases} \max\{-|S|(nc), -|S|(n - |S|)c - |S|b\}, & \text{if } S \subset N; \\ \max\{-n^2 c, -nb\}, & \text{if } S = N. \end{cases}$$

Here, $0 < c < b < nc$, and b is the cost of cleaning water before dumping while c is the cost of cleaning the water after dumping. We have seen that $C(0) \neq \emptyset$ since $\vec{x} = (-b, -b, \ldots, -b) \in C(0)$. Find the least core if $c = 1, b = 2, n = 3$.

6.24 Answer: The characteristic function becomes

$$v(S) = \begin{cases} -3, & \text{if } |S| = 1; \\ -6, & \text{if } |S| = 2; \end{cases}$$

and $v(123) = -6$.

The least core is

$$C(\varepsilon) = \{(x_1, x_2, x_3) \mid -3 \leq x_i + \varepsilon, i = 1, 2, 3, -6$$
$$\leq x_i + x_j + \varepsilon, i \neq j, x_1 + x_2 + x_3 = -6\}.$$

Analyzing the inequalities gives

$$-6 - \varepsilon \leq x_2 + x_3 \Rightarrow x_1 \leq \varepsilon, \quad -6 - \varepsilon \leq x_1 + x_3 \Rightarrow x_2 \leq \varepsilon;$$
$$-6 - \varepsilon \leq x_1 + x_2 \leq 2\varepsilon \Rightarrow \varepsilon \geq -2;$$
$$-6 - 2\varepsilon \leq x_1 + x_2 \leq 2\varepsilon \Rightarrow \varepsilon \geq -\frac{3}{2};$$
$$-9 - 3\varepsilon \leq x_1 + x_2 + x_3 = -6 \Rightarrow \varepsilon \geq -1;$$

and $\varepsilon^1 = -1$ is the first ε that makes $C(\varepsilon) \neq \emptyset$. Then $x_1 \geq -2, x_2 \geq -2, x_3 = -2$ implies that

$$C(-1) = \{(-2, -2, -2)\}.$$

is the least core and $\vec{x} = (-2, -2, -2)$ is the solution of our problem. Each factory should pay to clean the water before it is dumped in the river.

6.2 The Nucleolus

Problems

6.25 There are three types of planes (1, 2, and 3) that use an airport runway. Plane 1 needs a 100-yard runway, 2 needs a 150-yard runway, and 3 needs a 400-yard runway. The cost of maintaining a runway is equal to its length. Suppose this airport has one 400-yard runway used by all three types of planes and assume also that only one plane of each type will land at the airport on a given day. We want to know how much of the $400 cost should be allocated to each plane.

(a) Find the characteristic function.

6.25.a Answer: $v(\emptyset) = 0,\ v(1) = -100,\ v(2) = -150,\ v(3) = -400,\ v(12) = -150,$ $v(13) = -400,\ v(23) = -400,\ v(123) = -400.$

(b) Find the least core and show it has only one allocation.

6.25.b Answer: The least ε-core is given by

$$\begin{aligned} C(\varepsilon) = \{x_1 &\geq -100 - \varepsilon, x_2 \geq -150 - \varepsilon, x_3 \geq -400 - \varepsilon \\ &= x_1 + x_2 \geq -150 - \varepsilon, x_1 + x_3 \geq -400 - \varepsilon, \\ x_2 + x_3 &\geq -400 - \varepsilon, x_1 + x_2 + x_3 = -400\}. \end{aligned}$$

Next,

$$\begin{aligned} -400 - \varepsilon &\leq x_1 + x_3 = -400 - x_2 \Rightarrow x_2 \leq \varepsilon, \\ -400 - \varepsilon &\leq x_2 + x_3 = -400 - x_1 \Rightarrow x_1 \leq \varepsilon, \\ -150 - \varepsilon &\leq x_1 + x_2 \leq 2\varepsilon \Rightarrow \varepsilon \geq -50, \end{aligned}$$

and that gives us the first $\varepsilon^1 = -50$. This implies

$$x_1 \leq -50, x_2 \leq -50, x_1 + x_2 = -100 \Rightarrow x_1 = -50, x_2 = -50, x_3 = -300.$$

6.26 In a glove game with three players, player 1 can supply one left glove and players 2 and 3 can supply one right glove each. The value of a coalition is the number of paired gloves in the coalition.

(a) Find the characteristic function.

6.26.a Answer: $v(i) = 0, i = 1, 2, 3, v(12) = v(13) = v(123) = 1, v(23) = 0.$

(b) Find $C(0)$.

6.26.b Answer:

$$C(0) = \{x_i \geq 0, x_1 + x_2 \geq 1, x_1 + x_3 \geq 1, x_1 + x_2 + x_3 = 1\} = \{(1, 0, 0)\}$$

since $x_3 = 1 - x_1 - x_2 \leq 0$ implies $x_3 = 0$, and then $x_1 = 1, x_2 = 0$.

6.27 Consider the normalized characteristic function for a three-person game:

$$v(12) = \frac{4}{5}, v(13) = \frac{2}{5}, v(23) = \frac{1}{5}.$$

Find the core, the least core X^1, and the next least core X^2. X^2 will be the nucleolus.

6.27 Answer: We have

$$C(\varepsilon) = \{(x_1, x_2, x_3) \mid x_i \geq -\varepsilon, i = 1, 2, 3, x_1 + x_2 \geq \frac{4}{5} - \varepsilon,$$

$$x_1 + x_3 \geq \frac{2}{5} - \varepsilon, x_2 + x_3 \geq \frac{1}{5} - \varepsilon\}.$$

Checking all possible combinations of inequalities results in the first ε from the inequalities

$$\frac{4}{5} - \varepsilon \leq x_1 + x_2 = 1 - \xi_3 \Rightarrow -\varepsilon \leq x_3 \leq \frac{1}{5} + \varepsilon \Rightarrow \varepsilon \geq -\frac{1}{10},$$

and $\varepsilon^1 = -\frac{1}{10}$. Then, the first least core is

$$X^1 = C\left(-\frac{1}{10}\right) = \left\{x_3 = \frac{1}{10}, x_1 + x_2 = \frac{9}{10}, \frac{4}{10} \leq x_1, \frac{2}{10} \leq x_2\right\}.$$

Next, we calculate the excesses for $\vec{x} \in X^1$:

$$e(1, \vec{x}) = v(1) - x_1 = -x_1, \qquad e(12, \vec{x}) = \frac{4}{5} - x_1 - x_2 = -\frac{1}{10};$$

$$e(2, \vec{x}) = v(2) - x_2 = -x_2, \qquad e(13, \vec{x}) = \frac{2}{5} - x_1 - x_3 = \frac{3}{10} - x_1;$$

$$e(3, \vec{x}) = v(3) - x_3 = -x_3 = -\frac{1}{10}, \qquad e(23, \vec{x}) = \frac{1}{5} - x_2 - x_3 = \frac{1}{10} - x_2.$$

Thus, we may eliminate coalitions 3 and 12 since their excesses cannot be reduced further. The next least core is

$$X^2 = C(\varepsilon) = \left\{\left(x_1, x_2, \frac{1}{10}\right) \mid x_i \geq -\varepsilon, i = 1, 2, e(13, \vec{x}) \leq \varepsilon, e(23, \vec{x})\right.$$

$$\left. \leq \varepsilon, x_1 + x_2 = \frac{9}{10}\right\}$$

$$= \left\{\left(x_1, x_2, \frac{1}{10}\right) \mid x_i \geq -\varepsilon, i = 1, 2, \frac{3}{10} - x_1 \leq \varepsilon, \frac{1}{10} - x_2\right.$$

$$\left. \leq \varepsilon, x_1 + x_2 = \frac{9}{10}\right\}$$

We get

$$\frac{3}{10} - \varepsilon \leq x_1, \quad \frac{1}{10} - \varepsilon \leq x_2 \Rightarrow \frac{4}{10} - 2\varepsilon \leq x_1 + x_2 = \frac{9}{10} \Rightarrow \varepsilon \geq -\frac{1}{4}.$$

The first ε making $C(\varepsilon) \neq \emptyset$ is $\varepsilon^2 = -\frac{1}{4}$. Then we calculate the next least core is $X^2 = C(-\frac{1}{4}) = \{(\frac{11}{20}, \frac{7}{20}, \frac{2}{20})\}$. This contains only one point and is the nucleolus.

6.28 Find the fair allocation in the nucleolus for the three-person characteristic function game with

$$v(i) = 0, \quad i = 1, 2, 3,$$
$$v(12) = v(13) = 2, \quad v(23) = 10,$$
$$v(123) = 12.$$

6.28 Answer: The least core is the set $X^1 = C(-1) = \{x_1 = 1, x_2 + x_3 = 11, x_2 \geq 1, x_3 \geq 2\}$. The nucleolus is the single point $X^2 = \{(1, \frac{11}{2}, \frac{11}{2})\}$.

Since this is a three-person game, we may use the formulas in Theorem F.9 to get the nucleolus directly. First, we have to check if the core is empty. Since the normalized characteristic function is $v'(S) = \frac{v(S)}{12}$, we have

$$v'(12) + v'(13) + v'(23) = \frac{14}{12} \leq 2,$$

and by Proposition F.7 we know $C(0) \neq \emptyset$. Check the inequalities in Theorem F.9 to see that Case 13 holds:

$$v(123) \leq v(23) + 2v(13),$$
$$v(123) \leq v(23) + 2v(12),$$
$$v(123) + v(23) \geq 2(v(13) + v(12)),$$
$$y_1 = \frac{(v(123) - v(23))}{2} = \frac{12 - 10}{2} = 1,$$
$$y_2 = \frac{(v(123) + v(23) - 2(v(13) - v(12)))}{4} = \frac{12 + 10 - 2(2 - 2)}{2} = \frac{11}{2},$$
$$y_3 = \frac{(v(123) + v(23) + 2(v(13) - v(12)))}{4} = \frac{12 + 10 + 2(2 - 2)}{4} = \frac{11}{2}.$$

6.29 In Problem 6.12, we considered the problem in which companies can often get a better cash return if they invest larger amounts. There are three companies who may cooperate to invest money in a venture that pays a rate of return as follows:

Invested amount	Rate of return
0–1,000,000	4%
1,000,000–3,000,000	5%
>3,000,000	5.5%

Suppose Company 1 will invest $1,800,000, Company 2, $900,000, and Company 3, $400,000. This problem was considered earlier but with a different amount of cash for player 3. How should the net amount earned on the total investment be split among the three companies? Define an appropriate characteristic function. Find the nucleolus.

6.29 Answer: The characteristic function can be taken to be the interest earned on the investment:

$$v(1) = 90{,}000, \ v(2) = 36{,}000, \ v(3) = 16{,}000;$$
$$v(12) = 135{,}000, \ v(13) = 110{,}000 \ v(23) = 65{,}000;$$
$$v(123) = 170{,}500.$$

To get the least core we have

$$C(\varepsilon) = \{90 - \varepsilon \le x_1, 36 - \varepsilon \le x_2, 16 - \varepsilon \le x_3,$$
$$135 - \varepsilon \le x_1 + x_2, 110 - \varepsilon \le x_1 + x_3, 65 - \varepsilon \le x_2 + x_3$$
$$x_1 + x_2 + x_3 = 170.5\}.$$

Using the inequalities $90 - \varepsilon \le x_1 \le 105.5 - \varepsilon$, we get the first $\varepsilon^1 = -\frac{15.5}{2}$. Then $x_1 = 97.75$ and $x_2 + x_3 = 72.75$.

We need the next least core. We calculate the excesses for $\vec{x} \in X^1 = C(\varepsilon^1)$ to get

$$\begin{aligned}
e(1, \vec{x}) &= 90 - 97.75 = -70.75, & e(12, \vec{x}) &= 135 - 97.75 - x_2 \\
e(2, \vec{x}) &= 36 - x_2, & e(13, \vec{x}) &= 110 - 97.75 - x_3 \\
e(3, \vec{x}) &= 16 - x_3, & e(23, \vec{x}) &= 170.75 - 7.75
\end{aligned}$$

The next least core will work with coalitions $\{2, 3, 12, 13\}$ since only those coalitions have excesses which may be lowered. We have

$$X^2 = \{36 - x_2 \le \varepsilon, 16 - x_3 \le \varepsilon, 37.25 - x_2 \le \varepsilon, 12.25 - x_3 \le \varepsilon, x_2 + x_3 = 72.75\}.$$

The two inequalities $16 - x_3 \le \varepsilon, 37.25 - x_2 \le \varepsilon$ lead to $53.25 - 2\varepsilon \le x_2 + x_3 = 72.75$ and $\varepsilon^2 = -9.75$ as the first ε making $X^2 \ne \emptyset$. Then $37.25 + 9.75 = 47 \le x_2$, and $16 + 9.75 = 25.75 \le x_3$ imply that $x_2 = 47, x_3 = 25.75$.

The nucleolus is $x_1 = \frac{391}{4} = 97{,}750, x_2 = 47{,}000, x_3 = \frac{103}{4} = 25{,}750$.

It is interesting to compare this with the common assumption that the fair allocation should be that each player will get the amount of 170,500 proportional to the amount they invest. That would lead to the allocation $y_1 = \frac{90}{142} 170{,}500 = 108063.38, y_2 = \frac{36}{142} 170{,}500 = 43225.35$ and $y_3 = \frac{16}{142} 170{,}500 = 19211.27$. But, this does not take into account that player 3 is a very important investor. It is her money that pushes the grand coalition into the 5.5% rate of return. Without player 3 the most they could get is 5%. Consequently player 3 has to be compensated for this power. The proportional allocation doesn't do that.

6.30 There are three ambitious computer science students, named 1, 2, and 3, with a lucrative idea and they wish to start a business. None of the students can start a business on their own but any two of them can.
- If 1 and 2 form a business, their total salary will be 120,000.
- If 1 and 3 form a business, their total salary will be 100,000.

- If 2 and 3 form a business, their total salary will be 80,000.
- If all three go into business, together they will pay themselves a total salary of 150,000.

(a) Find an appropriate characteristic function and find the core.

6.30.a Answer: The characteristic function is

$$v(i) = 0, i = 1, 2, 3, \ v(12) = 120, \ v(13) = 100, \ v(23) = 80, \ v(123) = 150.$$

The core is $C(0) = \{(70, 50, 30)\}$.

(b) Suppose the 3 together will pay themselves 140,000. Find the least core.

6.30.b Answer: Now $v(123) = 140$. The core is empty since the normalized characteristic function is

$$v'(S) = \frac{v(S)}{140}, \Rightarrow v'(12) = \frac{6}{7}, \ v'(13) = \frac{5}{7}, \ v'(23) = \frac{4}{7}$$

and $v'(12) + v'(13) + v'(23) = \frac{15}{7} > 2$. Then Proposition F.7 tells us $C(0) = \emptyset$.

The least core is calculated in the usual way and we get $C(\frac{20}{3}) = \{(\frac{200}{3}, \frac{1400}{3}, \frac{80}{3})\}$. This is the nucleolus in this case.

(c) Find the nucleolus if they will pay themselves a total of 300,000.

6.30.c Answer: In this case, $v(123) = 300$ and it is easy to check the core is not empty. The first least core is

$$X^1 = C(-90) = \{(x_1, x_2, 90) \mid x_1 + x_2 = 210, 90 \le x_2 \le 110, 90 \le x_1 \le 130\}.$$

To find the next least core, we eliminate coalitions $S = 3, S = 12$, and calculate $X^2 = C(-100) = \{(110, 100, 90)\}$ which is the nucleolus. Interestingly, player 3 must be paid an additional 10 out of player 1 and 2's total to induce her to join the grand coalition.

6.31 Four doctors, Moe, Larry, Curly, and Shemp,[1] are in partnership and cover hours for each other. At most one doctor needs to be in the office to see patients at any one time. They advertise their office hours as follows:

(1)	Shemp	12:00–5:00
(2)	Curly	9:00–4:00
(3)	Larry	10:00–4:00
(4)	Moe	2:00–5:00

A coalition is an agreement by one or more doctors as to the times they will really be in the office to see everybody's patients. The characteristic function should be the amount of time saved by a given coalition. Note that $v(i) = 0$, and $v(1234) = 13$ hours. This problem is an example of the use of cooperative game theory in an scheduling problem and is an important application in many disciplines.

---

[1]This scheduling problem is adapted from an example due to Mesterton–Gibbons.

(a) Find the characteristic function for all coalitions.

6.31.a Answer: Let's draw an appointment schedule:

| 9 | 10 | 11 | 12 | 1 | 2 | 3 | 4 | 5 |

			Shemp	Shemp	Shemp	Shemp	Shemp
Curly	Curly	Curly	Curly	Curly	Curly	Curly	
	Larry	Larry	Larry	Larry	Larry	Larry	
					Moe	Moe	Moe

The characteristic function is the number of hours saved by a coalition. Single coalitions save nothing $v(i) = 0$, $i = 1, 2, 3, 4$, and

$$v(12) = 4, v(13) = 4, v(14) = 3, v(23) = 6, v(24) = 2, v(34) = 2,$$
$$v(123) = 10, v(124) = 7, v(134) = 7, v(234) = 8, v(1234) = 13.$$

For instance, $v(124) = 7$ since Shemp and Curly overlap 4 hours and Shemp and Moe overlap 3 hours.

(b) Find X^1, X^2. When you get to X^2, you should have the fair allocation in terms of hours saved.

6.31.b Answer: We will use either Maple or Mathematica to solve this problem. We show that the Nucleolus $= \{(\frac{13}{4}, \frac{33}{8}, \frac{33}{8}, \frac{3}{2})\}$ with units in hours.

The first least core is determined from $\varepsilon^1 = -\frac{3}{2}$,

$$X^1 = C\left(-\frac{3}{2}\right) = \left\{ x_1 + x_2 + x_3 = \frac{23}{2}, x_4 = \frac{3}{2}, \right.$$
$$x_1 + x_2 + x_3 + x_4 = 13, x_1 + x_2 + x_4 \geq \frac{17}{2},$$
$$x_2 + x_3 + x_4 \geq \frac{19}{2}, x_1 \geq \frac{3}{2}, x_2 \geq \frac{3}{2},$$
$$x_1 + x_2 \geq \frac{11}{2}, x_3 \geq \frac{3}{2}, x_1 + x_3 \geq \frac{11}{2},$$
$$x_2 + x_3 \geq \frac{15}{2}, x_1 + x_4 \geq \frac{9}{2}, x_2 + x_4 \geq \frac{7}{2},$$
$$\left. x_3 + x_4 \geq \frac{7}{2}, x_1 + x_3 + x_4 \geq \frac{17}{2} \right\}.$$

A decided allocation is $x_4 = \frac{3}{2}$. Next, to get X^2 first calculate the excesses $e(S, \vec{x})$ for $\vec{x} \in X^1$ and we see that we may eliminate coalitions $S = 4, S = 123$.

We then obtain $\varepsilon^2 = -\frac{7}{4}$, and $x_1 = \frac{13}{4}$. Again we calculate the excesses $e(S, \vec{x}), \vec{x} \in X^2 = C(-\frac{7}{4})$, and we determine we may eliminate coalitions $S = 1, S = 14, S = 1234$.

Finally, we calculate $\varepsilon^3 = -\frac{15}{8}$, $X^3 = C(\varepsilon^3)$ using only the coalitions $S = 2, 3$, 124, 134 and we get $X^3 = \{(\frac{13}{4}, \frac{33}{8}, \frac{33}{8}, \frac{3}{2})\}$. That is the nucleolus.

(c) Find the exact times of the day that each doctor will be in the office according to the allocation you found in X^3.

6.31.c Answer: The schedule is set up as follows:

1. Since Curly(2) starts at 9 AM, and works 7 hours, he saves 4.125 hours and works 2.875 hours. That means he works from 9 to 11:52.5 AM.

2. Larry(3) comes in after Curly. He saves 4.125 hours from the 6 he worked before cooperation and so he works 1.875 hours. So he starts at 11:52.5 AM and leaves at 1:45 PM.

3. Shemp(1) comes in after Larry. He saves 3.25 hours from the 5 he worked before. He must work 1.75 hours, which means he starts at 1:45 PM and leaves at 3:30 PM.

4. Moe(4) is the last to arrive. He saves 1.5 hours from the 3 hours he worked. He arrives at 3:30 PM and leaves at 5 PM, closing the office and turning out the lights.

6.32 Use software to solve the four-person game with unnormalized characteristic function

$$v(i) = 0, \quad i = 1, 2, 3, 4,$$
$$v(12) = v(34) = v(14) = v(23) = 1,$$
$$v(13) = \frac{3}{4}, \quad v(24) = 0,$$
$$v(123) = v(124) = v(134) = v(234) = 1,$$
$$v(1234) = 3.$$

6.32 Answer: For the first least core $\varepsilon^1 = -\frac{1}{2}$:

$$\text{Least core} = X^1 = C\left(-\frac{1}{2}\right) = \left\{ x_1 + x_2 = \frac{3}{2}, x_3 + x_4 = \frac{3}{2}, x_i \geq \frac{1}{2}, i = 1, 2, 3, 4, \right.$$
$$x_2 + x_3 \geq \frac{3}{2}, x_1 + x_4 \geq \frac{3}{2}, x_1 + x_3 \geq \frac{5}{4}, x_2 + x_4 \geq \frac{1}{2},$$
$$x_1 + x_2 + x_3 \geq \frac{3}{2}, x_1 + x_2 + x_4 \geq \frac{3}{2}, x_1 + x_3 + x_4 \geq \frac{3}{2},$$
$$\left. x_2 + x_3 + x_4 \geq \frac{3}{2}, x_1 + x_2 + x_3 + x_4 = 3 \right\}.$$

Next, X^2 has $\varepsilon^2 = 1$. X^3 has $\varepsilon^3 = 3$, and nucleolus $= \{(\frac{3}{4}, \frac{3}{4}, \frac{3}{4}, \frac{3}{4})\}$.

6.33 We have four players involved in a game to minimize their costs. We have transformed such games to savings games by defining $v(S) = \sum_{i \in S} c(i) - c(S)$, the total cost if each player is in a one-player coalition, minus the cost involved if players form the coalition S. Find the nucleolus allocation of costs for the four-player game with costs

$$c(1) = 7, c(2) = 6, c(3) = 4, c(4) = 5,$$
$$c(12) = 7.5, c(13) = 7, c(14) = 7.5, c(23) = 6.5, c(24) = 6.5, c(34) = 5.5,$$
$$c(123) = 7.5, c(124) = 8, c(134) = 7.5, c(234) = 7,$$
$$c(1234) = 8.5, c(\emptyset) = 0.$$

6.33 Answer: The characteristic function for the **savings game** is $v(\emptyset) = 0$, $v(i) = 0$, $v(1234) = 22 - 8.5$, and

$$v(12) = 13 - 7.5, v(13) = 11 - 7, v(14) = 12 - 7.5,$$
$$v(23) = 10 - 6.5, v(24) = 11 - 6.5, v(34) = 9 - 5.5,$$
$$v(123) = 17 - 7.5, v(124) = 18 - 8, v(134) = 16 - 7.5,$$
$$v(234) = 15 - 7.$$

The least core is

$$X^1 = C(-1.125) = \{x_1 + x_2 + x_3 + x_4 = 13.5, x_1 + x_2 + x_3 \geq 10.625,$$
$$x_1 + x_2 + x_4 \geq 11.125, x_1 + x_3 + x_4 \geq 9.625,$$
$$x_2 + x_3 + x_4 \geq 9.125, x_1 \geq 1.125, x_2 \geq 1.125, x_1 + x_2 \geq 6.625,$$
$$x_3 \geq 1.125, x_1 + x_3 \geq 5.125, x_2 + x_3 \geq 4.625,$$
$$x_4 \geq 1.125, x_1 + x_4 \geq 5.625, x_2 + x_4 \geq 5.625, x_3 + x_4 \geq 4.625\}.$$

This reduces to $X^1 = \{(\frac{35}{8}, \frac{31}{8}, \frac{19}{8}, \frac{23}{8})\} = \{(4.375, 3.875, 2.375, 2.875)\}$, so this is the nucleolus.

In the original terms of costs, we have

$$y_1 = c(1) - \frac{35}{8} = 2.625,$$
$$y_2 = c(2) - \frac{31}{8} = 2.125,$$
$$y_3 = c(3) - \frac{19}{8} = 1.625,$$
$$y_4 = c(4) - \frac{23}{8} = 2.125.$$

6.34 Show that in a cost savings game if we define the cost savings characteristic function $v(S) = \sum_{i \in S} c(i) - c(S)$, if $v(S)$ is superadditive, then $c(S)$ is subadditive, and conversely.

6.34 Answer: Let $S, T \subset N$, $S \cap T = \emptyset$. Then if c is subadditive $c(S \cup T) \leq c(S) + c(T)$ and

$$v(S \cup T) = \sum_{i \in S \cup T} c(i) - c(S \cup T)$$
$$= \sum_{i \in S} c(i) + \sum_{i \in T} c(i) - c(S \cup T)$$
$$\geq \sum_{i \in S} c(i) + \sum_{i \in T} c(i) - c(S) - c(T)$$
$$= v(S) + v(T),$$

which says v is superadditive. The reverse is also true since $c(S) = \sum_{i \in S} c(i) - v(S)$. Then,

$$
\begin{aligned}
c(S \cup T) &= \sum_{i \in S \cup T} c(i) - v(S \cup T) \\
&= \sum_{i \in S} c(i) + \sum_{i \in T} c(i) - v(S \cup T) \\
&\leq \sum_{i \in S} c(i) + \sum_{i \in T} c(i) - v(S) - v(T) \\
&= c(S) + c(T).
\end{aligned}
$$

6.3 The Shapley Value

Problems

6.35 The formula for the Shapley value is

$$
x_i = \sum_{S \in \Pi^i} [v(S) - v(S - i)] \frac{(|S| - 1)!(n - |S|)!}{n!}, \quad i = 1, 2, \ldots, n,
$$

where Π^i is the set of all coalitions $S \subset N$ containing i as a member (i.e., $i \in S$). Show that an equivalent formula is

$$
x_i = \sum_{T \in 2^N \setminus \Pi^i} [v(T \cup i) - v(T)] \frac{|T|!(n - |T| - 1)!}{n!}, \quad i = 1, 2, \ldots, n.
$$

6.35 Answer: We have $T \in 2^N \setminus \Pi^i$ if and only if $T \cup i \in \Pi^i$. Also, $|T \cup i| = |T| + 1$. Hence,

$$
\begin{aligned}
x_i &= \sum_{S \in \Pi^i} [v(S) - v(S - i)] \frac{(|S| - 1)!(n - |S|)!}{n!} \\
&= \sum_{T \in 2^N \setminus \Pi^i} [v(T \cup i) - v(T)] \frac{(|T| + 1 - 1)!(n - |T| - 1)!}{n!} \\
&= \sum_{T \in 2^N \setminus \Pi^i} [v(T \cup i) - v(T)] \frac{|T|!(n - |T| - 1)!}{n!}.
\end{aligned}
$$

6.36 Let v be a superadditive characteristic function for a simple game. Show that if $v(S) = 0$ and $A \subset S$ then $v(A) = 0$ and if $v(S) = 1$ and $S \subset A$ then $v(A) = 1$.

6.36 Answer: Since v is superadditive if $A \subset S$ and $v(S) = 0$, then

$$v(S) = 0 = v(A \cup (S \setminus A)) \geq v(A) + v(S \setminus A) \geq 0$$

and hence $v(A) = 0$ as well as $v(S \setminus A) = 0$. Similarly, if $S \subset B$ and $v(S) = 1$, then

$$v(B) = v(S \cup (B \setminus S)) \geq v(S) + v(B \setminus S) \geq 1.$$

Since the game is simple, we conclude that $v(B) = 1$.

6.37 Moe, Larry, and Curly have banded together to form a leg, back, and lip waxing business, LBLWax, Inc. The overhead to the business is 40K per year. Each stooge brings in annual business and incurs annual costs as follows: Moe-155K revenue, 40K costs; Larry-160K revenue, 35K costs; Curly-140K revenue, 38K costs. Costs include, wax, flame throwers, antibiotics, and so on. Overhead includes rent, secretaries, insurance, and so on. At the end of each year, they take out all the profit and allocate it to the partners.

(a) Find a characteristic function that can be used to determine how much each waxer should be paid.

6.37.a Answer: A reasonable characteristic function is

$$v(M) = 155 - 80 = 75, v(L) = 160 - 75 = 85, v(C) = 140 - 78 = 62,$$
$$v(ML) = 240 - 40 = 200, v(MC) = 217 - 40 = 177, v(LC) = 227 - 40 = 187,$$
$$v(MLC) = 115 + 125 + 102 - 40 = 302.$$

(b) Find the nucleolus.

6.37.b Answer: The Mathematica statements to find the least core are very simple:

$$c = \{75 - x_M \leq \varepsilon, 85 - x_L \leq \varepsilon, 62 - x_C \leq \varepsilon,$$
$$200 - x_M - x_L \leq \varepsilon, 177 - x_M - x_C \leq \varepsilon, 187 - x_L - x_C \leq \varepsilon,$$
$$x_M + x_L + x_C = 302\}.$$

Then

$$\text{Minimize } [\{\varepsilon, c\}, \{x_M, x_L, x_C\}]$$

Of course, this can also be done tediously by hand. The result is

$$\varepsilon^1 = -\frac{40}{3}, C(\varepsilon^1) = \left\{ \left(\frac{305}{3}, \frac{335}{3}, \frac{266}{3} \right) \right\}$$

and that's one fair way to divide the total of 302.

(c) Find the Shapley allocation.

6.37.c Answer: We calculate the Shapley value from the table:

Order of arrival	Moe	Larry	Curly
MLC	75	125	102
MCL	75	125	102
LMC	115	85	102
LCM	115	85	102
CML	115	125	62
CLM	115	125	62
Shapley	101.66	111.66	88.66

The Shapley allocation is the same as the nucleolus. Looks like waxing is lucrative.

6.38 In Problems 6.12 and 6.29, we considered the problem in which three investors can earn a greater rate of return if they invest together and pool their money. Find the Shapley values in both cases discussed in those exercises.

6.38 Answer: For Problem 6.12, the characteristic function is $v(1) = 90$, $v(2) = 36$, $v(3) = 12$, $v(12) = 135$, $v(13) = 105$, $v(23) = 60$, $v(123) = 150$. The Shapley value is from the table

Order of arrival	1	2	3
123	90	45	15
132	90	45	15
213	99	36	15
231	90	36	24
312	93	45	12
321	90	48	12
Shapley	$\frac{584}{6}$	$\frac{287}{6}$	$\frac{152}{6}$

We have $x_1 = 92$, $x_2 = 42.5$, $x_3 = 15.5$.

For Problem 6.29, the characteristic function is

$$v(1) = 90, v(2) = 36, v(3) = 16,$$
$$v(12) = 135, v(13) = 110, v(23) = 65,$$
$$v(123) = 170.5.$$

The Shapley Value is determined from the table

Order of arrival	1	2	3
123	90	45	35.5
132	90	60.5	20
213	99	36	35.5
231	105.5	36	29
312	94	60.5	16
321	105.5	49	16
Shapley	$\frac{552}{6}$	$\frac{255}{6}$	$\frac{93}{6}$

The resulting Shapley value is $x_1 = 97.33$, $x_2 = 47.83$, $x_3 = 25.33$.

6.39 Three chiropractors, Moe, Larry, and Curly, are in partnership and cover hours for each other. At most one chiropractor needs to be in the office to see patients at any one time. They advertise their office hours as follows:

(1)	Moe	2:00–5:00
(2)	Larry	11:00–4:00
(3)	Curly	9:00–1:00

A coalition is an agreement by one or more doctors as to the times they will really be in the office to see everyone's patients. The characteristic function should be the amount of time **saved** by a given coalition.

Find the Shapley value using the table of order of arrival.

6.39 Answer: Take the characteristic function representing the hours saved by a coalition,

$$v(M) = v(C) = v(L) = 0, v(MC) = 0, v(ML) = 2, v(CL) = 2, v(MCL) = 4.$$

Then the nucleolus is $C(-1) = \{(1, 1, 2)\}$. Thus, Moe and Curly both save 1 hour while Larry saves 2. Since Curly arrives first, he can now work 9-12. Then Larry works 12-3, and Moe works 3-5.

The Shapley allocation is the same as least core. For example,

$$x_M = (v(M) - v(\emptyset))\frac{(3-1)!}{(1-1)!}3!$$
$$+ (v(MC) - v(C) + v(ML) - v(L))\frac{(3-2)!}{(2-1)!}3!$$
$$+ (v(MCL) - v(CL))\frac{(3-3)!}{(3-1)!}3!$$
$$= 0 + 2\frac{1}{6} + 2\frac{2}{6} = 1$$

Here is the table giving the Shapley allocation:

Order of arrival	Moe	Larry	Curly
MLC	0	2	2
MCL	0	4	0
LMC	2	0	2
LCM	2	0	2
CLM	2	2	0
CML	0	4	0
Shapley	1	2	1

6.40 In the figure, three locations are connected as in the network. The numbers on the branches represent the cost to move one unit along that branch. The goal is to connect each location to the main trunk and to each other in the most economical way possible. The benefit of being connected to the main trunk is shown next to each location. Branches can be traversed in both directions. Take the minimal cost path to the main trunk in all cases.

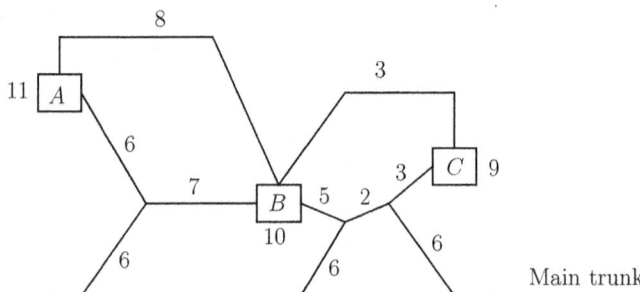

(a) Find an appropriate characteristic function and be sure it is superadditive.

6.40.a Answer: First the costs associated with connecting a coalition are as follows:

$$c(A) = 12, c(B) = 11, c(C) = 9;$$
$$c(AB) = 19, c(AC) = 20, c(BC) = 12, \quad c(ABC) = 20.$$

We take the characteristic function $v(S) = \sum_{i \in S} c(i) - c(S) + b(S)$, where $b(S)$ is the total benefit assigned to coalition S. We have

$$v(A) = 11, v(B) = 10, v(C) = 9,$$
$$v(AB) = 25, v(AC) = 21, v(BC) = 27,$$
$$v(ABC) = 42.$$

and it is easy to check that $v(S \cup T) \geq v(S) + v(T)$, $S \cap T = \emptyset$.

(b) Find the nucleolus.

6.40.b Answer: The first least core is $X^1 = C(-2)$ and we can eliminate coalitions $S = 1 = A$, $S = 23 = BC$. In addition $x_1 = 13 = AC$ and only x_2, x_3 are left to be determined.

The second least core is $X^2 = C(-4) = \{(13, 16, 13)\}$ and hence the nucleolus is $\vec{x} = (13, 16, 13)$.

(c) Find the Shapley value.

6.40.c Answer: The Shapley value is determined from the table with $A = 1, B = 2, C = 3$.

Order of arrival	1	2	3
123	11	14	17
132	11	21	10
213	15	10	17
231	15	10	17
312	12	21	9
321	15	18	9
Shapley	$\frac{79}{6}$	$\frac{94}{6}$	$\frac{79}{6}$

6.41 Suppose that the seller of an object (which is worthless to the seller) has two potential buyers who are willing to pay $100 or $130, respectively.

(a) Find the characteristic function and then the core and the Shapley value. Show that the Shapley value is **not** in the core.

6.41.a Answer: The characteristic function is $v(i) = 0, v(12) = 100, v(13) = 130,$ $v(23) = 0, v(123) = 130$. We calculate the core,

$$C(0) = \{x_i \geq 0, i = 1, 2, 3, 100 \leq x_1 + x_2, 130 \leq x_1 + x_3, 0 \leq x_2 + x_3,$$
$$x_1 + x_2 + x_3 = 130\}$$
$$= \{x_2 = 0, x_1 + x_3 = 130, x_1 \geq 0, x_2 \geq 0\}.$$

The Shapley value is $\vec{x} = \{(\frac{245}{3}, \frac{50}{3}, \frac{95}{3})\}$, and this point is not in $C(0)$ since $x_2 = \frac{50}{3} > 0$ and

$$x_1 + x_3 = \frac{340}{3} \neq 130.$$

The nucleolus of this game is $\{(115, 0, 15)\}$ which is quite different from the Shapley value. Under the nucleolus, of the 130 available if the players cooperate, the seller gets 115 for her object, and player 3 gets to keep 15 of her 130. She pays the seller 115. Under Shapley, the seller gets 81.67, player 3 pays a total of $130 - 31.67 = 98.33$ of which 81.67 goes to the seller to buy the object and 16.67 goes to player 2 to go away. Do you think this is what would actually happen?

(b) Show that the Shapley value is individually and group rational.

6.41.b Answer: To be individually rational we need $v(i) \leq x_i, i = 1, 2, 3$, which is clearly satisfied for the Shapley value. The group rational condition requires $\sum_{i=1}^{3} x_i = v(N)$. In our case,

$$x_1 + x_2 + x_3 = \frac{245 + 50 + 95}{3} = 130 = v(N).$$

6.42 Consider the characteristic function in Example 6.16 for the creditor–debtor problem. By writing out the table of the order of arrival of each player versus the benefit the player brings to a coalition when the player arrives as in Example 6.17, calculate the Shapley value.

6.42 Answer:

	Player A	Player B	Player C
ABC	25	65	10
ACB	25	65	10
BAC	50	40	10
BCA	50	40	10
CAB	35	65	0
CBA	50	50	0
Total	235	325	40

This gives the Shapley value $x_A = 39.17, x_B = 54.17, x_C = 6.67$.

6.43 Find the Shapley allocation for the three-person characteristic function game with

$$v(i) = 0, i = 1, 2, 3,$$
$$v(12) = v(13) = 2, v(23) = 10,$$
$$v(123) = 12.$$

6.43 Answer: Shapley value $= (\frac{4}{3}, \frac{16}{3}, \frac{16}{3})$.

6.44 Once again we consider the four doctors, Moe (4), Larry (3), Curly (2), and Shemp (1), and their problem to minimize the amount of hours they work as in Problem 6.31. The characteristic function is the number of hours saved by a coalition. We have $v(i) = 0$ and

$$v(12) = 4, v(13) = 4, v(14) = 3, v(23) = 6, v(24) = 2, v(34) = 2,$$
$$v(123) = 10, v(124) = 7, v(134) = 7, v(234) = 8, v(1234) = 13.$$

Find the Shapley allocation.

6.44 Answer: We may either do this by writing out the table of order of arrival or by using the formulas. The table with four players will have $4! = 24$ rows, so we use the formula instead.

To calculate the savings for Curly, we have $\Pi^2 = \{2, 12, 23, 24, 123, 124, 234, 1234\}$ and

$$
\begin{aligned}
x_2 &= \sum_{S \in \Pi^2} \frac{(|S| - 1)!(4 - |S|)!}{24} (v(S) - v(S - 2)) \\
&= \frac{2}{24}[4 + 6 + 2] + \frac{2}{24}[6 + 4 + 6] + \frac{1}{4}[6] \\
&= \frac{23}{6}.
\end{aligned}
$$

The other allocations are similar.

The Shapley value $= \{\frac{10}{3}, \frac{23}{6}, \frac{23}{6}, 2\}$. For comparison, the nucleolus was found to be $X^3 = \{(\frac{13}{4}, \frac{33}{8}, \frac{33}{8}, \frac{3}{2})\}$.

The hours of work for each player using the Shapley value are as follows:

1. Since Curly (2) starts at 9 AM, and works 7 hours, he saves 3.83 hours and works 3.166 hours. That means he works from 9 AM to 12:10 PM.
2. Larry (3) comes in after Curly. He saves 3.83 hours from the 6 he worked before cooperation and so he works 2.166 hours. Thus, he starts at 12:10 PM and leaves at 2:20 PM.
3. Shemp (1) comes in after Larry. He saves 3.33 hours from the 5 he worked before. He must work 1.66 hours, which means he starts at 2:20 PM and leaves at 4 PM.
4. Moe (4) is the last to arrive. He saves 2 hours from the 3 hours he worked. He arrives at 4 PM and leaves at 5 PM, closing the office and turning out the lights.

6.45 Garbage Game. Suppose that there are four property owners each with one bag of garbage that needs to be dumped on somebody's property (one of the four). Find the Shapley value for the garbage game with $v(N) = -4$ and $v(S) = -(4 - |S|)$.

6.45 Answer: The Shapley value for the game is $(-1, -1, -1, -1)$. To see why, since the game is symmetric clearly the formula for the Shapley value will give $x_1 = x_2 = x_3 = x_4$. Since $x_1 + x_2 + x_3 + x_4 = v(N) = -4$, we must have $x_1 = x_2 = x_3 = x_4 = -1$.

6.46 A farmer (player 1) owns some land that he values at \$100K. A speculator (player 2) feels that if she buys the land, she can subdivide it into lots and sell the lots for a total of \$150K. A home developer (player 3) thinks that he can develop the land and build homes that he can sell. So the land to the developer is worth \$160K.

(a) Find the characteristic function and the Shapley allocation.

6.46.a Answer: The characteristic function is $v(1) = 100$, $v(2) = v(3) = 0$, $v(12) = 150$, $v(13) = 160$, $v(23) = 0$, $v(123) = 160$.

The Shapley allocation is $\{(\frac{415}{3}, \frac{25}{3}, \frac{40}{3})\}$.

(b) Compare the Shapley allocation with the nucleolus allocation.

6.46.b Answer: The nucleolus is $\{(155, 0, 5)\}$. This reflects the fact that player 2 has no power at all because she is not the high bidder. The Shapley allocation grants player 2 the amount $\frac{25}{3}$ to reflect the fact that without player 2, player 1 has less negotiating power.

6.47 Find the Shapley allocation for the cost game in Problem 6.33.

6.47 Answer: Shapley value = $\{(3.91667, 3.66667, 2.75, 3.16667)\}$.

6.48 Consider the five player game with $v(S) = 1$ if $1 \in S$ and $|S| \geq 2$, $v(S) = 1$ if $|S| \geq 4$ and $v(S) = 0$ otherwise. Player 1 is called Mr BIG, and the others are called Peons. Find the Shapley value of this game. (Hint: Use symmetry to simplify.)

6.48 Answer: By symmetry, we should have the Shapley allocation $\vec{x} = (x_1, y, y, y, y)$ since all the peons are equal. Then since $v(12345) = 1$ and $x_1 + 4y = v(12345) = 1$, we only need to find x_1.

There are $2^{5-1} = 16$ coalitions that contain player 1. We have 4 two-player coalitions, 6 three player coalitions, 4 four-player coalitions, and

$$v(S) - v(S - 1) = \begin{cases} 0, & \text{if } S = 1; \\ 1, & \text{if } |S| = 2; \\ 1, & \text{if } |S| = 3; \\ 1, & \text{if } |S| = 4; \\ 0, & \text{if } S = N. \end{cases}$$

Thus, by the formula for the Shapley value,

$$\begin{aligned} x_1 &= \sum_{S \in \Pi^1} [v(S) - v(S - 1)] \frac{(|S| - 1)!(5 - |S|)!}{5!} \\ &= \frac{(2 - 1)!(5 - 2)!}{5!} \times 4 + \frac{(3 - 1)!(5 - 3)!}{5!} \times 6 + \frac{(|4| - 1)!(5 - |4|)!}{5!} \times 4 \\ &= \frac{1}{5} + \frac{1}{5} + \frac{1}{5} = \frac{3}{5}. \end{aligned}$$

Then $y = \frac{1 - x_1}{4} = \frac{1}{10}$. The Shapley value is then $\vec{x} = (\frac{3}{5}, \frac{1}{10}, \frac{1}{10}, \frac{1}{10}, \frac{1}{10})$.

6.49 Consider the glove game in which there are four players; player 1 can supply one left glove and players 2, 3, and 4 can supply one right glove each.

(a) Find $v(S)$ for this game if the value of a coalition is the number of paired gloves in the coalition.

6.49.a Answer: The characteristic function is

$$v(i) = 0, i = 1, 2, 3, 4, v(12) = v(13) = v(14) = 1, v(23) = v(24) = v(34) = 0,$$
$$v(123) = v(124) = v(134) = 2, v(234) = 0, v(1234) = 2.$$

(b) Find $C(0)$.

6.49.b Answer: The core is $C(0) = \{(2, 0, 0, 0)\}$. To see this,

$$C(0) = \{x_i \geq 0, i = 1, 2, 3, 1 \leq x_1 + x_2, 1 \leq x_1 + x_3, 1 \leq x_1 + x_4$$
$$2 \leq x_1 + x_2 + x_3, 2 \leq x_1 + x_2 + x_4, 2 \leq x_1 + x_3 + x_4,$$
$$x_1 + x_2 + x_3 + x_4 = 2\}.$$

Adding the 3-player coalitions, we see that

$$3x_1 + 2x_2 + 2x_3 + 2x_4 \geq 6 \Rightarrow x_1 \geq 2.$$

Then $x_1 + x_2 + x_3 + x_4 = 2 \Rightarrow x_1 = 2, x_2 = x_3 = x_4 = 0$. Thus $C(0) = \{(2, 0, 0, 0)\}$.

Since the core contains only one point, it is the nucleolus. Player 1 is allocated everything while players 2, 3 and 4 get nothing. Doesn't seem fair.

(c) What is the Shapley allocation?

6.49.c Answer: Players 2, 3, and 4 are symmetric so the Shapley allocation is of the form (x, y, y, y), where $x + 3y = 2$. By the formula for the Shapley value,

$$x = \sum_{S \in \Pi^1} [v(S) - v(S - 1)] \frac{(|S| - 1)!(4 - |S|)!}{4!}$$
$$= \frac{(2 - 1)!(4 - 2)!}{4!} \times 3 + \frac{(3 - 1)!(4 - 3)!}{4!} \times 6 + \frac{(|4| - 1)!(4 - |4|)!}{4!} \times 2$$
$$= \frac{1}{4} + \frac{1}{2} + \frac{1}{4} = \frac{5}{4}.$$

Then $y = \frac{2-x}{3} = \frac{1}{4}$. The Shapley value is then $\vec{x} = (\frac{5}{4}, \frac{1}{4}, \frac{1}{4}, \frac{1}{4})$.

(d) Now we change the problem a bit. Suppose there are two players, each with three gloves. Player 1 has two left-hand gloves and one right-hand glove; player 2 has two right-hand gloves and one left-hand glove. They can sell a pair of gloves for 10 dollars. How should they split the proceeds. Determine the nucleolus and the Shapley allocation.

6.49.d Answer: The characteristic function is $v(1) = v(2) = 10, v(12) = 30$. The core is

$$C(0) = \{x_1 \geq 10, x_2 \geq 10, x_1 + x_2 = 30\},$$

which has more than one point. That means we have to calculate the least core.

$$C(\varepsilon) = \{x_1 \geq 10 - \varepsilon, x_2 \geq 10 - \varepsilon, x_1 + x_2 = 30\}$$
$$= \{30 = x_1 + x_2 \geq 20 - 2\varepsilon, x_1 \geq 10 - \varepsilon, x_2 \geq 10 - \varepsilon\}.$$

Thus $\varepsilon \geq -5 \Rightarrow \varepsilon^1 = -5 \Rightarrow x_1 = x_2 = 15$ and $C(-5) = \{(15, 15)\}$, an even split. The Shapley allocation is

$$x_i = \frac{1}{2}[10 - 0] + \frac{1}{2}[30 - 10] = 15, \quad i = 1, 2,$$

which is the same as the nucleolus.

6.50 In Problem 6.25, we considered that there are three types of planes (1, 2, and 3) that use an airport. Plane 1 needs a 100-yard runway, 2 needs a 150-yard runway, and 3 needs a 400-yard runway. The cost of maintaining a runway is equal to its length. Suppose this airport has one 400-yard runway used by all three types of planes and assume that for the day under study, only one plane of each type lands at the airport. We want to know how much of the $400 cost should be allocated to each plane. We showed that the nucleolus is $C(\varepsilon = -50) = \{x_1 = x_2 = -50, x_3 = -300\}$.

(a) Find the Shapley cost allocation.

6.50.a Answer: The Shapley cost allocation is $x_1 = -\frac{200}{6}; x_2 = -\frac{350}{6}; x_3 = -\frac{1850}{6}$. This comes from the table

Order of arrival	Player 1	Player 2	Player 3
123	−100	−50	−250
132	−100	0	−300
213	0	−150	−250
231	0	−150	−250
312	0	0	−400
321	0	0	−400
Total	−200	−350	−1850

(b) Calculate the excesses for both the nucleolus and Shapley allocations.

6.50.b Answer: The excesses $e(S, \vec{x}) = v(S) - \vec{x}(S)$ are in the table.

Coalition	Excess Shapley	Excess nucleolus
1	−66.67	−50
2	−91.67	−100
3	−91.67	−100
12	−58.33	−50
13	−58.33	−50
23	−33.33	−50

Under Shapley the planes pay more in three cases and less in three cases than under the nucleolus, and by the same amounts.

6.51 A river has n pollution-producing factories dumping water into the river. Assume that the factory does not have to pay for the water it uses, but it may need to expend money to clean the water before it can use it. Assume the cost of a factory to clean polluted water before it can be used is proportional to the number of polluting factories. Let $c = $ cost per factory. Assume also that a factory may choose to clean the water it dumps into the river at a cost of b per factory. We take the inequalities $0 < c < b < nc$. The characteristic function is

$$v(S) = \begin{cases} \max\{|S|(-nc), |S|(-(n - |S|)c) - |S|b\}, & \text{if } S \subset N; \\ \max\{-n^2c, -nb\}, & \text{if } S = N. \end{cases}$$

Take $n = 5, b = 3, c = 2$. Find the Shapley allocation.

6.51 Answer: If $n = 5, b = 3, c = 2$, we have

$$v(S) = \begin{cases} \max\{|S|(-10), |S|(-(5 - |S|)2) - |S|3\}, & \text{if } S \subset N; \\ -15, & \text{if } S = N. \end{cases}$$

A direct calculation shows that

$$v(S) = \begin{cases} -10, & \text{if } |S| = 1; \\ -18, & \text{if } |S| = 2; \\ -21, & \text{if } |S| = 3; \\ -20, & \text{if } |S| = 4; \\ -15, & \text{if } S = N, |N| = 5. \end{cases}$$

Plugging into Shapley's formula gives $\vec{x} = (-3, -3, -3, -3, -3)$.

6.52 Suppose we have a game with characteristic function v that satisfies the property $v(S) + v(N - S) = v(N)$ for all coalitions $S \subset N$. These are called constant sum games.

(a) Show that for a two-person constant sum game, the nucleolus and the Shapley value are the same.

6.52.a Answer: Calculate the Shapley value from

Order of arrival	1	1
12	$v(1)$	$v(12) - v(1) = v(2)$
21	$v(12) - v(2) = v(1)$	$v(2)$
Shapley	$v(1)$	$v(2)$

and hence Shapley is $\vec{s} = (v(1), v(2))$.

Next, the least core is

$$C(\varepsilon) = \{v(1) - x_1 \le \varepsilon, v(2) - x_2 \le \varepsilon, x_1 + x_2 = v(12) = v(1) + v(2)\}.$$

But $v(1) + v(2) - 2\varepsilon \le x_1 + x_2 = v(1) + v(2) \Rightarrow \varepsilon \ge 0 \Rightarrow \varepsilon^1 = 0$ is the first ε that makes $C(\varepsilon) \ne \emptyset$. Then, if $\varepsilon = 0, x_1 - v(1) + x_2 - v(2) = 0$ implies $x_1 = v(1)$, and $x_2 = v(2)$. Thus $X^1 = C(0) = \{(v(1), v(2))\}$, the same as Shapley.

(b) Show that the nucleolus and the Shapley value are the same for a three person constant sum game.

6.52.b Answer: Let subscripts denote the coalition in v. The Shapley value is obtained from

Order of arrival	1	2	3
123	v_1	$v_{12} - v_1$	$v_{123} - v_{12} = v_3$
132	v_1	$v_{123} - v_{13} = v_2$	$v_{13} - v_1$
213	$v_{12} - v_2$	v_2	$v_{123} - v_{12} = v_3$
231	$v_{123} - v_{23} = v_1$	v_2	$v_{23} - v_2$
312	$v_{13} - v_3$	$v_{123} - v_{13} = v_2$	v_3
321	$v_{123} - v_{23} = v_1$	$v_{23} - v_3$	v_3
Shapley	$s_1 = \frac{4v_1 + v_{12} - v_2 + v_{13} - v_3}{6}$	$s_2 = \frac{4v_2 + v_{12} - v_1 + v_{23} - v_3}{6}$	$s_3 = \frac{4v_3 + v_{13} - v_1 + v_{23} - v_2}{6}$

Next, we need to calculate the nucleolus. We begin by normalizing the characteristic function to get $v'(i) = 0, i = 1, 2, 3, v'(123) = 1$, and, for example, with $S = 12$

$$v'(12) = \frac{v(12) - \sum_{k=1,2} v(k)}{v(N) - \sum_{k=1}^{3} v(k)} = \frac{v(123) - v(3) - v(1) - v(2)}{v(N) - \sum_{k=1}^{3} v(k)} = 1.$$

Hence, $v'(12) = v'(13) = v'(23) = 1$. But then it is easy to see directly, or using Proposition F.7 since $v'(12) + v'(13) + v'(23) = 3 > 2$, that $C(0) = \emptyset$.

Since the core is empty we may apply Theorem F.8 that gives a formula for the nucleolus. We have from Equation (F.1)

$$x_1' = \frac{v'(123) + v'(12) + v'(13) - 2v'(23)}{3} = \frac{1}{3}, x_2' = \frac{1}{3}, x_3' = \frac{1}{3}.$$

Now the unnormalized allocation in the nucleolus is

$$x_1 = x_1'\left(v(123) - \sum_{i=1}^{3} v(i)\right) + v(1) = \frac{v(123) - v(1) - v(2) - v(3) + 3v(1)}{3}.$$

Now note that the Shapley allocation is

$$s_1 = \frac{4v_1 + v_{12} - v_2 + v_{13} - v_3}{6} = \frac{4v_1 + v_{123} - v_3 - v_2 + v_{123} - v_2 - v_3}{6} = x_1.$$

Similarly, $s_2 = x_2, s_3 = x_3$ and hence the nucleolus and the Shapley allocation are the same.

(c) Check the result of the preceding part if $v(i) = i, i = 1, 2, 3, v(12) = 5$, $v(13) = 6$, $v(23) = 7$ and $v(123) = 8$.

6.52.c Answer: Direct computation shows that the nucleolus and Shapley value are both $(\frac{5}{3}, \frac{8}{3}, \frac{11}{3})$.

6.53 Three plumbers, Moe Howard (1), Larry Fine (2), and Curly Howard (3), work at the same company and at the same times. Their houses are located as in the figure and they would like to carpool to save money.[1] Once they reach the expressway, the distance to the company is 12 miles. They drive identical Chevy Corvettes, so each has the identical cost of driving to work of 1 dollar per mile. Assume that for any coalition, the route taken is always the shortest distance to pick up the passengers (doubling back may be necessary to pick someone up). It doesn't matter whose car is used. Only the direction from home to work is to be considered. Shemp is not to be considered in this part of the problem.

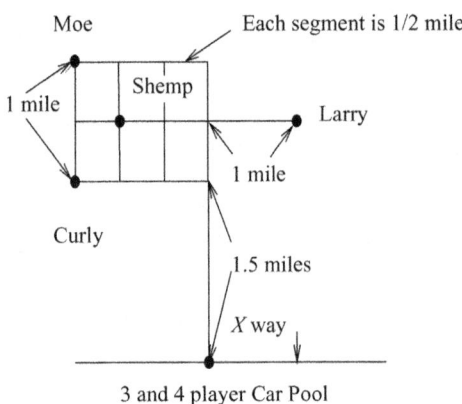

3 and 4 player Car Pool

Moe, Larry, and Curly Carpool to work.

(a) By considering $v(S) = \sum_{i \in S} c(i) - c(S)$, where $c(S)$ is the cost of driving to work if S forms a carpool, what is the characteristic function of this game.

6.53.a Answer: Since

$$c(1) = 16, c(2) = 15, c(3) = 15,$$
$$c(12) = 18, c(13) = 16, c(23) = 18, c(123) = 19,$$

we compute

$$v(i) = 0, v(12) = 13, v(13) = 15, v(23) = 12, v(123) = 27.$$

This characteristic function is superadditive and they all have an incentive to join the carpool.

(b) Find the least core.

6.53.b Answer: The inequalities we need to solve are as follows:

$$C(\varepsilon) = \{0 \le x_i \le -\varepsilon, 13 - \varepsilon \le x_1 + x_2, 15 - \varepsilon \le x_1 + x_3, 12 - \varepsilon \le x_2 + x_3$$
$$x_1 + x_2 + x_3 = 27\}.$$

Eliminating x_3 we get $x_2 \le 12 + \varepsilon, x_1 \le 15 + \varepsilon \Rightarrow 13 - \varepsilon \le x_1 + x_2 \le 27 + 2\varepsilon$. This leads to $\varepsilon^1 = -\frac{14}{3}$ and then $x_1 + x_2 = \frac{53}{3}, x_3 = \frac{28}{3}$.

---

[1] Based on a similar problem due to Mesterton–Gibbons.

Then $x_2 \leq 12 + \varepsilon^1 = \frac{22}{3}$, $x_1 \leq 15 + \varepsilon^1 = \frac{31}{3}$. Since $31 + 22 = 53$, we conclude $x_2 = \frac{22}{3}$, $x_1 = \frac{31}{3}$. Thus,

$$X^1 = C\left(-\frac{14}{3}\right) = \left\{\left(\frac{31}{3}, \frac{22}{3}, \frac{28}{3}\right)\right\}.$$

(c) Assuming that Moe is the driver, how much should Larry and Curly pay Moe if they all carpool?

6.53.c Answer: It seems reasonable that Larry and Curly should pay Moe the difference in the amount it costs them to drive on their own and the amount of the savings attributable to that player. We get Larry should pay Moe $c(2) - \frac{22}{3} = 15 - \frac{22}{3} = 7.66$ and Curly should pay Moe $c(3) - \frac{28}{3} = 5.66$.

(d) Find the Shapley allocation of savings for each player.

6.53.d Answer: The table for calculation of the Shapley value is given below.

Arrival/player	1	2	3
123	0	13	14
132	0	12	15
213	13	0	14
231	15	0	12
312	15	12	0
321	15	12	0
Total	58	49	55

The Shapley allocation is therefore $x_1 = \frac{58}{6}$, $x_2 = \frac{49}{6}$, $x_3 = \frac{55}{6}$. We get Larry should pay Moe $c(2) - \frac{49}{6} = 6.83$ and Curly should pay Moe $c(3) - \frac{55}{6} = 5.83$.

(e) The problem is the same as before but now Shemp Howard (4) wants to join the plumbers Howard, Fine and Howard at the company. Answer all of the questions posed above.

6.53.e Answer: Let Shemp be player 4. The costs are as follows:

$c(1) = 16, c(2) = 15, c(3) = 15, c(4) = 15,$
$c(12) = 18, c(13) = 16, c(14) = 16, c(23) = 18, c(24) = 17, c(34) = 16,$
$c(123) = 19, c(124) = 18, c(134) = 17, c(234) = 18,$
$c(1234) = 19.$

The corresponding cost savings are as follows:

$v(i) = 0, i = 1, 2, 3, 4,$
$v(12) = 13, v(13) = 15, v(14) = 15, v(23) = 12, v(24) = 13, v(34) = 14,$
$v(123) = 27, v(124) = 28, v(134) = 29, v(234) = 27,$
$v(1234) = 42.$

By going through similar calculations, we get

$$\text{Nucleolus} = C\left(-\frac{15}{4}\right) = \left\{\left(x_1 = \frac{45}{4}, x_2 = \frac{37}{4}, x_3 = \frac{41}{4}, x_4 = \frac{45}{4}\right)\right\}.$$

The Shapley value is

$$x_1 = \frac{133}{12}, x_2 = \frac{115}{12}, x_3 = \frac{125}{12}, x_4 = \frac{131}{12}.$$

Larry, Curly, and Shemp should pay Moe the difference in the amount it costs them to drive on their own and the amount of the savings attributable to that player:

1. Larry should pay Moe $c(2) - \frac{115}{12} = 5.42$ under Shapley and $c(2) - \frac{37}{4} = 5.75$ under nucleolus.
2. Curly should pay Moe $c(3) - \frac{125}{12} = 4.58$ under Shapley and $c(3) - \frac{41}{4} = 4.75$ under nucleolus.
3. Shemp should pay Moe $c(4) - \frac{131}{12} = 4.08$ under Shapley and $c(4) - \frac{45}{4} = 3.75$ under nucleolus.

6.54 An alternative to the Shapley–Shubik index is called the **Banzhaf–Coleman index**, which is the imputation $\vec{b} = (b_1, b_2, \ldots, b_n)$ defined by

$$b_i = \frac{|W^i|}{|W^1| + |W^2| + \cdots + |W^n|}, \quad i = 1, 2, \ldots, n,$$

where W^i is the set of coalitions that win with player i and lose without player i. We also refer to W^i as the coalitions for which player i is **critical** for passing a resolution. It is primarily used in **weighted voting systems** in which player i has w_i votes, and a resolution passes if $\sum_j w_j \geq q$, where q is known as the **quota**.

(a) Why is this a reasonable definition as an index of power.

6.54.a Answer: For player i,

$$b_i = \frac{|W^i|}{|W^1| + |W^2| + \cdots + |W^n|}$$

is the ratio of the number of times player i is critical to a resolution passing, to the total number of critical coalitions for the entire game. In other words, b_i is the fraction of all the critical votes provided by player i. That's why it's reasonable.

(b) Consider the four-player-weighted voting system in which a resolution passes if it receives at least 10 votes. Player 1 has 6 votes, player 2 has 5 votes, player 3 has 4 votes, and player 1 has 2 votes. Find the Shapley-Shubik index.

6.54.b Answer: The winning coalitions (winning with the player and losing without the player) for each player are as follows:

$$W^1 = \{12, 13, 123, 124, 134\},$$
$$W^2 = \{12, 124, 234\},$$
$$W^3 = \{13, 134, 234\},$$
$$W^4 = \{234\}.$$

Thus,

$$x_1 = \sum_{S \in W^1} \frac{(|S| - 1)!(n - |S|)!}{n!} = 2 \times \frac{2}{4!} + 3 \times \frac{2}{4!} = \frac{5}{12},$$

$$x_2 = 1 \times \frac{1}{12} + 2 \times \frac{1}{12} = \frac{3}{12},$$

$$x_3 = \frac{3}{12},$$

$$x_4 = \frac{1}{12}.$$

(c) Find the Banzhaf–Coleman index.

6.54.c Answer: We have the Banzhaf–Coleman index:

$$b_i = \frac{|W^i|}{|W^1| + |W^2| + \cdots + |W^n|}, i = 1, 2, \ldots, n.$$

Calculating, with $|W^1| = 5, |W^2| = 3, |W^3| = 3, |W^4| = 1,$

$$b_1 = \frac{5}{12}, b_2 = \frac{3}{12}, b_3 = \frac{3}{12}, b_4 = \frac{1}{12}.$$

This is the same as the Shapley–Shubik index.

6.55 Suppose a game has four players with votes 4, 2, 1, 1, respectively, for each player $i = 1, 2, 3, 4$. The quota is $q = 5$. Show that the Banzhaf–Coleman index for player 1 is more than twice the index for player 2 even though player 2 has exactly half the votes player 1 does.

6.55 Answer: One simple way to calculate the Banzhaf–Coleman index is to list for each coalition, the players who are critical for that coalition (critical means win with the player, lose without her). The table contains the following results:

Coalition	Votes	Critical players
12	6	1, 2
13	5	1, 3
14	5	1, 4
123	7	1
124	7	1
134	6	1
1234	8	1

Next, count up the number of times each player is critical for a coalition. For player 1 it is 7 times; player 2, 3, and 4, exactly 1 time each. The total number of winning coalitions for each critical player is 10. Hence,

$$b_1 = \frac{7}{10}, b_2 = b_3 = b_4 = \frac{1}{10}.$$

Player 1's power index is 7 times that of the other players even though she has only twice as many votes (as player 2).

6.56 The Senate of the 112th Congress has 100 members of whom 53 are Democrats and 47 are Republicans. Assume that there are three types of Democrats and three types of Republicans–Liberals, Moderates, Conservatives. Assume that these types vote as a block. For the Democrats, Liberals (1) have 20 votes, Moderates (2) have 25 votes, Conservatives (3) have 8 votes. Also, for Republicans, Liberals (4) have 2 votes, Moderates (5) have 15 votes, and Conservatives (6) have 30 votes. A resolution requires 60 votes to pass.

(a) Find the Shapley–Shubik index and the total power of the Republicans and Democrats.

6.56.a Answer: A straightforward computation using the Shapley formulas gives

$$x_1 = 21.67\%, \; x_2 = 25\%, \; x_3 = 5\%, \; x_4 = 1.67\%, \; x_5 = 16.67\%, \; x_6 = 30\%.$$

The total Democratic power is $x_1 + x_2 + x_3 = 51.67\%$ and Republican power is $x_4 + x_5 + x_6 = 48.33\%$.

(b) Find the Banzhaf–Coleman index.

6.56.b Answer: The Banzhaf–Coleman index is

$$b_1 = 21.15\%, \; b_2 = 25\%, \; b_3 = 5.77\%, \; b_4 = 1.92\%, \; b_5 = 17.31\%, \; b_6 = 28.85\%.$$

The total Democratic power is $b_1 + b_2 + b_3 = 51.92\%$ and Republican power is $b_4 + b_5 + b_6 = 48.08\%$.

Now this is a lot of work to do by hand, so here is the Maple code to do this. You can easily translate it into Mathematica.

```
>S:={1,2,3,4,5,6}:
>with(combinat):
>L:=powerset(S) minus {{}} :M:=convert(L list):
>K:=nops(M):P:=nops(S):
>v[1]:=20:v[2]:=25:v[3]:=8:v[4]:=2:v[5]:=15:v[6]:=30:
>q:=60:

>for k from 1 to K do
    tot[k] := 0;
    for i in M[k] do
    tot[k] := tot[k]+v[i]
  end do;
  tot[k]
  end do;

>for k from 1 to K do
    for i from 1 to P do
    crit[i, k] := 0:
    if tot[k] >= q and tot[k]-v[i] < q then
    crit[i, k] := 1
```

```
      end if
      end do:
   end do:

> for i from 1 to P do
      for k from 1 to K do
        if crit[i, k] > 0, and member(i, M[k]) then
            lprint(M[k], v[i], i, tot[k], crit[i, k]) ;
        end if
      end do;
   end do;

> for i from 1 to P do
      c[i] := 0:
      for k from 1 to K do
        if crit[i, k] > 0 and member(i, M[k]) then
        c[i] := c[i]+1;
        end if
      end do;
   lprint(c[i]);
end do;

> for i from 1 to P do
      b[i] := evalf(c[i]/add(c[i], i = 1 .. P))
   end do;
```

The result of this is that it finds all the wining coalitions for each player and calculates the Banzhaf–Coleman index and gives

$$b_1 = 0.2115, b_2 = 0.25, b_3 = 0.057, b_4 = 0.019, b_5 = 0.173, b_6 = 0.288.$$

(c) What happens if the Republican Moderate votes becomes 1, while the Republican Conservative votes becomes 44.

6.56.c Answer: The Shapley–Shubik index in this case is

$$x_1 = 16.67\%, x_2 = 16.67\%, x_3 = 0\%, x_4 = 0\%, x_5 = 0\%, x_6 = 66.67\%.$$

The total Democratic power is $x_1 + x_2 + x_3 = 33.34\%$ and Republican power is $x_4 + x_5 + x_6 = 66.67\%$.

The Banzhaf–Coleman index is found by running the code in the previous part (or doing it by hand) with votes

$$v[1] = 20, v[2] = 25, v[3] = 8, v[4] = 2, v[5] = 1, v[6] = 44.$$

We get the result

$$b_1 = 20\%, b_2 = 20\%, b_3 = 0\%, b_4 = 0\%, b_5 = 0\%, b_6 = 60\%.$$

The total Democratic power is $b_1 + b_2 + b_3 = 40\%$ and Republican power is $b_4 + b_5 + b_6 = 60\%$, even though the Democrats have a total of 53 votes.

In both indices, the Republicans, a minority in the Senate, have dominant control due to the conservative bloc. The conservatives in the Democratic party, and the moderates and liberals in the Republican party have no power at all.

6.4 Bargaining

Problems

6.57 A British game show involves a final prize. The two contestants may each choose either to Split the prize, or Claim the prize. If they Split the prize, they each get $\frac{1}{2}$; if they each Claim the prize, they each get 0. If one player Splits, and the other player Claims, the player who Claims the prize gets 1 and the other player gets 0. The game matrix is

$$\begin{bmatrix} (\frac{1}{2}, \frac{1}{2}) & (0, 1) \\ (1, 0) & (0, 0) \end{bmatrix}.$$

(a) Find the Nash bargaining solution without threats.

6.57.a Answer: The matrices are $A = \begin{bmatrix} \frac{1}{2} & 0 \\ 1 & 0 \end{bmatrix}$, $B^T = A$. Calculate easily that $value(A) = value(B^T) = 0$ and hence the security point is $(0, 0)$. The Nash bargaining problem is then

$$\text{Maximize } uv, \quad \text{subject to } u + v \leq 1, 0 \leq u \leq 1, 0 \leq v \leq 1.$$

By calculus, the solution is $\overline{u} = \frac{1}{2}, \overline{v} = \frac{1}{2}$. The bargained solution is that each contestant should Split. That seems fair and natural.

(b) Apparently, each player has a credible threat to always Claim the prize. Find the optimal threat strategies and the Nash solution as well as the combination of pure strategies that the players should agree to in the threat game.

6.57.b Answer: First, we calculate the threat security point. Since the Pareto-optimal boundary is $u + v = 1$, we have $v = -u + 1$ and so $m_p = -1, b = 1$. Then we calculate

$$-m_p A - B = A - B = \begin{bmatrix} 0 & -1 \\ 1 & 0 \end{bmatrix}, \text{ and } value(A - B) = 0, \text{ with a saddle point}$$

$X_t = (0, 1), Y_t = (0, 1)$. The threat security point is, therefore, $u^t = X_t A Y_t^T = 0$ and $v^t = X_t B Y_t^T = 0$, exactly the same as the security point we found before. Immediately we see that the threat bargaining solution is

$$\overline{u} = \frac{1}{-2m_p}(b + value(-m_p A - B)) = \frac{1}{2}, \overline{v} = \frac{1}{2}(b - value(-m_p A - B)) = \frac{1}{2},$$

the same solution as before. This means that in order to achieve the threat bargained solution, they should agree to always play row 1, column 1, or if they play many times, half the time they should play row 2, column 1, and half the time row 1, column 2. The expected payoffs are the same.

(c) If the game will only be played one time and one player announces that he will definitely Claim the prize and then split the winnings after the show is over, what must the other player do?

6.57.c Answer: If player 1 announces he will Claim the prize and split it after the show is over, player 2 has two choices.

  1. Player 2 can believe that player 1 will actually split the prize. In this case player 2 and player 1 both receive $\frac{1}{2}$.

  2. Player 2 does not believe player 1 will split the prize claimed. In that case, player 2 should threaten to claim the prize as well if player 1 does not agree to Split. If player 1 does not agree to Split, player 2 should carry out the threat. Either way, since player 2 doesn't believe player 1, player 2 ends up with either 0 or $\frac{1}{2}$, but if 0, then player 1 also gets 0, not 1.

6.58 Find the solution to the Nash bargaining problem for the game

$$\begin{bmatrix} (1,4) & (-1,-4) \\ (-4,-1) & (4,1) \end{bmatrix}.$$

6.58 Answer: The Pareto-optimal boundary will be $P = \{(u,v) \mid u + v = 5, 1 \le u \le 4\}$. We have then that $a = 1, b = 5$ and we consider the zero sum game with matrix $A - B$. The optimal threat strategies are the saddle point for this game. We calculate $X_t = (x, 1 - x), x = 0, 1, Y_t = (1, 0)$, and $v(A - B) = -3$. Hence, the optimal payoffs are as follows:

$$u^t = \frac{1}{2}(5 - 3) = 1, \qquad v^t = \frac{1}{2}(5 - (-3)) = 4.$$

Note that u^t is an endpoint of $[1, 4]$.

6.59 Find the Nash bargaining solution, the threat solution, and the KS solution to the battle of the sexes game with matrix

$$\begin{bmatrix} (4,2) & (2,-1) \\ (-1,2) & (2,4) \end{bmatrix}.$$

Compare the solutions with the solution obtained using the characteristic function approach.

6.59 Answer: The matrices are $A = \begin{bmatrix} 4 & 2 \\ -1 & 2 \end{bmatrix}$ and $B = \begin{bmatrix} 2 & -1 \\ 2 & 4 \end{bmatrix}$. A computation shows $value(A) = 2, value(B^T) = 2$ and the security point is $(2, 2)$. With $(2, 2)$ security point, the bargaining solution is $(3, 3)$ since that is the solution of the problem

$$\text{Maximize } (u - 2)(v - 2), \text{ subject to } u + v \le 6, 2 \le u \le 4, 2 \le v \le 4.$$

In fact take a derivative of $(u - 2)(6 - u - 2)$ and set to zero to see that $u = 3$.

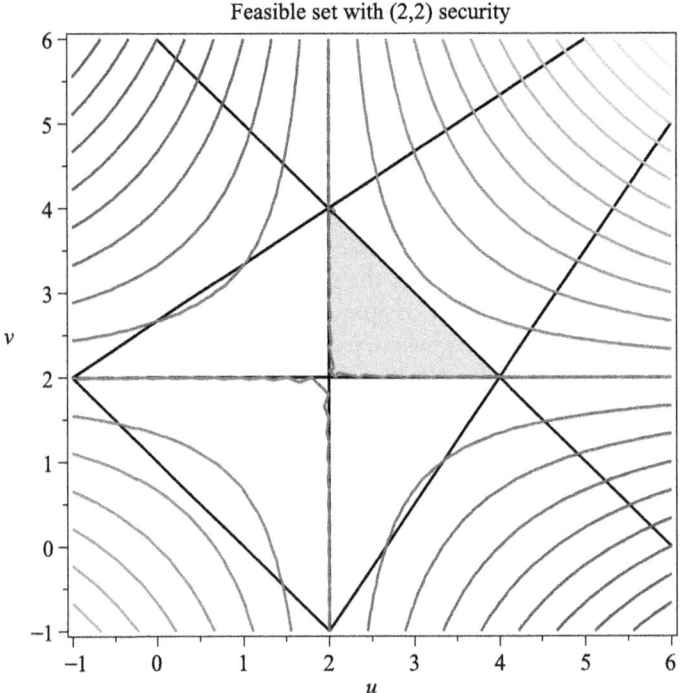

Feasible set with (2,2) security

Security point (2, 2).

Next, since the Pareto-optimal boundary is $u + v = 6$, we have $m_p = -1$, $b = 6$. We calculate $value(-m_p A - B)$ and the saddle strategies for that game to get $value(A - B) = 2$, $X_t = (1, 0)$, $Y_t = (1, 0)$.

The threat security point is, therefore, $(u^t, v^t) = (4, 2)$ with threat strategies $X_t = (1, 0) = Y_t$.

Next, we use the formulas

$$\overline{u} = \frac{(m_p u^t + v^t - b)}{2m_p} \quad \text{and} \quad \overline{v} = \frac{(m_p u^t + v^t + b)}{2}$$

with $b = 6$. The threat solution is $(\overline{u}, \overline{v}) = (4, 2)$ with both players threatening to play row 1, column 1.

The KS line for $(u^*, v^*) = (2, 2)$ is $v - 2 = k(u - 2)$, $k = \frac{\max v - 2}{\max u - 2} = 1$, or simplified to $v = u$. This intersects $u + v = 6$, the Pareto-optimal boundary, at $(3, 3)$. Therefore, $(3, 3)$ is the KS solution.

There is no KS line for $(u^*, v^*) = (4, 2)$ because it is on the edge of the Pareto line.

If we consider the cooperative game, the characteristic function is $v(1) = v(2) = 2$, $v(12) = 6$, which gives nucleolus and Shapley value $(3, 3)$ and matches with the solution for the $(2, 2)$ security point.

6.60 Find the Nash bargaining solution and the threat solution to the game with bimatrix

$$\begin{bmatrix} (-2, 5) & (-7, 3) & (3, 4) \\ (4, -3) & (6, 1) & (-6, -6) \end{bmatrix}.$$

Find the KS line and solution.

6.60 Answer: The matrices are

$$A = \begin{bmatrix} -2 & -7 & 3 \\ 4 & 6 & -6 \end{bmatrix} \quad \text{and} \quad \begin{bmatrix} 5 & 3 & 4 \\ -3 & 1 & -6 \end{bmatrix}$$

With security point $(u^*, v^*) = (value(A), value(B^T)) = (-\frac{12}{11}, 1)$, we get the bargaining solution $(\overline{u}, \overline{v}) = (3, 4)$. This occurs at the point of intersection of the 2 lines forming the Pareto-optimal boundary $v = -u + 7$, $v = (-\frac{1}{5})u + \frac{23}{5}$.

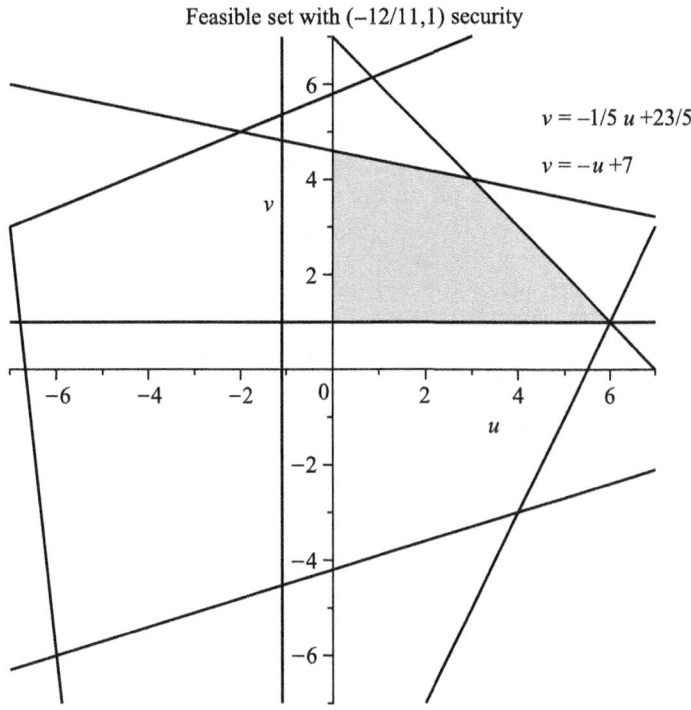

Feasible set with $(-12/11,1)$ security

$v = -1/5\, u + 23/5$

$v = -u + 7$

You can see this from the problem

$$\text{Maximize } \left(u + \frac{12}{11} \right)(v - 1) \text{ subject to } u + 5v \le 23, u + v \le 7$$

and $u \ge -\frac{12}{11}$, $v \ge 1$. Assuming the maximum is on the line $u + v = 7$, $3 \le u \le 6$, by calculus, we have $g'(u) = \frac{54}{11} - 2u = 0 \Rightarrow u = \frac{27}{11} < 3$. Hence, the maximum is achieved at $u = 3$, and then $v = 7 - u = 4$.

Next, since the Pareto-optimal boundary is $u + v = 7$, $3 \le u \le 6$ and $u + \frac{1}{5}v = \frac{23}{5}$, $-1 \le u \le 3$, we calculate for both $m_p = -1$, $b = 7$ and $m_p = -\frac{1}{5}$, $b = \frac{23}{5}$.

In the case $m_p = -1$, $b = 7$, we calculate $v(A - B) = 0$ and the threat strategies $X_t = (0, 1)$, $Y_t = (0, 0, 1)$. Then $u^t = X_t A Y_t^T = -6$, $v^t = X_t B Y_t^T = -6$ is our safety point. We next calculate with $b = 7$,

$$\bar{u} = \frac{(m_p u^t + v^t - b)}{2m_p} = \frac{7}{2} \quad \text{and} \quad \bar{v} = \frac{(m_p u^t + v^t + b)}{2} = \frac{7}{2}.$$

In the case $m_p = -\frac{1}{5}, b = \frac{23}{5}$, we calculate $v(\frac{1}{5}A - B) = \frac{1}{5}$ and the threat strategies $X_t = (0, 1), Y_t = (0, 1, 0)$. Then $u^t = X_t A Y_t^T = 6, v^t = X_t B Y_t^T = 1$ is our safety point. We next calculate with $b = \frac{23}{5}$,

$$\bar{u} = \frac{(m_p u^t + v^t - b)}{2m_p} = 12 \quad \text{and} \quad \bar{v} = \frac{(m_p u^t + v^t + b)}{2} = \frac{11}{5}.$$

However, since $\bar{u} = 12 > 3$, this is not on the line segment $u + \frac{1}{5}v = \frac{23}{5}, -1 \leq u \leq 3$. Therefore, this is not the threat solution.

We conclude that the threat solution is $\bar{u} = \bar{v} = \frac{7}{2}$, and the optimal threat strategies are $X_t = (0, 1), Y_t = (0, 0, 1)$.

For the KS line, we take $u^* = -\frac{12}{11}, v^* = 1$. The maximum possible feasible payoffs for each player is $a = 6$ for player I and $b = 4.818$ for player II, and so $k = \frac{b-v^*}{a-u^*} = 0.538$. The equation of the KS line is then $v - 1 = k(u + \frac{12}{11})$, and this intersects the Pareto-optimal boundary at the line segment $v = -u + 7$ and at the point $\bar{u} = 3.518, \bar{v} = 3.482$.

6.61 Find the Nash bargaining solution and the threat solution to the game with bimatrix

$$\begin{bmatrix} (-3, -1) & (0, 5) & (1, \frac{19}{4}) \\ (2, \frac{7}{2}) & (\frac{5}{2}, \frac{3}{2}) & (-1, -3) \end{bmatrix}.$$

Find the KS line and solution.

6.61 Answer: The matrices are

$$A = \begin{bmatrix} -3 & 0 & 1 \\ 2 & \frac{5}{2} & -1 \end{bmatrix}, \qquad B = \begin{bmatrix} -1 & 5 & \frac{19}{4} \\ \frac{7}{2} & \frac{3}{2} & -3 \end{bmatrix}.$$

We have $value(A) = -\frac{1}{7}, value(B^T) = \frac{19}{8}$. That is our safety point. The Pareto-optimal boundary has three line segments:

$$\begin{cases} \frac{1}{4}u + v = 5, & \text{if } 0 \leq u \leq 1; \\ \frac{2}{4}u + v = 6, & \text{if } 1 \leq u \leq 2; \\ 2u + \frac{1}{2}v = \frac{23}{4}, & \text{if } 2 \leq u \leq \frac{5}{2}. \end{cases}$$

The Nash problem is

$$\text{Maximize } \left(u + \frac{1}{7}\right)\left(v - \frac{19}{8}\right)$$

subject to $(u, v) \in S$. The part of the Pareto-optimal boundary for this problem is the line segment $\frac{5}{4}u + v = 6$, $1 \leq u \leq 2$. Using calculus, we find

$$\overline{u} = \frac{193}{140}, \quad \overline{v} = \frac{479}{112}.$$

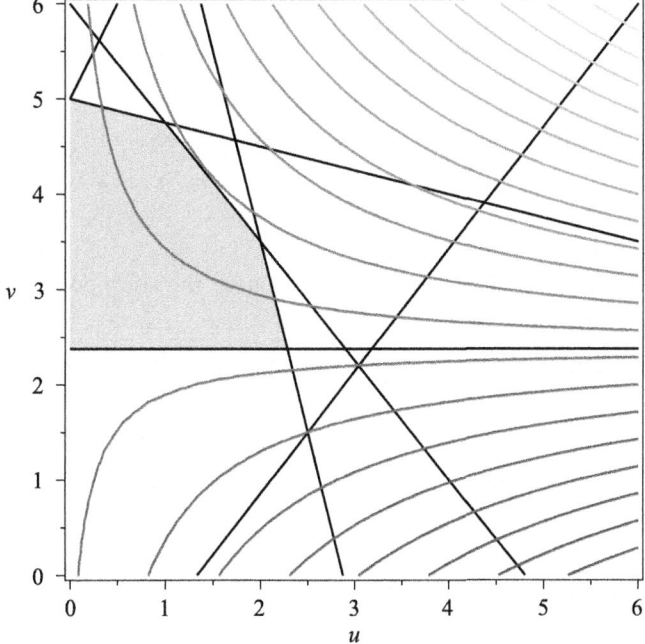

Next, we consider the threat solution. We have to find the threat strategies for all three line segments.

1. $v = -\frac{1}{4}u + 5$, $0 \leq u \leq 1$. Then $m_p = -\frac{1}{4}$, $b = 5$, and

$$value\left(\frac{1}{4}A - B\right) = -\frac{487}{236}, \; X_t = \left(\frac{17}{59}, \frac{42}{59}\right), \; Y_t = \left(\frac{33}{59}, \frac{26}{59}, 0\right),$$

and

$$u^t = X_t A Y_t^T = 1.097, \; v^t = X_t B Y_t^T = 2.34, \Rightarrow \overline{u} = 5.87, \overline{v} = 3.53.$$

Since $5.87 \notin [0, 1]$, this is not the threat solution.

2. $v = -\frac{5}{4}u + 6$, $1 \leq u \leq 2$. Then $m_p = -\frac{5}{4}$, $b = 6$, and

$$value\left(\frac{5}{4}A - B\right) = -1, \; X_t = (0, 1), \; Y_t = (1, 0, 0).$$

Then

$$u^t = X_t A Y_t^T = 2, \; v^t = X_t B Y_t^T = \frac{7}{2} \Rightarrow \overline{u} = 2, \overline{v} = \frac{7}{2}.$$

This is in the range. Let's check the final segment.

3. $v = -4u + \frac{23}{2}, 2 \leq u \leq \frac{5}{2}$. In this case, $m_p = -4, b = \frac{23}{2}$, and

$$value(4A - B) = -\frac{115}{126}, X_t = \left(\frac{22}{63}, \frac{41}{63}\right), Y_t = \left(\frac{1}{63}, 0, \frac{62}{63}\right).$$

The safety point is then

$$u^t = X_t A Y_t^T = -0.294, v^t = -0.258 \Rightarrow \overline{u} = 1.32, \overline{v} = 6.206.$$

Since $1.32 \notin [2, \frac{5}{2}]$, this too is not the threat solution.

We conclude that the threat solution is $\overline{u} = 2, \overline{v} = \frac{7}{2}$ and player 1 threatens to always play the second row; player 2 threatens to use the third column unless they both come to their senses.

Finally, we determine the KS solution for the safety point $u^* = -\frac{1}{7}, v^* = \frac{19}{8}$. We have $a = \max_{(u,v) \in S} u = \frac{73}{32}, b = \max_{(u,v) \in S} v = 5$. Then

$$v = v^* + k(u - u^*) = \frac{19}{8} + \frac{196}{181}\left(u + \frac{1}{7}\right)$$

is the KS line, and this intersects the Pareto-optimal boundary through the segment $v = -\frac{5}{4}u + 6$ at the point $\overline{u} = 1.487, \overline{v} = 4.140$, and that is the KS solution.

6.62 Consider the sequential bargaining problem. Suppose that each player has their own discount factor $\delta_1, \delta_2, 0 < \delta_i < 1$. Find the subgame perfect equilibrium for each player assuming the bargaining has three stages and ends, as well as assuming the stages could continue forever.

6.62 Answer: The tree is shown in the following figure.

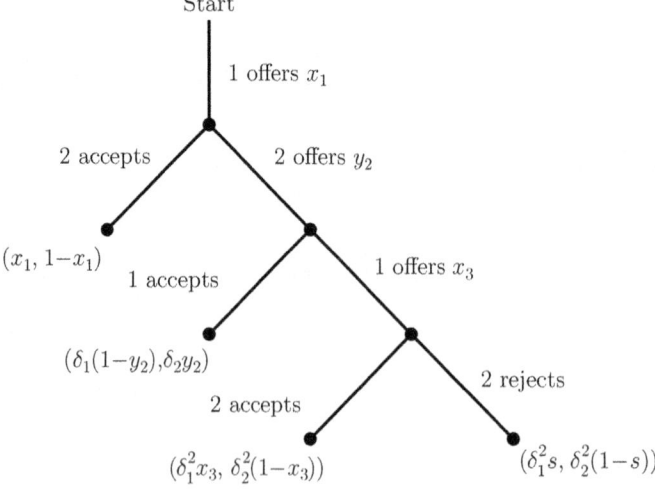

Bargaining with differing discount factors.

The subgame perfect equilibria for each player assuming a possibly infinite number of stages is given by

Player 1

1. At the start of the game the first offer for player 1 should be $x_1^* = 1 - \delta_2(1 - \delta_1 s)$.

2. If player 2 rejects x_1^*, player 1 will accept player 2's offer of y_2 if $y_2 \geq \delta_1 s$ and otherwise reject player 2's offer.

3. If player 1 rejects player 2's stage 2 offer and the final stage is reached, player 1 should offer $x_3^* = s$.

Player 2

1. At the start of the game if the first offer by player 1 is x_1, then player 2 should accept the offer if $x_1 \leq 1 - \delta_2(1 - \delta_1 s)$ and otherwise reject it.

2. If player 2 rejects player 1's first offer, player 2 should counter offer $y_2 = \delta_1 s$.

3. In the final stage, if player 1 rejects player 2's stage 2 offer, player 2 should accept player 1's counter of s.

The payoffs to each player are $x_1^* = 1 - \delta_2(1 - \delta_1 s)$, $1 - x_1^* = \delta_2(1 - \delta_1 s)$, respectively.

To determine s, we note that the conditions in stage 3 are the same as at the beginning of the game. This means that s should satisfy $s = 1 - \delta_2(1 - \delta_1 s)$. Solving for s, we get

$$s = \frac{1 - \delta_2}{1 - \delta_1 \delta_2} \Rightarrow x_1^* = \frac{1 - \delta_2}{1 - \delta_1 \delta_2}, 1 - x_1^* = \frac{\delta_2(1 - \delta_1)}{1 - \delta_1 \delta_2}.$$

6.63 You want to buy a seller's condo. You look up what the seller paid for the condo 2 years ago and find that she paid \$305,000. You figure that she will absolutely not sell below this price. The seller has listed the condo at \$350,000. Assume that the sequential bargaining problem will go at most 3 rounds before everyone is fed up with the process and the deal falls apart. Take the discount factors for each player to be $\delta = 0.99$. What should the offers be at each stage? What if the process could go on indefinitely?

6.63 Answer: The spread is $350 - 305 = 45K$. If the process ends with at most three stages, player 1 should offer $305 + 0.99(45) = 349.55$. If the process could go on indefinitely, player 1 should offer $305 + 0.5025(45) = 327.625$, because $x_1 = 1 - 0.99 + 0.99^2 \frac{1}{1+0.99}$.

6.64 The Nash solution also applies to payoff functions with a continuum of strategies. For example, suppose that two investors are bargaining over a piece of real estate and they have payoff functions $u(x, y) = x + y$, while $v(x, y) = x + \sqrt{y}$, with $x, y \geq 0$, and $x + y \leq 1$. Both investors want to maximize their own payoffs. The bargaining solution with safety point $u^* = 0$, $v^* = 0$ (because both players get zero if negotiations break down) is given by the solution of the problem

$$\text{Maximize } (u(x, y) - u^*)(v(x, y) - v^*) = (x + y)(x + \sqrt{y})$$
$$\text{subject to } x, y \geq 0, x + y \leq 1.$$

Solve this problem to find the Nash bargaining solution.

6.64 Answer: Use calculus or Maple to get $x^* = \frac{3}{4}$, $y^* = \frac{1}{4}$, and then $\overline{u} = 1, \overline{v} = \frac{5}{4}$.

6.65 A classic bargaining problem involves a union and management in contract negotiations. If management hires $w \geq 0$ workers, the company produces $f(w)$ revenue units, where f is a continuous, increasing function. The maximum number of workers who are represented by the union is W. A person who is not employed by the company gets a payoff $p_0 \geq 0$, which is either unemployment benefits or the pay at another job. In negotiations with the union, the firm agrees to the pay level p and to employ $0 \leq w \leq W$ workers. We may consider the payoff functions as

$$u(p, w) = f(w) - pw \quad \text{to the company}$$

and

$$v(p, w) = pw + (W - w)p_0 \quad \text{to the union.}$$

Assume the safety security point is $u^* = 0$ for the company and $v^* = Wp_0$ for the union.

(a) What is the nonlinear program to find the Nash bargaining solution?

6.65.a Answer: The Nash problem is

$$\text{Maximize } (u - u^*)(v - v^*) = (f(w) - pw)(pw + (W - w)p_0 - Wp_0)$$
$$= (f(w) - pw)(p - p_0)w$$

with $(p, w) \in S$, where

$$S = \{(u, v) \mid u \geq u^*, v \geq v^*\}$$
$$= \{(u, v) \mid u \geq 0, v \geq Wp_0\}$$
$$= \{(p, w) \mid f(w) - pw \geq 0, p \geq p_0, 0 \leq w \leq W\}.$$

(b) Assuming an interior solution (which means you can find the solution by taking derivatives), show that the solution (p^*, w^*) of the Nash bargaining solution satisfies

$$f'(w^*) = p_0 \quad \text{and} \quad p^* = \frac{w^* p_0 + f(w^*)}{2w^*}.$$

6.65.b Answer: Set $h(p, w) = (f(w) - pw)(p - p_0)w$. We have

$$\frac{\partial h}{\partial p} = w((p_0 - 2p)w + f(w)) = 0$$
$$\frac{\partial h}{\partial w} = (p - p_0)(f(w) + w(f'(w) - 2p)) = 0.$$

Solving the first equation for p gives

$$p = \frac{wp_0 + f(w)}{2w}.$$

Substitute this p into the equation $h_w = 0$ to get

$$f(w) + w\left(f'(w) - \frac{wp_0 + f(w)}{w}\right) = f(w) + wf'(w) - p_0w - f(w)$$
$$= w(f'(w) - p_0) = 0,$$

which implies $f'(w) = p_0$.

(c) Find the Nash bargaining solution for $f(w) = \ln(w + a) + b, a > 0, \frac{1}{a} > p_0, b > -\ln a$.

6.65.c Answer: Using the formulas from the previous part, $p_0 = f'(w^*) = \frac{1}{w^*+a}$ that implies $w^* = \frac{1}{p_0} - a > 0$ and

$$p^* = \frac{wp_0 + f(w)}{2w} = \frac{p_0(ap_0 - \ln(\frac{1}{p_0} - a) - b - 1)}{2ap_0 - 2}.$$

Review Problems

True or False or Fill in the Blank or Complete the Statement.

6.66 A Nash equilibrium in a zero sum game is not the same as a saddle point.

6.66 Answer: False.

6.67 The least core is always nonempty.

6.67 Answer: False

6.68 If the least core has only one allocation then that allocation is fair in the sense that it minimizes . . . .

6.68 Answer: . . . the maximum dissatisfaction.

6.69 In order for $v(S)$ to be a characteristic function it must satisfy three conditions. What are they?

6.69 Answer:
1. $v(\emptyset) = 0$
2. $v(N) \geq \sum_{i=1}^n v(i)$
3. $v(S \cup T) \geq v(S) + v(T), \forall S, T \subset N, S \cap T = \emptyset$.

6.70 In order for a vector $\vec{x} = (x_1, \ldots, x_n)$ to be an imputation it must satisfy two conditions. What are they?

6.70 Answer:
1. $x_i \geq v(i), i = 1, 2, \ldots, n$. This is individual rationality.
2. $\sum_{i=1}^n x_i = v(N)$. This is group rationality.

6.71 In order for a pair of strategies X^*, Y^* to be a Nash equilibrium for the game (A, B) it must satisfy two conditions. What are they?

6.71 Answer: $E_I(X^*, Y^*) \geq E_I(X, Y^*)$, and $E_{II}(X^*, Y^*) \geq E_{II}(X, Y^*)$ for evey strategy $X \in S_n$ and $Y \in S_m$.

6.72 The core of a cooperative game is the set . . . .

6.72 Answer:

$$C(0) = \{\vec{x} = (x_1, \ldots, x_n) \mid v(S) - \vec{x}(S) \leq 0, \ \forall \emptyset \neq S \subsetneq N\}$$

where $\sum_{i=1}^{n} x_i = v(N)$.

6.73 Is the following statement correct? If not, correct it. Equality of Payoffs in a two-person nonzero sum game says that $E_1(i, Y^*) = E_1(j, Y^*)$ for all rows i, j if (X^*, Y^*) is a Nash equilibrium.

6.73 Answer: Equality of payoffs says that if $(X^* = (x_1, \ldots, x_n), Y^*)$ is a Nash equilibrium and $x_i > 0, x_j > 0$ then $E_1(i, Y^*) = E_1(j, Y^*)$.

6.74 In the Nash bargaining solution, the problem reduces to

$$\text{Maximize } g(u, v) := (u - value(A))(v - value(B))$$
$$\text{subject to } (u, v) \in S, u \geq value(A), v \geq value(B).$$

6.74 Answer: False

6.75 The Pareto-optimal boundary must have a negative slope because . . . .

6.75 Answer: . . . if the slope was positive it would mean that both players could increase their payoffs and remain in the feasible set.

6.76 The optimal threat strategies (X_t, Y_t) in the Nash bargaining solution are given as the saddle point for the zero sum game with matrix $-m_p A - B$, where m_p is . . . .

6.76 Answer: . . . the slope of the line which is the Pareto-optimal boundary.

CHAPTER SEVEN

Evolutionary Stable Strategies and Population Games

7.1 Evolution

Problems

7.1 In the currency game (Example 7.2), derive the same result we obtained but using the equivalent definition of ESS: X^* is an evolutionary stable strategy if for every strategy $X = (x, 1 - x)$, with $x \neq x^*$, there is some $p_x \in (0, 1)$, which depends on the particular choice x, such that

$$u(x^*, px + (1 - p)x^*) > u(x, px + (1 - p)x^*), \quad \text{for all } 0 < p < p_x.$$

Find the value of p_x in each case an ESS exists.

7.1 Answer: Let's look at the equilibrium $X_1 = (1, 0)$. We need to show that for $x \neq 1$, $u(1, px + (1 - p)) > u(x, px + (1 - p))$ for some p_x, and for all $0 < p < p_x$. Now $u(1, px + (1 - p)) = 1 - p + px$, and $u(x, px + (1 - p)) = p + x - 3px + 2px^2$. In order for X_1 to be an ESS, we need $1 > 2p(x - 1)^2$, which implies $0 < p < 1/(2(x - 1)^2)$. So, for $0 \leq x < 1$, we can take $p_x = 1/(2(x - 1)^2)$ and the ESS requirement will be satisfied. Similarly, the equilibrium $X_2 = (0, 1)$, can be shown to be an ESS. For $X_3 = (\frac{1}{2}, \frac{1}{2})$, we have

$$u\left(\frac{1}{2}, px + \frac{1 - p}{2}\right) = \frac{1}{2} \text{ and } u\left(x, px + \frac{1 - p}{2}\right) = \frac{1}{2} + \frac{p}{2} - 2px + 2px^2.$$

In order for X_3 to be an ESS, we need

$$\frac{1}{2} > \frac{1}{2} + \frac{p}{2} - 2px + 2px^2,$$

which becomes $0 > 2p(x - \frac{1}{2})^2$, for $0 < p < p_x$. This is clearly impossible, so X_3 is not an ESS.

Solutions Manual to Accompany Game Theory: An Introduction, Second Edition. E.N. Barron.
© 2013 John Wiley & Sons, Inc. Published 2013 by John Wiley & Sons, Inc.

7.2 It is possible that there is an economy that uses a dominant currency in the sense that the matrix becomes

I/II	Euros	Dollars
Euros	(1, 1)	(0, 0)
Dollars	(0, 0)	(2, 2)

Find all Nash equilibria and determine which are ESSs.

7.2 Answer: There are three Nash equilibria $X_1 = Y_1 = (1, 0)$, $X_2 = Y_2 = (0, 1)$, and the mixed $X_3 = Y_3 = (\frac{2}{3}, \frac{1}{3})$. The first two are ESSs. For X_3, $u(\frac{2}{3}, \frac{2}{3}) = \frac{2}{3}$, $u(x, \frac{2}{3}) = \frac{2}{3}$. Is $u\left(\frac{2}{3}, x\right) = \frac{2}{3} > u(x, x)$? No, because $\frac{2}{3} > x^2 + 2(1 - x)^2$ is false for all $0 < x < 1$.

7.3 The decision of whether or not each of two admirals should attack an island was studied in Problem 4.16. The analysis resulted in the following matrix.

Fr/Brit	11	12
11	$-\frac{5}{2}, -\frac{5}{2}$	$\frac{7}{4}, \frac{1}{2}$
12	$\frac{1}{2}, \frac{7}{4}$	$\frac{5}{4}, \frac{7}{4}$

Note that this is a symmetric game. There are three Nash equilibria:
1. $X_1 = Y_1 = (\frac{1}{7}, \frac{6}{7})$, payoff French $= \frac{8}{7}$, British $= \frac{8}{7}$.
2. $X_2 = (1, 0)$, $Y_2 = (0, 1)$, payoff French $= \frac{1}{2}$, British $= \frac{7}{4}$.
3. $X_3 = (0, 1)$, $Y_3 = (1, 0)$, payoff French $= \frac{7}{4}$, British $= \frac{1}{2}$.
Determine which, if any of these are ESSs.

7.3 Answer: The pure Nash equilibria (X_2, Y_2), (X_3, Y_3) are not symmetric and hence cannot be ESSs. We determine whether or not X_1 is an ESS.

We calculate for $x \neq \frac{1}{7}$,

$$u\left(\frac{1}{7}, p\,x + (1 - p)\frac{1}{7}\right) = \frac{8}{7} + \frac{p}{28}(5 - 35x)$$

and

$$u\left(x, p\,x + (1 - p)\frac{1}{7}\right) = \frac{8}{7} + \frac{p}{28}(3 - 7x - 98x^2).$$

The question is whether (for small enough $0 < p < 1$) we have

$$\frac{8}{7} + \frac{p}{28}(5 - 35x) > \frac{8}{7} + \frac{p}{28}(3 - 7x - 98x^2) \Leftrightarrow 98x^2 - 28x + 2 > 0.$$

Since $98x^2 - 28x + 2 = 2(7x - 1)^2 > 0$ for any $x \neq \frac{1}{7}$, we conclude that $X_1 = (\frac{1}{7}, \frac{6}{7})$ is indeed an ESS. Consequently, in a series of naval battles between the British and French, strong navys will always attack, while weak navys will attack with probability $\frac{1}{7}$.

7.4 Analyze the Nash equilibria for a version of the prisoner's dilemma game:

I/II	Confess	Deny
Confess	$(4, 4)$	$(1, 6)$
Deny	$(6, 1)$	$(1, 1)$

7.4 Answer: The only symmetric (nonstrict) Nash is $(X^* = (0, 1), X^*)$. Then $u(0, 0) = 1$, $u(x, 0) = 1$, $u(x, x) = -2x^2 + 5x + 1$, and $u(0, x) = 5x + 1$. Hence, $u(0, 0) = 1 = u(x, 0)$ and $u(x, x) < u(0, x)$, for any $0 < x \le 1$. This means that $X^* = (0, 1)$ is an ESS.

7.5 Determine the Nash equilibria for rock-paper-scissors with matrix

I/II	Rock	Paper	Scissors
Rock	$(2, 2)$	$(-1, 1)$	$(1, -1)$
Paper	$(1, -1)$	$(2, 2)$	$(-1, 1)$
Scissors	$(-1, 1)$	$(1, -1)$	$(2, 2)$

There are three pure Nash equilibria and four mixed equilibria (all symmetric). Determine which are evolutionary stable strategies, if any, and if an equilibrium is not an ESS, show how the requirements fail.

7.5 Answer: The Nash equilibria and their payoffs are shown in the following table; they are all symmetric.

X^*	$u(X^*, X^*)$
$(1, 0, 0)$	2
$(0, 1, 0)$	2
$(0, 0, 1)$	2
$(\frac{3}{4}, \frac{1}{4}, 0)$	$\frac{5}{4}$
$(\frac{1}{4}, 0, \frac{3}{4})$	$\frac{5}{4}$
$(0, \frac{3}{4}, \frac{1}{4})$	$\frac{5}{4}$
$(\frac{1}{3}, \frac{1}{3}, \frac{1}{3})$	$\frac{2}{3}$

For $X^* = (1, 0, 0)$, you can see this is an ESS because it is strict. Consider next $X^* = (\frac{3}{4}, \frac{1}{4}, 0)$. Since $u(Y, X^*) = \frac{5}{4}(y_1 + y_2) - y_3/2$, the set of best response strategies is $Y = (y, 1 - y, 0)$. Then $u(Y, Y) = 4y^2 - 4y + 2$ and $u(X^*, Y) = -\frac{1}{4} + 2y$. Since it is **not** true that $u(Y, Y) < u(X^*, Y)$, for all best responses $Y \ne X^*$, X^* is not an ESS.

7.6 In Problem 3.18, we considered the game of the format to be used for radio stations WSUP and WHAP. The game matrix is

WSUP/WHAP	RB	EM	AT
RB	25, 25	50, 30	50, 20
EM	30, 50	15, 15	30, 20
AT	20, 50	20, 30	10, 10

Determine which, if any of the three Nash equilibria are ESSs.

7.6 Answer: First notice that this is a symmetric game since $B = A^T$. The three Nash equilibria are $X_1 = (1, 0, 0)$, $Y_1 = (0, 1, 0)$, $X_2 = (0, 1, 0)$, $Y_2 = (1, 0, 0)$, and $X^* = (\frac{7}{8}, \frac{1}{8}, 0) = Y^*$. This last is the only symmetric Nash and so is the only candidate to be an ESS.

To do the calculations we take advantage of the fact that the third row and column are dominated and may be dropped. The strategy we examine is then $X^* = (\frac{7}{8}, \frac{1}{8})$ and the matrix is $A = \begin{bmatrix} 25 & 50 \\ 30 & 15 \end{bmatrix}$. Set

$$u(x, y) = (x, 1 - x)A \begin{bmatrix} y \\ 1 - y \end{bmatrix}.$$

We calculate

$$u\left(\frac{7}{8}, \frac{7}{8}\right) = \frac{225}{8}, \quad \text{and} \quad u\left(x, \frac{7}{8}\right) = \frac{225}{8}.$$

We are in the case $u(X^*, X^*) = u(X, X^*)$. We have to check if $u(X^*, X) > u(X, X)$ for all $X \neq X^*$.

Next,

$$u\left(\frac{7}{8}, x\right) = \frac{365}{8} - 20x, \quad \text{and} \quad u(x, x) = 15 + 50x - 40x^2$$

and the question is if $\frac{365}{8} - 20x > 15 + 50x - 40x^2$, for all $x \neq \frac{7}{8}$. By algebra, we see that

$$\frac{365}{8} - 20x - (15 + 50x - 40x^2) = \frac{5}{8}(7 - 8x)^2 > 0, \quad \forall \, 0 \leq x \leq 1, x \neq \frac{7}{8}.$$

We answer the question affirmatively and we have shown

$$u(X^*, X^*) = u(X, X^*) \Rightarrow u(X^*, X) > u(X, X), \quad \forall \, X \neq X^*.$$

Thus, $X^* = (\frac{7}{8}, \frac{1}{8})$ is indeed an ESS. In the long run, both WSUP and WHAP will play RB 87.5% of the time and EM 12.5% of the time.

7.7 Consider a game with matrix $A = \begin{bmatrix} a & 0 \\ 0 & b \end{bmatrix}$. Suppose that $ab \neq 0$.

    (a) Show that if $ab < 0$, then there is exactly one ESS. Find it.

7.7.a Answer: There is a unique Nash, strict and symmetric ESS $X^* = (0, 1)$ if $a < 0$, $b > 0$, and $X^* = (1, 0)$ if $b < 0$, $a > 0$.

    (b) Suppose that $a > 0$, $b > 0$. Then there are three symmetric Nash equilibria. Show that the Nash equilibria which are evolutionary stable are the pure ones and that the mixed Nash is not an ESS.

7.7.b Answer: There are three Nash equilibria, all symmetric, Nash equilibria $=$ $(1, 0)$, $(0, 1)$, X, where $X = \left(\frac{b}{(a+b)}, \frac{a}{(a+b)} \right)$. Both $(1, 0)$, $(0, 1)$ are strict, $(1, 0)$, $(0, 1)$ are both evolutionary stable. The mixed X is not an ESS since

$$E(1, 1) = a > \frac{ab}{(a + b)} = E(X, 1).$$

(c) Suppose that $a < 0$, $b < 0$. Show that this game has two pure nonsymmetric Nash equilibria and one symmetric mixed Nash equilibrium. Show that the mixed Nash is an ESS.

7.7.c Answer: There are two strict asymmetric Nash Equilibria, and one symmetric Nash Equilibrium given by $X = (\frac{b}{(a+b)}, \frac{a}{(a+b)})$. However, now X is an ESS since

$$E(X, Y) = cay_1 + (1 - c)by_2 = \frac{ab}{(a + b)}, \quad \text{where } c = \frac{b}{(a + b)},$$

and for every strategy

$$Y \neq X, \quad E(Y, Y) = ay_1^2 + by_2^2 < \frac{ab}{(a + b)} = E(X, Y),$$

so X is an ESS.

7.8 Verify that $X^* = (x^*, 1 - x^*)$ is an ESS if and only if (X^*, X^*) is a Nash equilibrium and $u(x^*, x) > u(x, x)$ for every $X = (x, 1 - x) \neq X^*$ that is a best response to X^*.

7.8 Answer: If X^* is an ESS then directly from the definition $u(X^*, X^*) > u(Y, X^*)$ for all $Y \neq X^*$ or $u(X^*, X^*) = u(Y, X^*)$. In either case $u(X^*, X^*) \geq u(Y, X^*)$ for all $Y \neq X^*$, which means that (X^*, X^*) is a Nash equilibrium.

If (X^*, X^*) is a Nash equilibrium but is not strict, then there is some strategy $Y = (y, 1 - y) \neq X^*$ which is a best response to X^*, and $u(y, x^*) = u(x^*, x^*)$. Now X^* is an ESS if and only if

$$(1 - p)[u(x^*, x^*) - u(x, x^*)] > p[u(x, x) - u(x^*, x)] \tag{7.1}$$

for all $x \neq x^*$, for all $0 < p < p_x$, for some $0 < p_x < 1$. Thus for the best response Y, if $u(y, x^*) = u(x^*, x^*)$, the ESS condition is equivalent to $u(y, y) < u(x^*, y)$ since the left side of (7.1) is zero with $x = y$.

7.2 Population Games

Problems

7.9 Consider a game in which a seller can be either honest or dishonest and a buyer can either inspect or trust (the seller). One game model of this is the matrix $A = \begin{bmatrix} 3 & 2 \\ 4 & 1 \end{bmatrix}$, where the rows are inspect and trust, and the columns correspond to dishonest and honest.

(a) Find the replicator dynamics for this game.

7.9.a Answer: The replicator equation in simplified form is

$$\frac{dp}{dt} = p(1 - p)(1 - 2p).$$

(b) Find the Nash equilibria and determine which, if any, are ESSs.

7.9.b Answer: The three Nash equilibria are $X_1 = (\frac{1}{2}, \frac{1}{2}) = Y_1$, and the two nonsymmetric Nash points $((0, 1), (1, 0))$ and $((1, 0), (0, 1))$. So only X_1 is a possible ESS. It is not hard to show directly that X_1 is an ESS but we can also use the fact from that if $a_{11} \neq a_{21}$ and $a_{22} \neq a_{12}$, then there must be an ESS. Since the nonsymmetric Nash equilibria cannot be ESSs, the only possibility is X_1. Therefore, X_1 is an ESS.

(c) Analyze the stationary solutions for stability.

7.9.c Answer: From the following figure you can see that $(p_1(t), p_2(t)) \to (\frac{1}{2}, \frac{1}{2})$ as $t \to \infty$ and conclude that (X_1, X_1) is an ESS. Verify directly using the stability theorem that it is asymptotically stable.

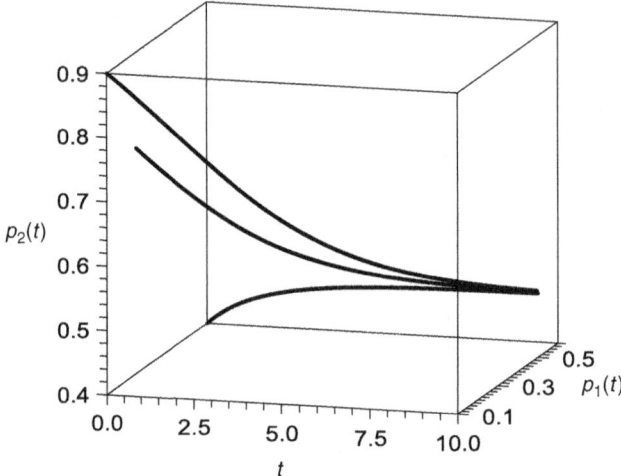

The figure shows trajectories starting from three different initial points. In the three-dimensional figure, you can see that the trajectories remain in the plane $p_1 + p_2 = 1$. The Maple commands used to find the stationary solutions, check their stability, and produce the graph are as follows:

```
> restart:with(DEtools):with(plots):with(LinearAlgebra):
> A:=Matrix([[3,2],[4,1]]); X:=<x1,x2>;
> Transpose(X).A.X;
> s:=expand(%);
> L:=A.X; f:=(x1,x2)->L[1]-s;g:=(x1,x2)->L[2]-s;
> solve({f(x1,x2)=0,g(x1,x2)=0},[x1,x2]);
> q:=diff(f(x1,x2),x1)+diff(g(x1,x2),x2);
> with(VectorCalculus):Jacobian(<f(x1,x2),g(x1,x2)>,[x1,x2]);
> j:=simplify(%);
> a1:=subs(x1=1/2,x2=1/2,j);b1:=subs(x1=1/2,x2=1/2,q);
```

```
> Determinant(a1);
> DEplot3d({D(p1)(t)=p1(t)*(3*p1(t)+2*p2(t)
                    -3*p1(t)^2-6*p1(t)*p2(t)-p2(t)^2),
        D(p2)(t)=p2(t)*(4*p1(t)+p2(t)
                    -3*p1(t)^2-6*p1(t)*p2(t)-p2(t)^2) },
        {p1(t),p2(t)}, t=0..10,
        [[p1(0)=0.1,p2(0)=0.9],[p1(0)=0.6,p2(0)=.4],
        [p1(0)=1/4,p2(0)=3/4]],scene=[t,p1(t),p2(t)],
        stepsize=.1,linecolor=t);
```

For the problem, `Determinant(a1)=5>0`, and `b1=-6<0`, so the stability Theorem G.6 allows us to conclude that $(\frac{1}{2}, \frac{1}{2})$ is asymptotically stable.

7.10 Analyze all stationary solutions for the game with matrix $A = \begin{bmatrix} 1 & 1 \\ 1 & 3 \end{bmatrix}$.

7.10 Answer: The replicator equation becomes

$$\frac{dp_1}{dt} = -2p_1(1 - p_1)^2 \equiv f(p_1).$$

The stationary solutions are $p_1 = 0, 1$. The Nash equilibria of this game are $X_1 = Y_1 = (0, 1)$ and $X_2 = Y_2 = (1, 0)$. Now $f'(p_1) = -2(1 - p_1)^2 + 4(1 - p_1)$ and $f'(0) = -2 < 0$, but $f'(1) = 0$. The asymptotically stable solution is $X_1 = (0, 1)$, so that X_1 is an ESS, which you may verify directly.

7.11 Consider the symmetric game with matrix

$$A = \begin{bmatrix} 2 & 1 & 5 \\ 5 & 1 & 0 \\ 1 & 4 & 0 \end{bmatrix}.$$

(a) Find the one and only Nash equilibrium.

7.11.a Answer: The unique Nash equilibrium is $X = (\frac{15}{44}, \frac{20}{44}, \frac{9}{44}) = Y$, and it is symmetric. We obtain this from the equality of payoffs theorem or Problem 3.25. We calculate using

$$Y^{*T} = \frac{A^{-1}J_3^T}{J_3 A^{-1} J_3^T}$$

and get $Y^* = \left(\frac{15}{44}, \frac{20}{44}, \frac{9}{44}\right)$. Observe that if we want to use the formula for X^* in Problem 3.25, we must use $B = A^T$.

(b) Determine whether the Nash equilibrium is an ESS.

7.11.b Answer: We check that with $u(X, Y) = X A Y^T$

$$u(X^*, X^*) = \frac{95}{44} = u(X, X^*)$$

for any $X \in S_3$. Now we check if $u(X^*, X) > u(X, X)$ for all $X \in S_3$. We have

$$u(X^*, X) - u(X, X) = \frac{75}{44} - \frac{50}{11}x_1 + 4x_1^2 - \frac{45}{11}x_2 + 4x_1x_2 + 3x_2^2.$$

The easiest way to check if this is >0 is to graph the function, or to minimize the function. We'll do both. First, we have

$$f(x_1, x_2) = \frac{75}{44} - \frac{50}{11}x_1 + 4x_1^2 - \frac{45}{11}x_2 + 4x_1x_2 + 3x_2^2$$

and the problem becomes

$$\text{Minimize } f(x_1, x_2), 0 \le x_1 \le 1, 0 \le x_2 \le 1, x_1 + x_2 \le 1.$$

We can do this by calculus or with Mathematica. The minimum is 0 attained at $x_1 = \frac{15}{44}, x_2 = \frac{20}{44}$. Thus, $f(x_1, x_2) > 0$ at any point except at the Nash equilibrium X^*. This tells us that X^* is indeed an ESS. Here is the graph of f showing it is positive except for the minimum point.

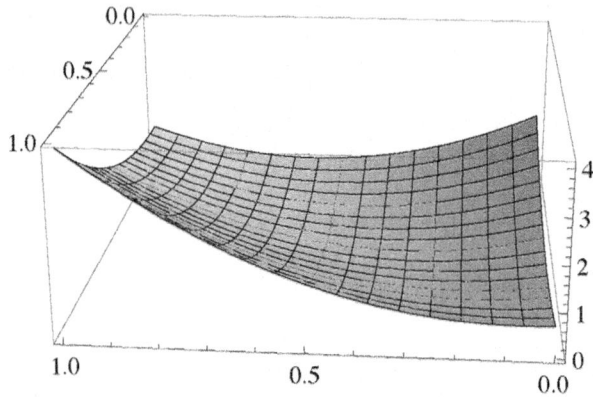

(c) Reduce the replicator equations to two equations and find the stationary solutions. Check for stability using the stability Theorem G.6.

7.11.c Answer: The replicator equations are as follows:

$$\frac{dp_i}{dt} = p_i(E(i, \pi) - E(\pi, \pi)), \quad i = 1, 2, 3.$$

Using $p_3 = 1 - p_1 - p_2$ and a lot of algebra, we get the replicator equations for p_1, p_2 as follows:

$$\frac{dp_1(t)}{dt} = p_1(t)(-9p_1(t) - 8p_2(t) + 5 + 4p_1(t)^2 + 4p_1(t)p_2(t) + 3p_2(t))^2,$$

$$\frac{dp_2(t)}{dt} = p_2(t)(-p_1(t) - 3p_2(t) + 4p_1(t)^2 + 4p_1(t)p_2(t) + 3p_2(t))^2.$$

Then $p_1 = \frac{15}{44}$, $p_2 = \frac{20}{44}$ is a stationary solution. To check stability, we use Theorem G.6. These calculations may be done by hand but Mathematica is way easier. Here are the commands to do this:

$$f[p_1, p_2] = p_1 \left(-9p_1 - 8p_2 + 5 + 4p_1^2 + 4p_1 p_2 + 3p_2\right)^2,$$

$$g[p_1, p_2] = p_2 \left(-p_1 - 3p_2 + 4p_1^2 + 4p_1 p_2 + 3p_2\right)^2,$$

$$s[p_1, p_2] = D[f[p_1, p_2], p_1] + D[g[p_1, p_2], p_2],$$

$$s[15/44, 20/44] = -0.230517 < 0,$$

$$t[p_1, p_2] = D[\{f[p_1, p_2], g[p_1, p_2]\}, \{\{p_1, p_2\}\}],$$

$$Det[t[15/44, 20/44]] = 1.90447 > 0.$$

The third line calculates $f_{p_1} + g_{p_2}$, and the fifth line calculates the determinant of the Jacobian matrix given in the fourth line. By the Theorem G.6, we conclude that $p_1 = \frac{15}{44}$, $p_2 = \frac{20}{44}$ is asymptotically stable.

The remaining stationary solutions are $(p_1 = 0, p_2 = 1)$, $(p_1 = 0, p_2 = 0)$, $(p_1 = 1, p_2 = 0)$. The convergence to $(\frac{15}{44}, \frac{20}{44}, \frac{9}{44})$ is shown in the figure.

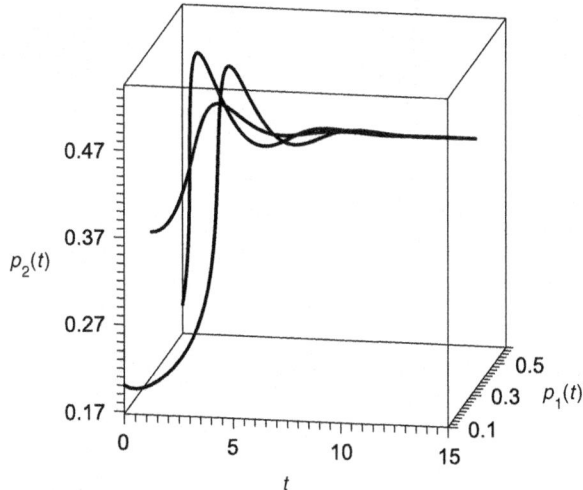

7.12 Consider the symmetric game with matrix

$$A = \begin{bmatrix} 1 & 1 & 0 \\ 1 & 0 & 1 \\ 0 & 0 & 0 \end{bmatrix}.$$

Show that $X = (1, 0, 0)$ is an ESS that is asymptotically stable for Equation (G.3).

7.12 Answer: The set of best response strategies to X^* consists of all strategies $Y = (y_1, y_2, y_3)$ that maximize $Y A X^{*T} = y_1 + y_2$. The maximum is achieved when $y_1 + y_2 = 1$ and $y_3 = 0$. Thus any best response to X^* are the strategies in which $Y = (y, 1 - y, 0), 0 \le y \le 1$. Then $u(Y, Y) = Y A Y^{*T} = -y^2 + 2y, u(X^*, Y) = 1$, and $-y^2 + 2y < 1$ for all $0 \le y < 1$. Consequently, by definition, X^* is an ESS and hence it must be asymptotically stable.

To see asymptotic stability using the replicator equations, we have

$$\frac{dp_1}{dt} = -p_1\left(-p_1 + p_1^2 + p_1 p_2 - p_2^2\right),$$

$$\frac{dp_2}{dt} = -p_2\left(-1 + 2p_2 + p_1^2 + p_1 p_2 - p_2^2\right).$$

If we check the conditions of Theorem G.6, we see that it is inconclusive because $f_{p_1} + g_{p_2} = -1 < 0$ but the determinant of the Jacobian at $(1, 0)$ is zero. Nevertheless, since we know $X^* = (1, 0, 0)$ is an ESS, we also know that $(p_1 = 1, p_2 = 0)$ is asymptotically stable. Here is the picture.

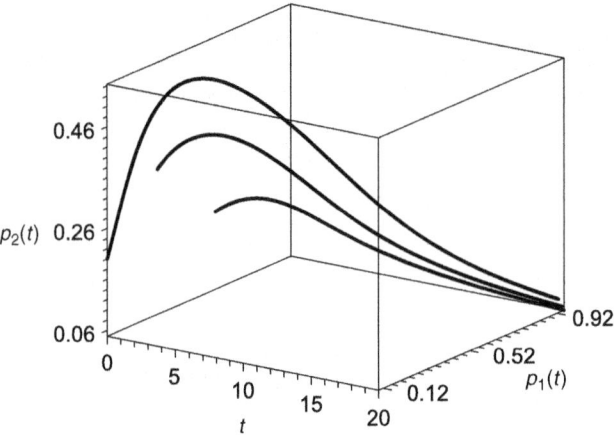

7.13 The simplest version of the rock-paper-scissors game has matrix

$$A = \begin{bmatrix} 0 & -1 & 1 \\ 1 & 0 & -1 \\ -1 & 1 & 0 \end{bmatrix}.$$

(a) Show that there is one and only one completely mixed Nash equilibrium but it is not an ESS.

7.13.a Answer: $X^* = (\frac{1}{3}, \frac{1}{3}, \frac{1}{3})$ is the unique symmetric Nash equilibrium. But since $u(Y, Y) = 0, u(Y, X^*) = 0$, and $u(X^*, Y) = 0$, it will **not** be true that $u(Y, Y) < u(X^*, Y)$ for all Y that is a best response to X^*.

(b) Show that the statement in the first part is still true if you replace 0 in each row of the matrix by $a > 0$.

7.13.b Answer: It still is true that $X^* = (\frac{1}{3}, \frac{1}{3}, \frac{1}{3})$ is the unique symmetric Nash equilibrium. Now we calculate for any $Y = (y_1, y_2, 1 - y_1 - y_2)$:

$$u(Y, Y) = YAY^T = a\left(1 + 2y_1^2 + (2y_1 + 2y_2)(-1 + y_2)\right),$$

$$u(Y, X^*) = \frac{a}{3},$$

$$u(X^*, Y) = \frac{a}{3},$$

$$u(X^*, X^*) = \frac{a}{3}.$$

Hence, we are in the condition $u(X^*, X^*) = u(Y, X^*)$ and we have to check if $u(X^*, Y) > u(Y, Y)$ for every $Y \neq X^*$. However,

$$\text{Minimum } [u(X^*, Y) - u(Y, Y)]$$

is $-\frac{8a}{3} < 0$. Thus, it is not true that X^* is an ESS.

(c) Analyze the stability of the stationary points for the replicator dynamics for $a > 0$ and Equation (G.3) by reducing to two equations and using the stability Theorem G.6.

7.13.c Answer: The reduced replicator equations are as follows:

$$\frac{\mathrm{d}p_1}{\mathrm{d}t} = f(p_1, p_2)$$
$$= -p_1 \left(-1 + p_1 + 2p_2 + a \left(1 + 2p_1^2 + 2p_2(-1 + p_2) + p_1(-3 + 2p_2)\right)\right),$$

$$\frac{\mathrm{d}p_2}{\mathrm{d}t} = g(p_1, p_2)$$
$$= -p_2 \left(1 - 2p_1 - p_2 + a \left(1 - 3p_2 + 2 \left(p_1^2 + p_1(-1 + p_2) + p_2^2\right)\right)\right).$$

Then

$$(f_{p_1} + g_{p_2}) \left(\frac{1}{3}, \frac{1}{3}\right) = \frac{2a}{3} > 0$$

and $\det(J(f, g))(\frac{1}{3}, \frac{1}{3}) = \frac{1}{3} + \frac{a^2}{9} > 0$. According to G.6, X^* is unstable. This is illustrated in the following figure in the case when $a = 0$.

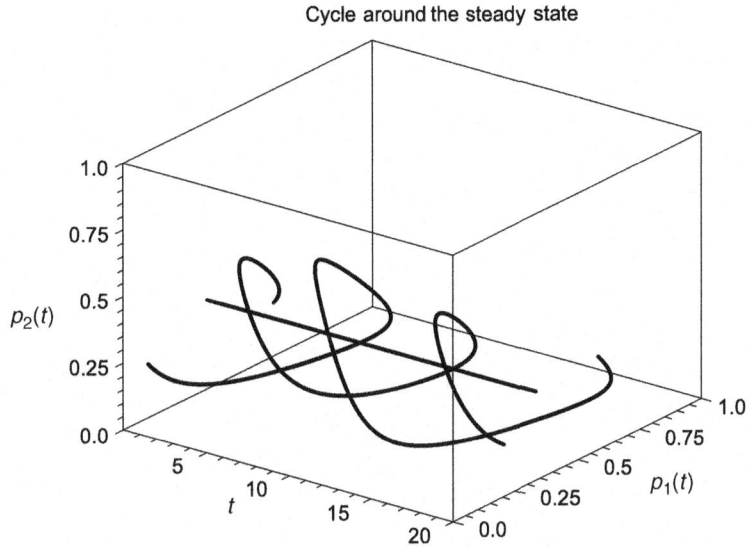

Cycle around the steady state

The case when $a = 0$, $X^* = \left(\frac{1}{3}, \frac{1}{3}, \frac{1}{3}\right)$ is not ESS.

The case when $a = 2$ is shown in the following figure.

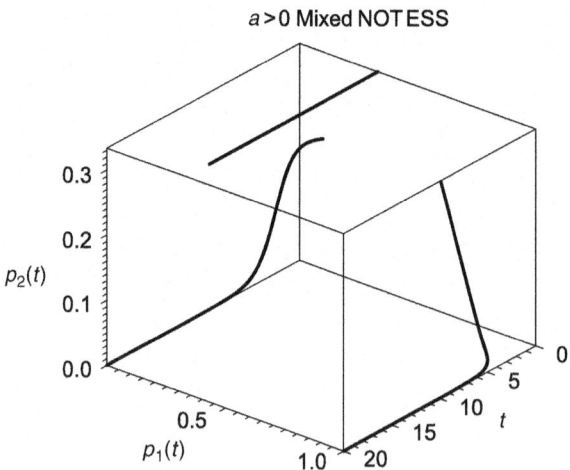

The case $a = 2$, $X^* = \left(\frac{1}{3}, \frac{1}{3}, \frac{1}{3}\right)$ is not ESS.

You can see that unless you start exactly at X^*, the trajectories converge to one of the pure Nash equilibria.

7.14 Find the frequency dynamics (G.3) for the game

$$A = \begin{bmatrix} 0 & 3 & 1 \\ 3 & 0 & 1 \\ 1 & 1 & 1 \end{bmatrix}.$$

Find the steady-state solutions and investigate their stability, then reach a conclusion about the Nash equilibria and the ESSs.

7.14 Answer: There is only one symmetric Nash equilibrium given by $X^* = \left(\frac{1}{2}, \frac{1}{2}, 0\right)$. The frequency dynamics become

$$\frac{dp_1(t)}{dt} = p_1\left(2p_2 - p_1 - 4p_1p_2 + p_1^2 + p_2^2\right),$$
$$\frac{dp_2(t)}{dt} = p_2\left(2p_1 - p_2 - 4p_1p_2 + p_1^2 + p_2^2\right).$$

The stationary solution $p_1 = \frac{1}{2}$, $p_2 = \frac{1}{2}$ is asymptotically stable because $f_{p_1}\left(\frac{1}{2}, \frac{1}{2}\right) + g_{p_2}\left(\frac{1}{2}, \frac{1}{2}\right) = -2 < 0$ and $\det J\left(\frac{1}{2}, \frac{1}{2}\right) = \frac{3}{4} > 0$. We conclude that X^* is asymptotically stable. Even though we have shown it is stable, doesn't mean it is an ESS. We show that directly.

Calculate the best response strategies from

$$\max_Y Y A X^{*T} = \max_Y \left(\frac{1}{2}(y_1 + y_2) + 1\right).$$

Any best response is of the form $Y = (y, 1 - y, 0)$. Next we check if $u(X^*, Y) > u(Y, Y)$ for any such best response $Y \neq X^*$. We have

$$u(X^*, Y) = \frac{3}{2}, \quad u(Y, Y) = 6y - 6y^2$$

Since the maximum of $6y - 6y^2$ is $\frac{3}{2}$ achieved at $y = \frac{1}{2}$, we conclude that indeed $u(X^*, Y) > u(Y, Y)$ as long as $Y \neq X^*$. That means X^* is an ESS.

7.15 Consider the symmetric game with matrix $A = \begin{bmatrix} -a & 1 & -1 \\ -1 & -a & 1 \\ 1 & -1 & -a \end{bmatrix}$, where $a > 0$.

Show that $X^* = (\frac{1}{3}, \frac{1}{3}, \frac{1}{3})$ is an ESS for any $a > 0$. Compare with Problem 7.13.b.

7.15 Answer: The difference between this problem and the matrix in Problem 7.13.b is that we are taking the negative of the matrix. We consider

$$A = \begin{bmatrix} -a & 1 & -1 \\ -1 & -a & 1 \\ 1 & -1 & -a \end{bmatrix} = - \begin{bmatrix} a & -1 & 1 \\ 1 & a & -1 \\ -1 & 1 & a \end{bmatrix}.$$

It still is true that $X^* = (\frac{1}{3}, \frac{1}{3}, \frac{1}{3})$ is the unique symmetric Nash equilibrium. We see this by calculating

$$X^* = \frac{J_3 A^{-1}}{J_3 A^{-1} J_3^T} = \left(\frac{1}{3}, \frac{1}{3}, \frac{1}{3} \right).$$

Now we calculate for any $Y = (y_1, y_2, 1 - y_1 - y_2)$:

$$u(Y, Y) = Y A Y^T = -a \left(1 + 2y_1^2 + (2y_1 + 2y_2)(-1 + y_2) \right),$$
$$u(Y, X^*) = -\frac{a}{3},$$
$$u(X^*, Y) = -\frac{a}{3},$$
$$u(X^*, X^*) = -\frac{a}{3}.$$

Again the condition $u(X^*, X^*) = u(Y, X^*)$ holds and we have to check if $u(X^*, Y) > u(Y, Y)$ for every $Y \neq X^*$. We consider,

$$\text{Minimum } [u(X^*, Y) - u(Y, Y)]$$

over all strategies Y and calculate that this minimum is zero, attained only when $Y = X^*$. Thus, in this case X^* is an ESS. Here is the figure for the replicator dynamics.

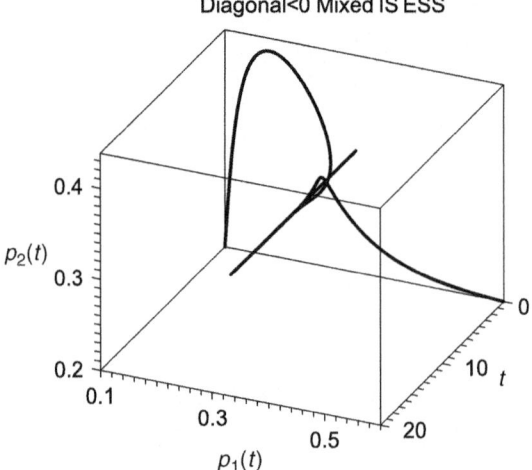

Diagonal<0 Mixed IS ESS

When diagonal< 0, $X^* = \left(\frac{1}{3}, \frac{1}{3}, \frac{1}{3}\right)$ is ESS.

7.16 Consider the symmetric game with matrix $A = \begin{bmatrix} 2 & 1 & 5 \\ 5 & \alpha & 0 \\ 1 & 4 & 3 \end{bmatrix}$, where α is a real number. This problem shows that a stable equilibrium of the replicator dynamics need not be an ESS.

(a) First consider the case $-8 < \alpha < 8.5$. Find the symmetric Nash equilibrium and check if it is an ESS.

7.16.a Answer: To find the completely mixed Nash equilibrium we use Problem 3.25. We have with $B = A^T$,

$$X^* = \frac{J_3 B^{-1}}{J_3 B^{-1} J_3^T} = \left(2 + \frac{55}{\alpha - 36}, \frac{11}{36 - \alpha}, \frac{\alpha + 8}{36 - \alpha}\right)$$

and

$$Y^* = \frac{A^{-1} J_3^T}{J_3 A^{-1} J_3^T} = \left(2 + \frac{55}{\alpha - 36}, \frac{11}{36 - \alpha}, \frac{\alpha + 8}{36 - \alpha}\right)^T.$$

Now these are legitimate strategies if and only if $-8 < \alpha < 8.5$ as you can readily check by solving the inequalities

$$0 < 2 + \frac{55}{\alpha - 36} < 1, 0 < \frac{11}{36 - \alpha} < 1, 0 < \frac{\alpha + 8}{36 - \alpha} < 1.$$

This is the only symmetric mixed Nash and there are no pure Nash equilibria in this case.

Now we check if X^* is an ESS. We calculate

$$u(X^*, X^*) = X^* A X^{*T} = \frac{85 + \alpha}{36 - \alpha},$$

$$u(X, X^*) = X A X^{*T} = \frac{85 + \alpha}{36 - \alpha},$$

and

$$u(X^*, X) - u(X, X)$$

$$= \frac{1}{36 - \alpha} \left(\alpha^2 x_2^2 + (1 - 6x_1 + 6x_2)^2 - \alpha \left((-2 + x_1)^2 - 2(9 + x_1)x_2 + 37x_2^2 \right) \right)$$

We minimize this over all $(x_1, x_2), 0 \le x_1, x_2 \le 1, x_1 + x_2 \le 1$. If this minimum is zero X^* will be an ESS. In this case it is not zero. We conclude that in the case $-8 < \alpha < 8.5$, the unique mixed Nash equilibrium is not an ESS.

(b) Take $\alpha = 2$. Analyze the mixed Nash equilibrium in this case and determine if it is asymptotically stable.

7.16.b Answer: Let $\pi = (p_1, p_2, 1 - p_1, p_2)$. The replicator equations for this game are as follows:

$$\frac{dp_1}{dt} = f(p_1, p_2) = p_1 (2 + p_1^2 - p_2(2 + p_2) - p_1(3 + 2p_2)),$$

$$\frac{dp_2}{dt} = g(p_1, p_2) = p_2 \left(-3 + p_1(5 + p_1) + 4p_2 - 2p_1 p_2 - p_2^2 \right).$$

When $\alpha = 2$ we have $X^* = (0.38235, 0.3235, 0.29411)$ and we calculate

$$f_{p_1} + g_{p_2}(0.38235, 0.3235) = -0.264 < 0$$

and

$$\det \begin{bmatrix} f_{p_1} & f_{p_2} \\ g_{p_1} & g_{p_2} \end{bmatrix} (0.38235, 0.3235) = 1.237 > 0.$$

We conclude that X^* is asymptotically stable, even though it is not an ESS.

The Main Definitions and Theorems

Appendix A: Matrix Two-Person Games

Definition A.1 *A matrix game with matrix $A_{n \times m} = (a_{ij})$ has the lower value*

$$v^- \equiv \max_{i=1,\dots,n} \min_{j=1,\dots,m} a_{ij}.$$

and the upper value

$$v^+ \equiv \min_{j=1,\dots,m} \max_{i=1,\dots,n} a_{ij},$$

*The lower value v^- is the smallest amount that player I is guaranteed to receive (v^- is player I's gain floor), and the upper value v^+ is the guaranteed greatest amount that player II can lose (v^+ is player II's loss ceiling). The **game has a value** if $v^- = v^+$, and we write it as $v = v(A) = v^+ = v^-$. This means that the smallest max and the largest min must be equal and the row and column i^*, j^* giving the payoffs $a_{i^*,j^*} = v^+ = v^-$ are **optimal**, or a **saddle point in pure strategies**.*

Definition A.2 *We call a particular row i^* and column j^* a **saddle point in pure strategies** of the game if*

$$a_{ij^*} \le a_{i^* j^*} \le a_{i^* j}, \quad \text{for all rows } i = 1, \dots, n \text{ and columns } j = 1, \dots, m. \quad \text{(A.1)}$$

Theorem A.3 *Let $f : C \times D \to \mathbb{R}$ be a continuous function. Let $C \subset \mathbb{R}^n$ and $D \subset \mathbb{R}^m$ be convex, closed, and bounded. Suppose that $x \mapsto f(x, y)$ is concave and $y \mapsto f(x, y)$ is convex. Then*

$$v^+ = \min_{y \in D} \max_{x \in C} f(x, y) = \max_{x \in C} \min_{y \in D} f(x, y) = v^-.$$

Solutions Manual to Accompany Game Theory: An Introduction, Second Edition. E.N. Barron.
© 2013 John Wiley & Sons, Inc. Published 2013 by John Wiley & Sons, Inc.

Definition A.4 *A **mixed strategy** is a vector (or $1 \times n$ matrix) $X = (x_1, \ldots, x_n)$ for player I and $Y = (y_1, \ldots, y_m)$ for player II, where*

$$x_i \geq 0, \sum_{i=1}^{n} x_i = 1 \quad and \quad y_j \geq 0, \sum_{j=1}^{m} y_j = 1.$$

The components x_i represent the probability that row i will be used by player I, so $x_i = Prob(I \text{ uses row } i)$, and y_j the probability column j will be used by player II, that is, $y_j = Prob(II \text{ uses row } j)$. Denote the set of mixed strategies with k components by

$$S_k \equiv \left\{ (z_1, z_2, \ldots, z_k) \mid z_i \geq 0, i = 1, 2, \ldots, k, \sum_{i=1}^{k} z_i = 1 \right\}.$$

In this terminology, a mixed strategy for player I is any element $X \in S_n$ and for player II any element $Y \in S_m$. A pure strategy $X \in S_n$ is an element of the form $X = (0, 0, \ldots, 0, 1, 0, \ldots, 0)$, which represents always playing the row corresponding to the position of the 1 in X.

Definition A.5 *Given a choice of mixed strategy $X \in S_n$ for player I and $Y \in S_m$ for player II, chosen independently, the **expected payoff** to player I of the game is*

$$E(X, Y) = \sum_{i=1}^{n} \sum_{j=1}^{m} a_{ij} Prob(I \text{ uses } i \text{ and } II \text{ uses } j)$$

$$= \sum_{i=1}^{n} \sum_{j=1}^{m} a_{ij} Prob(I \text{ uses } i) P(II \text{ uses } j)$$

$$= \sum_{i=1}^{n} \sum_{j=1}^{m} x_i a_{ij} y_j = X A Y^T.$$

In a zero sum two-person game, the expected payoff to player II would be $-E(X, Y)$. The independent choice of strategy by each player justifies the fact that

$$Prob(I \text{ uses } i \text{ and } II \text{ uses } j) = Prob(I \text{ uses } i) P(II \text{ uses } j).$$

Definition A.6 *A **saddle point in mixed strategies** is a pair (X^*, Y^*) of probability vectors $X^* \in S_n, Y^* \in S_m$, which satisfies*

$$E(X, Y^*) \leq E(X^*, Y^*) \leq E(X^*, Y), \quad \forall \, (X \in S_n, Y \in S_m).$$

*If player I decides to use a strategy other than X^* but player II still uses Y^*, then I receives an expected payoff smaller than that obtainable by sticking with X^*. A similar statement holds for player II. So (X^*, Y^*) is an **equilibrium** in this sense.*

Theorem A.7   *For any $n \times m$ matrix A, we have*

$$v^+ = \min_{Y \in S_m} \max_{X \in S_n} XAY^T = \max_{X \in S_n} \min_{Y \in S_m} XAY^T = v^-.$$

*The common value is denoted $v(A)$, or $value(A)$, and that is the **value of the game**. In addition, there is at least one saddle point $X^* \in S_n$, $Y^* \in S_m$ so that*

$$E(X, Y^*) \le E(X^*, Y^*) = v(A) \le E(X^*, Y), \quad \text{for all } X \in S_n, Y \in S_m.$$

Theorem A.8   *Let $A = (a_{ij})$ be an $n \times m$ game with value $v(A)$. Let w be a real number. Let $X^* \in S_n$ be a strategy for player I and $Y^* \in S_m$ be a strategy for player II.*

(a) *If $w \le E(X^*, j) = X^* A_j = \sum_{i=1}^n x_i^* a_{ij}$, $j = 1, \ldots, m$, then $w \le v(A)$.*

(b) *If $w \ge E(i, Y^*) = {}_i A \, Y^{*T} = \sum_{j=1}^m a_{ij} y_j^*$, $i = 1, 2, \ldots, n$, then $w \ge v(A)$.*

(c) *If $E(i, Y^*) = {}_i A \, Y^{*T} \le w \le E(X^*, j) = X^* A_j, i = 1, 2, \ldots, n, \ j = 1, 2, \ldots, m$, then $w = v(A)$ and (X^*, Y^*) is a saddle point for the game.*

(d) *If $v(A) \le E(X^*, j)$ for all columns $j = 1, 2, \ldots, m$, then X^* is optimal for player I. If $v(A) \ge E(i, Y^*)$ for all rows $i = 1, 2, \ldots, n$, then Y^* is optimal for player II.*

(e) *A strategy X^* for player I is optimal (i.e., part of a saddle point) if and only if $v(A) = \min_{1 \le j \le m} E(X^*, j)$. A strategy Y^* for player II is optimal if and only if $v(A) = \max_{1 \le i \le n} E(i, Y^*)$.*

Proposition A.9   *The value of the game, $v(A)$, and the optimal strategies $X^* = (x_1^*, \ldots, x_n^*)$ for player I and $Y^* = (y_1^*, \ldots, y_m^*)$ for player II must satisfy the system of equations $E(i, Y^*) = v(A)$ for each row with $x_i^* > 0$ and $E(X^*, j) = v(A)$ for every column j with $y_j^* > 0$. In particular, if $x_i > 0, x_k > 0$, then $E(i, Y^*) = E(k, Y^*)$. Similarly, if $y_j^* > 0, y_\ell^* > 0$, then $E(X^*, j) = E(X^*, \ell)$.*

Definition A.10   *A mixed strategy X^* for player I is a **best response strategy** to the strategy Y for player II if it satisfies*

$$\max_{X \in S_n} E(X, Y) = \max_{X \in S_n} \sum_{i=1}^n \sum_{j=1}^m x_i^* a_{ij} y_j = E(X^*, Y).$$

A mixed strategy Y^* for player II is a best response strategy to the strategy X for player I if it satisfies

$$\min_{Y \in S_m} E(X, Y) = \min_{Y \in S_m} \sum_{i=1}^n \sum_{j=1}^m x_i a_{ij} y_j^* = E(X, Y^*).$$

A.1 Properties of Optimal Strategies

1. A number v is the value of the game and and (X, Y) is a saddle point if and only if $E(i, Y) \leq v \leq E(X, j), i = 1, \ldots, n, j = 1, \ldots, m$.

2. If X is a strategy for player I and $value(A) \leq E(X, j), j = 1, \ldots, m$, then X is optimal for player I.

   If Y is a strategy for player II and $value(A) \geq E(i, Y), i = 1, \ldots, m$, then Y is optimal for player II.

3. If Y is optimal for II and $y_j > 0$, then $E(X, j) = value(A)$ for any optimal mixed strategy X for I. Similarly, if X is optimal for I and $x_i > 0$, then $E(i, Y) = value(A)$ for any optimal Y for II. In symbols,

$$y_j > 0 \Rightarrow E(X, j) = v(A), \quad \text{and} \quad x_i > 0 \Rightarrow E(i, Y) = v(A).$$

   Thus, if any optimal mixed strategy for a player has a strictly positive probability of using a row or a column, then that row or column played against any optimal opponent strategy will yield the value. This result is also called the **Equilibrium Theorem**.

4. If X is any optimal strategy for player I and $E(X, j) > value(A)$ for some column j, then for any optimal strategy Y for player II, we must have $y_j = 0$. Player II would never use column j in any optimal strategy for player II. Similarly, if Y is any optimal strategy for player II and $E(i, Y) < value(A)$, then any optimal strategy X for player I must have $x_i = 0$. If row i for player I gives a payoff when played against an optimal strategy for player II strictly below the value of the game, then player I would never use that row in any optimal strategy for player I. In symbols, if $(X = (x_i), Y = (y_j))$ is optimal, then

$$E(X, j) > v(A) \Rightarrow y_j = 0, \quad \text{and} \quad E(i, Y) < v(A) \Rightarrow x_i = 0.$$

5. If for any optimal strategy Y for player II, $y_j = 0$, then there is an optimal strategy X for player I so that $E(X, j) > value(A)$. If for any optimal strategy X for I, $x_i = 0$, then there is an optimal strategy Y for II so that $E(i, Y) < value(A)$. This is the converse statement to property 4.

6. If player I has more than one optimal strategy, then player I's set of optimal strategies is a convex, closed, and bounded set. Also, if player II has more than one optimal strategy, then player II's set of optimal strategies is a convex, closed, and bounded set.

Appendix B: Solution Methods for Matrix Games

Theorem B.1 *In the 2×2 game with matrix A,* **assume that there are no pure optimal strategies**. *If we set*

$$x^* = \frac{a_{22} - a_{21}}{a_{11} - a_{12} - a_{21} + a_{22}}, \qquad y^* = \frac{a_{22} - a_{12}}{a_{11} - a_{12} - a_{21} + a_{22}},$$

then $X^* = (x^*, 1 - x^*)$, $Y^* = (y^*, 1 - y^*)$ are optimal mixed strategies for players I and II, respectively. The value of the game is

$$v(A) = E(X^*, Y^*) = \frac{a_{11}a_{22} - a_{12}a_{21}}{a_{11} - a_{12} - a_{21} + a_{22}}.$$

Theorem B.2 *Assume that*
  1. $A_{n \times n}$ has an inverse A^{-1};
  2. $J_n A^{-1} J_n^T \neq 0$;
  3. $v(A) \neq 0$.
  Set $X = (x_1, \ldots, x_n)$, $Y = (y_1, \ldots, y_m)$, and

$$v \equiv \frac{1}{J_n A^{-1} J_n^T}, \qquad Y^T = \frac{A^{-1} J_n^T}{J_n A^{-1} J_n^T}, \qquad X = \frac{J_n A^{-1}}{J_n A^{-1} J_n^T}.$$

If $x_i \geq 0$, $i = 1, \ldots, n$ and $y_j \geq 0$, $j = 1, \ldots, n$, we have that $v = v(A)$ is the value of the game with matrix A and (X, Y) is a saddle point in mixed strategies.

Definition B.3 *A game is* **completely mixed** *if every saddle point consisting of strategies $X = (x_1, \ldots, x_n) \in S_n$, $Y = (y_1, \ldots, y_m) \in S_m$ satisfies the property $x_i > 0$, $i = 1, 2, \ldots, n$ and $y_j > 0$, $j = 1, 2, \ldots, m$. Every row and every column is used with positive probability.*

Solutions Manual to Accompany Game Theory: An Introduction, Second Edition. E.N. Barron.
© 2013 John Wiley & Sons, Inc. Published 2013 by John Wiley & Sons, Inc.

B.1 Linear Programming Methods

B.1.1 METHOD 1

$$\text{Player I's program} = \begin{cases} \text{Minimize } z_{\text{I}} = \mathbf{p}\, J_n^T = \sum_{i=1}^{n} p_i, \quad J_n = (1, 1, \ldots, 1) \\ \text{subject to } \mathbf{p}\, A \geq J_m, \qquad\qquad\quad \mathbf{p} \geq 0. \end{cases}$$

The set of constraints is

$$\mathbf{p}\, A \geq J_m \iff \mathbf{p} \cdot A_j \geq 1,\, j = 1, \ldots, m.$$

Also $\mathbf{p} \geq 0$ means $p_i \geq 0,\, i = 1, \ldots, n$.

Once we solve player I's program, we will have in our hands the optimal $\mathbf{p} = (p_1, \ldots, p_n)$ that minimizes the objective $z_{\text{I}} = \mathbf{p}\, J_n^T$. The solution will also give us the minimum objective z_{I}, labeled z_{I}^*.

Unwinding the formulation back to our original variables, we find the optimal strategy X for player I and the value of the game as follows:

$$\boxed{value(A) = \frac{1}{\sum_{i=1}^{n} p_i} = \frac{1}{z_{\text{I}}^*} \quad \text{and} \quad x_i = p_i\, value(A).}$$

$$\text{Player II's program} = \begin{cases} \text{Maximize } z_{\text{II}} = \mathbf{q}\, J_m^T, J_m = (1, 1, \ldots, 1) \\ \text{subject to } A\, \mathbf{q}^T \leq J_n^T,\ \mathbf{q} \geq 0. \end{cases}$$

Player II's problem is the **dual** of player I's. At the conclusion of solving this program we are left with the optimal maximizing vector $\mathbf{q} = (q_1, \ldots, q_m)$ and the optimal objective value z_{II}^*. We obtain the optimal mixed strategy for player II and the value of the game from

$$\boxed{value(A) = \frac{1}{\sum_{j=1}^{m} q_j} = \frac{1}{z_{\text{II}}^*} \quad \text{and} \quad y_j = q_j\, value(A).}$$

B.1.2 METHOD 2

Player I

Choose a mixed strategy $X^* = (x_1^*, \ldots, x_n^*)$ so as to

$$\text{Maximize } v$$

subject to the constraints

$$\sum_{i=1}^{n} a_{ij} x_i^* = X^* A_j = E(X^*, j) \geq v, \quad j = 1, \ldots, m;$$

and

$$\sum_{i=1}^{n} x_i^* = 1, \; x_i \geq 0, \quad i = 1, \ldots, n.$$

Player II
    Choose a strategy $Y^* = (y_j{}^*)$ so as to

$$\text{Minimize } v$$

subject to the constraints

$$\sum_{j=1}^{m} a_{ij} y_j^* = {}_i A Y^{*T} = E(i, Y^*) \leq v, \quad i = 1, \ldots, n;$$

and

$$\sum_{j=1}^{m} y_j^* = 1, \; y_j \geq 0, \quad j = 1, \ldots, m.$$

Appendix C: Two-Person Nonzero Sum Games

Definition C.1   *A pair of mixed strategies $(X^* \in S_n, Y^* \in S_m)$ is a Nash equilibrium if $E_I(X, Y^*) \leq E_I(X^*, Y^*)$ for every mixed $X \in S_n$ and $E_{II}(X^*, Y) \leq E_{II}(X^*, Y^*)$ for every mixed $Y \in S_m$. If (X^*, Y^*) is a Nash equilibrium we denote by $v_A = E_I(X^*, Y^*)$ and $v_B = E_{II}(X^*, Y^*)$ as the optimal payoff to each player. Written out with the matrices, (X^*, Y^*) is a Nash equilibrium if*

$$E_I(X^*, Y^*) = X^* A Y^{*T} \geq X A Y^{*T} = E_I(X, Y^*), \quad \text{for every } X \in S_n,$$

$$E_{II}(X^*, Y^*) = X^* B Y^{*T} \geq X^* B Y^{T} = E_{II}(X^*, Y), \quad \text{for every } Y \in S_m.$$

Definition C.2   *A strategy $X^0 \in S_n$ is a **best response strategy** to a given strategy $Y^0 \in S_m$ for player II, if*

$$E_I(X^0, Y^0) = \max_{X \in S_n} E_I(X, Y^0).$$

*Similarly, a strategy $Y^0 \in S_m$ is a **best response strategy** to a given strategy $X^0 \in S_n$ for player I, if*

$$E_{II}(X^0, Y^0) = \max_{Y \in S_m} E_{II}(X^0, Y).$$

Definition C.3   *Consider the bimatrix game with matrices (A, B). The **safety value** for player I is $value(A)$. The **safety value** for player II in the bimatrix game is $value(B^T)$.*
   *If A has the saddle point (X^A, Y^A), then X^A is called the **maxmin strategy** for player I.*
   *If B^T has saddle point (X^{B^T}, Y^{B^T}), then X^{B^T} is the **maxmin strategy** for player II.*

Theorem C.4   **(Equality of Payoffs Theorem)** *Suppose that*

$$X^* = (x_1, x_2, \ldots, x_n), \qquad Y^* = (y_1, y_2, \ldots, y_m)$$

is a Nash equilibrium for the bimatrix game (A, B).

Solutions Manual to Accompany Game Theory: An Introduction, Second Edition. E.N. Barron.
© 2013 John Wiley & Sons, Inc. Published 2013 by John Wiley & Sons, Inc.

For any row k that has a positive probability of being used, $x_k > 0$, we have $E_I(k, Y^*) = E_I(X^*, Y^*) \equiv v_I$.

For any column j that has a positive probability of being used, $y_j > 0$, we have $E_{II}(X^*, j) = E_{II}(X^*, Y^*) \equiv v_{II}$. That is,

$$x_k > 0 \Rightarrow E_I(k, Y^*) = v_I,$$
$$y_j > 0 \Rightarrow E_{II}(X^*, j) = v_{II}.$$

Definition C.5   *The **best response sets** for each player are defined as*

$$BR_I(Y) = \{X \in S_n \mid E_I(X, Y) = \max_{p \in S_n} E_I(p, Y)\},$$
$$BR_{II}(X) = \{Y \in S_m \mid E_{II}(X, Y) = \max_{t \in S_m} E_I(X, t)\}.$$

Theorem C.6   *(Lemke–Howson)Consider the two-person game with matrices (A, B) for players I and II. Then, $(X^* \in S_n, Y^* \in S_m)$ is a Nash equilibrium if and only if they satisfy, along with scalars p^*, q^* the nonlinear program:*

$$\max_{X, Y, p, q} \; XAY^T + XBY^T - p - q$$

subject to

$$AY^T \le pJ_n^T$$
$$B^T X^T \le qJ_m^T \qquad \text{(equivalently } XB \le qJ_m)$$
$$x_i \ge 0, y_j \ge 0, \quad XJ_n = 1 = YJ_m^T,$$

where $J_k = (1\ 1\ 1\ \cdots\ 1)$ is the $1 \times k$ row vector consisting of all 1s. In addition, $p^* = E_I(X^*, Y^*)$ and $q^* = E_{II}(X^*, Y^*)$.

Proposition C.7   *Consider the nonzero sum game $(A_{n \times n}, B_{n \times n})$ and suppose that A^{-1} and B^{-1} exist. If $J_n A^{-1} J_n^T \ne 0, J_n B^{-1} J_n^T \ne 0$ and*

$$X^* = \frac{J_n B^{-1}}{J_n B^{-1} J_n^T}, \qquad Y^{*T} = \frac{A^{-1} J_n^T}{J_n A^{-1} J_n^T}$$

are legitimate strategies, then (X^*, Y^*) is a Nash equilibrium, and

$$v_I = \frac{1}{J_n A^{-1} J_n^T} = E_I(X^*, Y^*), \qquad v_{II} = \frac{1}{J_n B^{-1} J_n^T} = E_{II}(X^*, Y^*).$$

Proposition C.8   *If A, B are 2×2 matrices, set*

$$A^* = \begin{bmatrix} a_{22} & -a_{12} \\ -a_{21} & a_{11} \end{bmatrix} \quad \text{and} \quad B^* = \begin{bmatrix} b_{22} & -b_{12} \\ -b_{21} & b_{11} \end{bmatrix}.$$

Assuming both rows and columns are played with positive probability,

$$X^* = \left(\frac{b_{22} - b_{21}}{b_{11} - b_{12} + b_{22} - b_{21}}, \frac{b_{11} - b_{12}}{b_{11} - b_{12} + b_{22} - b_{21}} \right) = \frac{J_n B^*}{J_n B^* J_n^T},$$

and

$$Y^{*T} = \left(\frac{a_{22} - a_{12}}{a_{11} - a_{21} + a_{22} - a_{12}}, \frac{a_{11} - a_{21}}{a_{11} - a_{21} + a_{22} - a_{12}} \right) = \frac{A^* J_n^T}{J_n A^* J_n^T}$$

(X^*, Y^*) is a Nash equilibrium if they are strategies, and then

$$v_I = \frac{\det(A)}{J_n A^* J_n^T} = E_I(X^*, Y^*), \qquad v_{II} = \frac{\det(B)}{J_n B^* J_n^T} = E_{II}(X^*, Y^*).$$

Definition C.9   *A distribution $P = (p_{ij})$ is a **correlated equilibrium** if*

$$\sum_{j=1}^{m} a_{ij} p_{ij} \geq \sum_{j=1}^{m} a_{qj} p_{ij}$$

for all rows $i = 1, 2, \ldots, n$, $q = 1, 2, \ldots, n$, and

$$\sum_{i=1}^{n} b_{ij} p_{ij} \geq \sum_{i=1}^{n} b_{ir} p_{ij}$$

for all columns $j = 1, 2, \ldots, m$, $r = 1, 2, \ldots, m$.

Definition C.10   *The social welfare payoff of a game is the maximum sum of each individual payoff. That is,*

$$\text{Maximum } a_{ij} + b_{ij}, \quad i = 1, 2, \ldots, n, \quad j = 1, 2, \ldots, m.$$

*The social welfare of a **pure strategy** (i^*, j^*) is $a_{i^* j^*} + b_{i^* j^*}$.*
*The expected social welfare of a **mixed pair** (X, Y) is*

$$\sum_{i=1}^{n} \sum_{j=1}^{m} (a_{ij} + b_{ij}) x_i y_j = E_I(X, Y) + E_{II}(X, Y).$$

*The expected social welfare of a **distribution** P is*

$$\sum_{i=1}^{n} \sum_{j=1}^{m} (a_{ij} + b_{ij}) p_{ij} = E_P(I) + E_P(II).$$

LP Problem for a Correlated Equilibrium C.11

$$Maximize \sum_{i,j} p_{i,j}(a_{i,j} + b_{i,j})$$

over variables $P = (p_{ij})$ subject to

$$\sum_j a_{ij} p_{ij} \geq \sum_j a_{qj} p_{ij}$$

for all rows $i = 1, 2, \ldots, n, q = 1, 2, \ldots, n$, and

$$\sum_{i=1}^n b_{ij} p_{ij} \geq \sum_{i=1}^n b_{ir} p_{ij}$$

for all columns $j = 1, 2, \ldots, m, r = 1, 2, \ldots, m.$

$$\sum_{i,j} p_{i,j} = 1$$

$$p_{i,j} \geq 0.$$

Definition C.12   *Given a collection of payoff functions*

$$(u_1(q_1, \ldots, q_n), \ldots, u_n(q_1, \ldots, q_n))$$

for an n-person nonzero sum game, where the q_i is a pure or mixed strategy for player $i =$ 1, 2, \ldots, n, we say that $(q_1^*, \ldots, q_n^*)$ is Pareto-optimal if there does not exist any other strategy for any of the players that makes that player better off, that is, increases her or his payoff, without making other players worse off, namely, decreasing at least one other player's payoff.

Definition C.13   *A Nash equilibrium is* **payoff-dominant** *if it is Pareto-optimal compared to all other Nash equilibria in the game.*

C.1 Summary of Methods for Finding Mixed Nash Equilibria

The methods we have for finding the mixed strategies for nonzero sum games are recapped here.

1. Equality of payoffs. Suppose that we have mixed strategies $X^* = (x_1, \ldots, x_n)$ and $Y^* = (y_1, \ldots, y_m)$. For any rows $k_1, k_2, \ldots$ that have a positive probability of being used, the expected payoffs to player I for using any of those rows must be equal: $E_I(k_r, Y^*) = E_I(k_s, Y^*) = E_I(X^*, Y^*)$. You can find Y^* from these equations. Similarly, for any columns j that have a positive probability of being used, we have $E_{II}(X^*, j_r) = E_{II}(X^*, j_s) = E_{II}(X^*, Y^*)$. You can find X^* from these equations.

2. You can use the calculus method directly by computing

$$f(x_1, \ldots, x_{n-1}, y_1, \ldots, y_{m-1}) = \left(x_1, \ldots, x_{n-1}, 1 - \sum_{i=1}^{n-1} x_i \right) A \begin{bmatrix} y_1 \\ y_2 \\ \vdots \\ 1 - \sum_{j=1}^{m-1} y_j \end{bmatrix}$$

and then

$$\frac{\partial f}{\partial x_i} = 0, \quad i = 1, 2, \ldots, n-1.$$

This will let you find Y^*. Next, compute

$$g(x_1, \ldots, x_{n-1}, y_1, \ldots, y_{m-1}) = \left(x_1, \ldots, x_{n-1}, 1 - \sum_{i=1}^{n-1} x_i \right) B \begin{bmatrix} y_1 \\ y_2 \\ \vdots \\ 1 - \sum_{j=1}^{m-1} y_j \end{bmatrix},$$

and then,

$$\frac{\partial g}{\partial y_j} = 0, \quad j = 1, 2, \ldots, m-1.$$

From these you will find X^*.

3. You can use the system of equations to find interior Nash points given by

$$\sum_{j=1}^{m} y_j [a_{kj} - a_{nj}] = 0, \qquad k = 1, 2, \ldots, n-1,$$

$$\sum_{i=1}^{n} x_i [b_{is} - b_{im}] = 0, \qquad s = 1, 2, \ldots, m-1.$$

$$x_n = 1 - \sum_{i=1}^{n-1} x_i, \quad y_m = 1 - \sum_{j=1}^{m-1} y_j.$$

4. If you are in the 2×2 case, or if you have square invertible $n \times n$ matrices you may use the formulas derived in Problem 3.25.

5. In the 2×2 case, you can find the rational reaction sets for each player and see where they intersect. This gives all the Nash equilibria including the pure ones.

6. Use the nonlinear programming method: set up the objective, the constraints, and solve. Use the option `initialpoint` to modify the starting point the algorithm uses to find additional Nash points.

Appendix D: Games in Extensive Form: Sequential Decision Making

Definition D.1 *A subgame perfect equilibrium for an extensive form game is a Nash equilibrium whose restriction to any subgame is also a Nash equilibrium of this subgame.*

Theorem D.2 *Any finite tree game in extensive form has at least one subgame perfect equilibrium.*

Theorem D.3 *(1) Any finite game in extensive form with perfect information has at least one subgame perfect equilibrium in pure strategies.*

(2) Suppose all payoffs at all terminal nodes are distinct for any player. Then there is one and only one subgame perfect equilibrium.

(3) If a game has no proper subgames, then every Nash equilibrium is subgame perfect.

Solutions Manual to Accompany Game Theory: An Introduction, Second Edition. E.N. Barron.
© 2013 John Wiley & Sons, Inc. Published 2013 by John Wiley & Sons, Inc.

Appendix E: N-Person Nonzero Sum Games and Games with a Continuum of Strategies

Definition E.1   *A collection of strategies $q^* = (q_1^*, \ldots, q_n^*) \in Q_1 \times \cdots \times Q_N$ is a pure Nash equilibrium for the game with payoff functions $\{u_i(q_1, \ldots, q_n)\}$, $i = 1, \ldots, N$, if for each player $i = 1, \ldots, N$, we have*

$$u_i(q_1^*, \ldots, q_{i-1}^*, q_i^*, q_{i+1}^*, \ldots, q_N^*)$$
$$\geq u_i(q_1^*, \ldots, q_{i-1}^*, q_i, q_{i+1}^*, \ldots, q_N^*), \text{ for all } q_i \in Q_i.$$

A short hand way to write this is $u_i(q_i^*, q_{-i}^*) \geq u_i(q_i, q_{-i}^*)$ for all $q_i \in Q_i$, where q_{-i} refers to all the players except the ith.

Definition E.2   *Given payoff functions $u_i(q_1, \ldots, q_n)$, a best response of player k, written $q_k = BR_k(q_1, \ldots, q_{k-1}, q_{k+1}, \ldots, q_n)$ is a value that satisfies*

$$u_k(q_1, \ldots, q_n) = \max_{t \in Q_k} u_k(q_1, \ldots, q_{k-1}, t, q_{k+1}, \ldots, q_n)$$

or $u_k(q_k, q_{-k}) = \max_{t \in Q_k} u_k(t, q_{-k})$. In other words, q_k provides the maximum payoff for player k, given the values of the other player's $q_i's$.

In general, we say that $q_k = BR_k(q_{-k}) \in \arg\max_x u_k(q_k, q_{-k})$[1] where we use the notation

$$q_{-k} = (q_1, \ldots, q_{k-1}, q_{k+1}, \ldots, q_n) = \text{ all players except } k,$$

$$(q_k, q_{-k}) = (q_1, \ldots, q_{k-1}, q_k, q_{k+1}, \ldots, q_n) = \text{ all players including } k.$$

Theorem E.3   *Let $Q_1 \subset \mathbb{R}^n$ and $Q_2 \subset \mathbb{R}^m$ be compact and convex sets.*

1. *Suppose that the payoff functions $u_i : Q_1 \times Q_2 \to \mathbb{R}$, $i = 1, 2$, satisfy*

   (a) *u_1 and u_2 are continuous;*

---

[1]The arg max of a function is the set of points where the maximum is attained.

Solutions Manual to Accompany Game Theory: An Introduction, Second Edition. E.N. Barron.
© 2013 John Wiley & Sons, Inc. Published 2013 by John Wiley & Sons, Inc.

(b) $q_1 \mapsto u_1(q_1, q_2)$ *is concave for each fixed* q_2;

(c) $q_2 \mapsto u_2(q_1, q_2)$ *is concave for each fixed* q_1.

Then, there is a Nash equilibrium for (u_1, u_2).

2. *Let* $Q_i \subset \mathbb{R}^{n_i}$ *be convex and compact. If we have N payoff functions* $u_i : Q_1 \times Q_2 \times \cdots \times Q_N \to \mathbb{R}$, $i = 1, 2, \ldots, N$, *which are continuous and* $q_i \mapsto u_i(q_i, q_{-i})$ *is concave, then there is a Nash equilibrium* $(q_1^*, \ldots, q_N^*)$.

Appendix F: Cooperative Games

Definition F.1 *Let 2^N denote the set of all possible coalitions for the players N. If $S = \{i\}$ is a coalition containing the single member i, we simply denote S by i.*
   Any function $v : 2^N \to \mathbb{R}$ satisfying

$$v(\emptyset) = 0 \quad and \quad v(N) \geq \sum_{i=1}^{n} v(i)$$

is a characteristic function (of an n-person cooperative game).

Definition F.2 *Let x_i be a real number for each $i = 1, 2, \ldots, n$, with $\sum_{i=1}^{n} x_i \leq v(N)$. A vector $\vec{x} = (x_1, \ldots, x_n)$ is an **imputation** if*

- $x_i \geq v(i)$ (**individual rationality**);

- $\sum_{i=1}^{n} x_i = v(N)$ (**group rationality**).

*Each x_i represents the share of the value of $v(N)$ received by player i. The imputation $\vec{x}$ is also called a **payoff vector** or an **allocation**, and we will use these words interchangeably.*

Definition F.3 *Let $S \subset N$ be a coalition and let $\vec{x} \in X$. The **excess** of coalition $S \subset N$ for imputation $\vec{x} \in X$ is defined by*

$$e(S, x) = v(S) - \sum_{i \in S} x_i.$$

It is the amount by which the rewards allocated to the coalition S differs from the benefits associated with S.

Solutions Manual to Accompany Game Theory: An Introduction, Second Edition. E.N. Barron.
© 2013 John Wiley & Sons, Inc. Published 2013 by John Wiley & Sons, Inc.

The **core of the game** *is*

$$C(0) = \{\vec{x} \in X \mid e(S, \vec{x}) \leq 0, \ \forall S \subset N\} = \{\vec{x} \in X \mid v(S) \leq \sum_{i \in S} x_i, \ \forall S \subset N\}.$$

The core is the set of allocations so that each coalition receives at least the rewards associated with that coalition. The core may be empty.

   *The ε-***core***, for $-\infty < \varepsilon < +\infty$, is*

$$C(\varepsilon) = \{\vec{x} \in X \mid e(S, \vec{x}) \leq \varepsilon, \ \forall S \subset N, S \neq N, S \neq \emptyset\}.$$

Let $\varepsilon^1 \in (-\infty, \infty)$ be the first ε for which $C(\varepsilon) \neq \emptyset$. The **least core***, labeled X^1, is $C(\varepsilon^1)$. It is possible for ε^1 to be positive, negative, or zero.*

Definition F.4   *If we have two imputations $\vec{x} \in X, \vec{y} \in X$, and a nonempty coalition $S \subset N$, then $\vec{x}$* **dominates** *$\vec{y}$ (for the coalition S) if $x_i > y_i$ for all members $i \in S$, and $\vec{x}(S) = \sum_{i \in S} x_i \leq v(S)$. If $\vec{x}$ dominates $\vec{y}$ for the coalition S, we write $\vec{x} \succ_S \vec{y}$.*

Theorem F.5   *The core of a game is the set of all undominated imputations for the game; that is,*

$$C(0) = \{\vec{x} \in X \mid \text{there is no } \vec{z} \in X \text{ and } S \subset N \text{ such that } \vec{z} \succ_S \vec{x}\}.$$
$$= \{\vec{x} \in X \mid \text{there is no } \vec{z} \in X \text{ and } S \subset N \text{ such that}$$
$$z_i > x_i, \forall \ i \in S, \text{ and } \sum_{i \in S} z_i \leq v(S)\}.$$

Proposition F.6   *Let $\delta_i = v(N) - v(N - i)$. Then, $C(0) = \emptyset$ if $\sum_{i=1}^{n} \delta_i < v(N)$.*

Proposition F.7   *We take $N = \{1, 2, 3\}$ and a characteristic function in normalized form*

$$v(i) = v(\emptyset) = 0, \quad i = 1, 2, 3, \quad v(123) = 1,$$
$$v(12) = a_{12}, \quad v(13) = a_{13}, \quad v(23) = a_{23}.$$

(We have $0 \leq a_{ij} \leq 1$.) For the three-person cooperative game with normalized characteristic function v, we have $C(0) \neq \emptyset$ if and only if

$$a_{12} + a_{13} + a_{23} \leq 2.$$

Theorem F.8   *(Leng and Parlar) Empty Core. Suppose we have the three-player cooperative game with players $N = \{1, 2, 3\}$ and characteristic function $v(S), S \subset N$ with $v(i) = 0, i = 1, 2, 3, v(S) \geq 0$. If $C(0) = \emptyset$ the nucleolus allocation is given by*

$$x_i = \frac{v(N) + v(ij) + v(ik) - 2v(jk)}{3}, \quad i, j, k = 1, 2, 3 \text{ and } i \neq j \neq k. \quad \text{(F.1)}$$

Theorem F.9 *(Leng and Parlar) Nonempty Core. Let $\vec{x}^* = (x_1, x_2, x_3)$ denote the imputation in the nucleolus of a three-player cooperative game with nonempty core. Assume that $v(i) = 0$, $i = 1, 2, 3$, $v(S) \geq 0$, $C(0) \neq \emptyset$. Then*

1. *if $i, j = 1, 2, 3$ are such that $v(N) \geq 3v(ij)$, $i \neq j \Rightarrow x_1 = x_2 = x_3 = \dfrac{v(N)}{3}$.;*

2. *if $i, j, k = 1, 2, 3, i \neq j \neq k$ are such that*

$$3v(ij) \geq v(N) \begin{cases} \geq v(ij) + 2v(ik), & \text{if } i, j, k = 1, 2, 3, i \neq j \neq k; \\ \geq v(ij) + 2v(jk), & \text{if } i, j, k = 1, 2, 3, i \neq j \neq k; \end{cases}$$

then $x_i = x_j = \dfrac{v(N) + v(ij)}{4}$, $x_k = \dfrac{v(N) - v(ij)}{2}$.

3. *If $i, j, k = 1, 2, 3, i \neq j \neq k$ are such that*

$$v(ij) \geq v(ik) \text{ and } v(N) \begin{cases} \leq v(ij) + 2v(ik), & \text{if } i, j, k = 1, 2, 3, i \neq j \neq k; \\ \geq v(ij) + 2v(jk), & \text{if } i, j, k = 1, 2, 3, i \neq j \neq k; \end{cases}$$

then $x_i = \dfrac{v(ij) + v(ik)}{2}$, $x_j = \dfrac{v(N) - v(ik)}{2}$, $x_k = \dfrac{v(N) - v(ij)}{2}$.

4. *If for $i, j, k = 1, 2, 3, i \neq j \neq k$*

$$v(N) + v(ij) \geq 2(v(ik) + v(jk)) \quad \text{and} \quad v(N) \begin{cases} \leq v(ij) + 2v(ik), \\ \leq v(ij) + 2v(jk), \end{cases}$$

then

$$x_i = \frac{v(N) + v(ij) + v(ik) - 2(v(ik) - v(jk))}{4},$$

$$x_j = \frac{v(N) + v(ij) + 2(v(jk) - v(ik))}{4},$$

$$x_k = \frac{v(N) - v(ij)}{2}.$$

5. *If for $i, j, k = 1, 2, 3, i \neq j \neq k$*

$$v(N) + v(ij) \leq 2(v(ik) + v(jk)),$$

then

$$x_i = \frac{v(N) + v(ij) + v(ik) - 2v(jk)}{3},$$

$$x_j = \frac{v(N) + v(ij) + v(jk) - 2v(ik)}{3},$$

$$x_k = \frac{v(N) + v(ik) + v(jk) - 2v(ij)}{3}.$$

Definition F.10   *An allocation $\vec{x} = (x_1, \ldots, x_n)$ is called the* **Shapley value** *if*

$$x_i = \sum_{\{S \in \Pi^i\}} [v(S) - v(S - i)] \frac{(|S| - 1)!(|N| - |S|)!}{|N|!}, \quad i = 1, 2, \ldots, n,$$

where Π^i is the set of all coalitions $S \subset N$ containing i as a member (i.e., $i \in S$), $|S|$ = number of members in S, and $|N| = n$.

Definition F.11   *Suppose that we are given a characteristic function $v(S)$ that satisfies that for every $S \subset N$, either $v(S) = 0$ or $v(S) = 1$. This is called a* **simple game**. *If $v(S) = 1$, the coalition S is said to be a* **winning coalition**. *If $v(S) = 0$, the coalition S is said to be a* **losing coalition**. *Let*

$$W^i = \{S \in \Pi^i \mid v(S) = 1, v(S - i) = 0\},$$

denote the set of coalitions who win with player i and lose without player i. These are the winning coalitions for which player i is **critical**.

Definition F.12   *The Pareto-optimal boundary of the feasible set is the set of payoff points in which no player can improve his payoff without at least one other player decreasing her payoff.*

Definition F.13   *The* **status quo payoff point**, *or* **safety point**, *or* **security point** *in a two-person game is the pair of payoffs (u^*, v^*) that each player can achieve if there is no cooperation between the players.*

Theorem F.14   *Let the set of feasible points for a bargaining game be nonempty and convex, and let $(u^*, v^*) \in S$ be the security point. Consider the nonlinear programming problem*

$$Maximize\ g(u, v) := (u - u^*)(v - v^*)$$
$$subject\ to\ (u, v) \in S, u \geq u^*, v \geq v^*.$$

Assume that there is at least one point $(u, v) \in S$ with $u > u^*$, $v > v^*$. Then there exists one and only one point $(\overline{u}, \overline{v}) \in S$ that solves this problem, and this point is the unique solution of the bargaining problem $(\overline{u}, \overline{v}) = f(S, u^*, v^*)$ that satisfies the axioms $1 - 6$. If, in addition, the game satisfies the symmetry assumption, then the conclusion of axiom 6 tells us that $\overline{u} = \overline{v}$.

Appendix G: Evolutionary Stable Strategies and Population Games

Definition G.1 *Given the symmetric game* (A, A^T), *pure strategy* i^* *is evolutionary stable (ESS) if there is a* $0 < p^* < 1$ *so that*

$$(1 - p)a_{i^*i^*} + p\, a_{i^*j} > (1 - p)a_{j,i^*} + p\, a_{jj}, \text{for any} j \neq i^* \text{and all } 0 < p < p^* \quad \text{(G.1)}$$

Definition G.2 *A strategy* X^* *is an ESS against (deviant strategy) strategies* $X_1, \ldots, X_s$ *if either of (1) or (2) hold:*

(1) $u(x^*, x^*) > u(x_k, x^*)$, *for each* $k = 1, 2, \ldots, s$,

(2) *for any* x_k *such that* $u(x^*, x^*) = u(x_k, x^*)$,
   we must have $u(x^*, x_j) > u(x_k, x_j)$, *for all* $j = 1, 2, \ldots, s$.

Definition G.3 *A strategy* $X^* = (x^*, 1 - x^*)$ *is an evolutionary stable strategy if for every strategy* $X = (x, 1 - x)$, *with* $x \neq x^*$, *there is some* $p_x \in (0, 1)$, *which depends on the particular choice* x, *such that*

$$u(x^*, px + (1 - p)x^*) > u(x, px + (1 - p)x^*), \text{ for all } 0 < p < p_x. \quad \text{(G.2)}$$

G.1 Population Games

G.1.1 THE REPLICATOR EQUATIONS

$$\frac{dp_i(t)}{dt} = p_i(t)\left[E(i, \pi(t)) - E(\pi(t), \pi(t))\right] \quad \text{(G.3)}$$

$$= p_i(t)\left[\sum_{k=1}^{n} a_{i,k} p_k(t) - \pi(t) A\pi(t)^T\right], \quad i = 1, 2, \ldots, n,$$

Solutions Manual to Accompany Game Theory: An Introduction, Second Edition. E.N. Barron.
© 2013 John Wiley & Sons, Inc. Published 2013 by John Wiley & Sons, Inc.

or, equivalently,

$$\frac{dp_i(t)}{p_i(t)} = \left[\sum_{k=1}^{n} a_{i,k} p_k(t) - \pi(t) A \pi(t)^T \right] dt. \tag{G.4}$$

Theorem G.4 *Suppose that you have a system of differential equations*

$$\frac{d\pi}{dt} = f(\pi(t)), \pi = (p_1, \ldots, p_n). \tag{G.5}$$

Assume that $f : \mathbb{R}^n \to \mathbb{R}^n$ and $\partial f / \partial p_i$ are continuous. Then for any initial condition $\pi(0) = \pi_0$, there is a unique solution up to some time $T > 0$.

Definition G.5 *A steady state (or stationary, or equilibrium, or fixed-point) solution of the system of ordinary differential equations (G.5) is a constant vector π^* that satisfies $f(\pi^*) = 0$. It is (locally) **stable** if for any $\varepsilon > 0$ there is a $\delta > 0$ so that every solution of the system with initial condition π_0 satisfies*

$$|\pi_0 - \pi^*| < \delta \Rightarrow |\pi(t) - \pi^*| < \varepsilon, \quad \forall t > 0.$$

*A stationary solution of Equation (G.5) is (locally) **asymptotically stable** if it is locally stable and if there is $\rho > 0$ so that*

$$|\pi_0 - \pi^*| < \rho \Rightarrow \lim_{t \to \infty} |\pi(t) - \pi^*| = 0.$$

The set

$$B_{\pi^*} = \{\pi_0 \mid \lim_{t \to \infty} \pi(t) = \pi^*\}$$

*is called the **basin of attraction** of the steady state π^*. Here, $\pi(t)$ is a trajectory through the initial point $\pi(0) = \pi_0$. If every initial point that is possible is in the basin of attraction of π^*, we say that the point π^* is **globally asymptotically stable**.*

Theorem G.6 *A steady-state solution (p_1^*, p_2^*) of the system*

$$\frac{dp_1(t)}{dt} = f(p_1(t), p_2(t)),$$

$$\frac{dp_2(t)}{dt} = g(p_1(t), p_2(t)),$$

is asymptotically stable if

$$f_{p_1}(p_1^*, p_2^*) + g_{p_2}(p_1^*, p_2^*) < 0$$

and

$$\det J(p_1^*, p_2^*) = \det \begin{bmatrix} f_{p_1}(p_1^*, p_2^*) & f_{p_2}(p_1^*, p_2^*) \\ g_{p_1}(p_1^*, p_2^*) & g_{p_2}(p_1^*, p_2^*) \end{bmatrix} > 0.$$

If either $f_{p_1} + g_{p_2} > 0$ or $\det J(p_1^*, p_2^*) < 0$, the steady-state solution $\pi^* = (p_1^*, p_2^*)$ is unstable (i.e., not stable).

Theorem G.7 *In any 2×2 game, a strategy $X^* = (x_1^*, x_2^*)$ is an ESS if and only if the system (G.3) has $p_1^* = x_1^*$, $p_2^* = x_2^*$ as an asymptotically stable-steady state.*

Theorem G.8 *If X^* is an ESS, then X^* is an asymptotically stable stationary solution of (G.3). In addition, if X^* is completely mixed, then it is globally asymptotically stable.*

G.2 Properties of an ESS

1. **If X^* is an ESS, then (X^*, X^*) is a Nash equilibrium.** Why? Because if it isn't a Nash equilibrium, then there is a player who can find a strategy $Y = (y, 1 - y)$ such that $u(y, x^*) > u(x^*, x^*)$. Then, for all small enough $p = p_y$, we have

   $$p(u(x^*, y) - u(y, y)) + (1 - p)(u(x^*, x^*) - u(y, x^*)) < 0, \quad \forall\, 0 < p < p_y,$$

   This is a contradiction of the definition (Definition G.3) of ESS. One consequence of this is that **only the symmetric Nash equilibria** of a game are candidates for ESSs.

2. **If (X^*, X^*) is a strict Nash equilibrium, then X^* is an ESS.** Why? Because if (X^*, X^*) is a strict Nash equilibrium, then $u(x^*, x^*) > u(y, x^*)$ for any $y \neq x^*$. But then, for every small enough p, we would obtain

   $$pu(x^*, y) + (1 - p)u(x^*, x^*) > pu(y, y) + (1 - p)u(y, x^*), \text{ for } 0 < p < p_y,$$

   for some $p_y > 0$ and all $0 < p < p_y$. But this defines X^* as an ESS according to (Definition G.3).

3. **A symmetric Nash equilibrium X^* is an ESS for a symmetric game if and only if $u(x^*, y) > u(y, y)$ for every strategy $y \neq x^*$ that is a best response strategy to x^*.** Recall that $Y = (y, 1 - y)$ is a best response to $X^* = (x^*, 1 - x^*)$ if

   $$v(x^*, y) = X^* A^T Y^T = Y A X^{*T} = u(y, x^*) = \max_{W \in S_2} u(w, x^*) = \max_{W \in S_2} W A X^{*T}$$

   because of symmetry. In short, $u(y, x^*) = \max_{0 \leq w \leq 1} u(w, x^*)$.

4. **A symmetric 2×2 game with $A = (a_{ij})$ and $a_{11} \neq a_{21}, a_{12} \neq a_{22}$, must have an ESS.**

Index

Solutions Manual to Accompany Game Theory: An Introduction, Second Edition. E.N. Barron.
© 2013 John Wiley & Sons, Inc. Published 2013 by John Wiley & Sons, Inc.